DOLPHIN
SOCIETIES

Early days of dolphin research. Editors Kenneth S. Norris and Karen Pryor on the deck of a fishing boat in Hawaii in 1966, creating an instrument belt for a trained dolphin to wear during open ocean diving tests. Odds and ends from the hardware store were used to make a belt with a rachet on it, which was supposed to measure how much the animal's chest compressed during a dive. Unfortunately, the dolphin got so much smaller in circumference, even during relatively shallow dives, that it usually came back with the belt around its tail. (Photo by Henry Grozhinsky for LIFE magazine.)

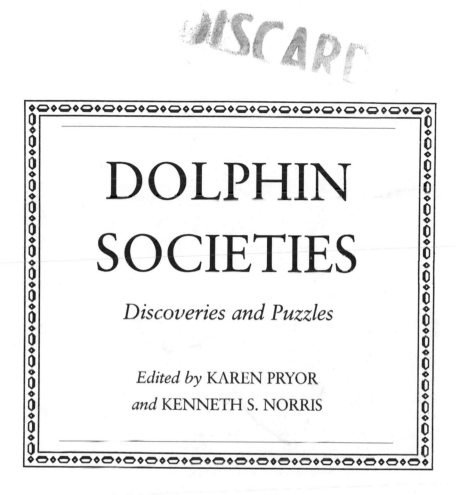

DOLPHIN

SOCIETIES

Discoveries and Puzzles

Edited by KAREN PRYOR

and KENNETH S. NORRIS

UNIVERSITY OF CALIFORNIA PRESS

BERKELEY   LOS ANGELES   OXFORD

University of California Press
Berkeley and Los Angeles, California

University of California Press, Ltd.
Oxford, England

© 1991 by
The Regents of the University of California

Library of Congress Cataloging-in-Publication Data

Dolphin societies : discoveries and puzzles / edited by Karen Pryor
   and Kenneth S. Norris.
      p.   cm.
   Includes bibliographical references.
   ISBN 0-520-06717-7 (alk. paper)
   1. Dolphins—Behavior.   2. Social behavior in animals.   I. Pryor,
Karen, 1932–   .   II. Norris, Kenneth S. (Kenneth Stafford)
QL737.C432D653   1991
599.5'3—dc20   90-35670
                                                                          CIP

Printed in the United States of America
1   2   3   4   5   6   7   8   9

The paper used in this publication meets the minimum requirements of American
National Standard for Information Sciences—Permanence of Paper for Printed
Library Materials, ANSI Z39.48-1984. ∞

CONTENTS

Acknowledgments   vii

Introduction • Karen Pryor and Kenneth S. Norris   1

PART I. FIELD STUDIES

Essay: Looking at Wild Dolphin Schools • Kenneth S. Norris   7

Chapter One. Herd Structure, Hunting, and Play: Bottlenose
   Dolphins in the Black Sea • Edited by V. M. Bel'kovich   17

Chapter Two. Dolphin Movement Patterns: Information from
   Radio and Theodolite Tracking Studies • Bernd Würsig,
   Frank Cipriano, and Melany Würsig   79

Chapter Three. The Feeding Ecology of Killer Whales (*Orcinus
   orca*) in the Pacific Northwest • Frederic L. Felleman,
   James R. Heimlich-Boran, and Richard W. Osborne   113

Chapter Four. The Interactions between Killer Whales and Boats
   in Johnstone Strait, B.C. • Susan Kruse, with Introductory
   Comments by Kenneth S. Norris   149

Chapter Five. Social Structure in Spotted Dolphins (*Stenella
   attenuata*) in the Tuna Purse Seine Fishery in the Eastern
   Tropical Pacific • Karen Pryor and Ingrid Kang Shallenberger   161

Chapter Six. The Role of Long-Term Study in Understanding
   the Social Structure of a Bottlenose Dolphin Community
   • Randall S. Wells   199

Chapter Seven. Using Aerial Photogrammetry to Study Dolphin
   School Structure • Michael D. Scott and Wayne L. Perryman   227

PART II. LABORATORY STUDIES

Essay: Mortal Remains: Studying Dead Animals • Karen Pryor   245

Chapter Eight. Some New and Potential Uses of Dental Layers in Studying Delphinid Populations • Albert C. Myrick, Jr.   251

Chapter Nine. An Overview of the Changes in the Role of a Female Pilot Whale with Age • Helene Marsh and Toshio Kasuya   281

Essay: Some Thoughts on Grandmothers • Kenneth S. Norris and Karen Pryor   287

PART III. CAPTIVE STUDIES: THE KEY TO UNDERSTANDING WILD DOLPHINS

Essay: Looking at Captive Dolphins • Kenneth S. Norris   293

Chapter Ten. Changes in Aggressive and Sexual Behavior between Two Male Bottlenose Dolphins (*Tursiops truncatus*) in a Captive Colony • Jan Östman   305

Chapter Eleven. Use of a Telemetry Device to Identify which Dolphin Produces a Sound • Peter Tyack   319

Essay: The Domestic Dolphin • Karen Pryor   345

Chapter Twelve. What the Dolphin Knows, or Might Know, in Its Natural World • Louis M. Herman   349

Chapter Thirteen. Dolphin Psychophysics: Concepts for the Study of Dolphin Echolocation • Patrick W. B. Moore   365

Afterword: Dolphin Politics and Dolphin Science • Karen Pryor and Kenneth S. Norris   383

Notes on Contributors   387

Index   391

ACKNOWLEDGMENTS

In addition to the reviewers acknowledged by the authors of individual chapters, the editors wish to thank the following people who read and commented on all or part of the manuscript for this book. Their perceptive critiques were invaluable.

Michael Bigg, Ph.D., Nanaimo Marine Laboratory, British Columbia

Donald Griffin, Ph.D., Rockefeller University, New York, New York

Erich Klinghammer, Ph.D., Purdue University, Battle Ground, Indiana

William Perrin, Ph.D., Southwest Fisheries Center, National Marine Fisheries Service, La Jolla, California

Sam Ridgway, D.V.M., Ph.D., Naval Ocean Systems Center, San Diego, California

William Schevill, Woods Hole Oceanographic Institution, Woods Hole, Massachusetts

Michael Schulman, Ph.D., Psychologist, New York, New York

William Watkins, Ph.D., Woods Hole Oceanographic Institution, Woods Hole, Massachusetts

INTRODUCTION

Karen Pryor and Kenneth S. Norris

We had three purposes in organizing this book. First, we wanted to report on the considerable progress currently being made in research on dolphins in the wild. Second, we wanted to show that we learn most about cetaceans when we study them both in the wild and in captivity; captive animals offer us understanding that cannot be acquired at a distance, and such understanding is fundamental to caring about cetaceans. Finally, we hope to inspire both the scientist and the general reader by the wide variety of imaginative ways in which a single difficult problem is being tackled: how to study the behavior and social structure of animals that are constantly on the move and mostly invisible.

We editors, Karen Pryor and Kenneth Norris, have been studying dolphins for 25 years and 38 years, respectively. We have watched the field grow from a time when not more than a dozen people were actively involved in dolphin research to now, when hundreds are contributing to the field, and marine mammalogy has been dignified with the formation of a scientific society and the publication of a journal. When we started out, scientists were sometimes hard put even to identify the species of dolphins they saw swimming in the ocean; now, many species are being studied at sea and in captivity all over the world. A sampling of this new work is reported here.

Outside the marine mammal scientific community, several misconceptions about dolphin research are still widely held. The lay public's perception of dolphins sometimes differs widely from that of serious students. Legitimate workers, who have plodded along seeking true things just as

1

other scientists do, unavoidably work in the midst of a din of publicity and speculation that has turned the dolphin into a mythic beast, a sort of floating hobbit. In this book, we trust we demonstrate that dolphin research has come a long way beyond first experiments in "communication" and that the truth about cetaceans is more interesting than fanciful speculations.

A second misconception, common in the academic community, is that dolphins are so inaccessible that they are not subjects for fruitful investigation. Many graduate students have been forced to drop ideas of doing marine mammal research because supervisory professors deemed it impractical. Dolphin research is certainly difficult and often expensive; it is undoubtedly easier to work on field mice. Nevertheless, marine mammal work is possible and worthwhile. We hope that a look at our cross section of problem solvers will give confidence to others who want to work in this field.

A third misconception, held unconsciously even by some marine mammalogists, is that all dolphins are *Tursiops;* that the familiar bottlenose dolphin, good old Flipper, the animal most often kept in captivity, is the model for the whole clan. In fact, the behavior and organization of the thirty-odd species of dolphins and small whales seem to be as rich and varied as their appearance and geographic distributions. Here, you will encounter a pleasant variety of species, from dainty little spinner dolphins in the tropics to big black pilot whales in chilly northern seas. You will meet killer whales that live in small, closed societies and spinner dolphins that seem to live in large, fluctuating groups, spotted dolphins in male-dominant groups, pilot whales apparently in family clusters without adult males but containing very old females, and dusky dolphins whose social organization changes with their choice of prey.

We hope that these studies will also provide some evidence to refute those who propagandize against keeping dolphins in captivity or who maintain that studies of captive dolphins have no application to understanding of dolphins in the wild. Our cooperative captives show us whole levels and kinds of information we simply have not been able to approach in the wild, from bioacoustics to cognition. As you will see in the contributions of Peter Tyack and Jan Östman, captive studies can throw a lot of light on what goes on in nature; and from Louis Herman and Patrick Moore, you will learn of captive dolphins that apparently relish their working partnerships with us.

Finally, this book examines an aspect of science that is not often exposed to view: the scientific imagination. The good scientist asks questions; the creative scientist asks new questions and finds new ways to tease out answers from a mass of inscrutable information. The papers herein range from a young student's first publication to essays by senior scientists summarizing decades of work and thinking; yet each one is like a mystery story. The scientist saw a puzzle, conceived of a way to tackle it, and came up with new conclusions. Albert Myrick started with a huge collection of dolphin teeth; Helene Marsh and Toshio Kasuya faced daunting piles of pilot whale corpses; Michael Scott had to make sense of aerial photographs in which thousands of dolphins appeared as not much more than dots. Each of them—and our other contributors—asked good questions, invented ways of answering them, and discovered, through exercise of the scientific imagination and a great deal of work, something new about dolphin societies.

In the case of every author, this work is ongoing. You are not seeing completion here but science in progress. We cross the paths of these workers and thinkers as we might pass through a school of dolphins in the ocean, enjoying an exciting glimpse of a group of separate individuals, moving purposefully in the same direction, destination unknown.

PART I

Field Studies

Essay

LOOKING AT WILD DOLPHIN SCHOOLS

Kenneth S. Norris

Rather like trying to perceive motion from inside a moving airplane, it is hard to detect a revolution from within. But in the last fifty years, there has been a revolution in our knowledge of the biology of live dolphins. Most of what we know about dolphin sounds, dolphin physiology, and the dolphin mind, and virtually *everything* we know about the doings of wild schools, has been learned in that time.

The first captive dolphin community was established at Saint Augustine, Florida, in the late 1930s. Observers soon saw that these were complex mammals, emphatically not fish, and that out there in the ocean they had managed to evolve societies that seemed as complex as, say, those of primates or social carnivores.

Dolphins at sea seemed all but unreachable. In vast tracts of ocean, even the species of dolphins present were not fully known. As late as the middle 1960s, for example, the dolphins of tropical seas were the most mysterious of faunas on the planet. We called the pigmy killer whale (*Feresa attenuata*) the "rarest large mammal on earth," because until one was collected alive in Hawaii in 1964, all we knew of it was derived from a couple of skulls in museums. More abundant, but no better understood, were the many species of slim oceanic dolphins of the genus

Stenella. Since then, the pigmy killer whale has been observed around
the world at several locations. It is, indeed, a fairly abundant little killer
whale–like animal; and the *Stenellas* are known to comprise a group of
species that occur in tens of millions in the tropical current systems of
all oceans.

But knowing an animal exists still does not tell us much about how
that wild dolphin lives; and the prospects of learning anything more
were daunting. How could one hope to learn anything about the behavior of animals that only appeared now and then, flirted with a vessel,
and were gone into the vast ocean?

The first attempts to understand wild dolphins were acoustic. In the
late 1940s, William Schevill and Barbara Lawrence made the first at-sea
recording of a cetacean's sounds. Their subject was the beluga, or white
whale. They worked with a stenographic machine and a primitive hydrophone. Belugas, they found, were noisy animals. This simple discovery demonstrated that one could record and study dolphin sounds; it
did not unravel much about what the sounds meant. That task became
a principal activity of field-oriented cetologists in the decade of the
1950s. It is a task still actively pursued today.

Schevill and his colleague William Watkins have been central figures
in the endeavor to record wild dolphin sounds, followed by many other
workers from around the world. The task, as Schevill and Watkins ad-
monished us, may sound simple, but it is not. The listening scientist
must exercise the most careful judgment about what animal is actually
making the sound one is recording. Because sounds travel well and far
underwater, it is often difficult to pinpoint the phonating animal or even
to assign the sound to a species if more than one kind of dolphin is
somewhere in the area. This problem is exacerbated because nearly
everyone listens through just one hydrophone at a time; there is no way,
in such a circumstance, to determine the source of the sound.

The observations of the Saint Augustine captive school made it clear
that whole societies of mammals were out there in the sea, and understanding of sounds alone would never reveal more than a single facet of
their lives. Further, it was clear that only certain intimate things about
dolphin lives could be learned from captive schools. Animals confined
in pools, even big ones, obviously were not able to carry out all of their
normal life patterns. The wild dolphin's long traverses along beaches
and around rocky headlands were gone, its feeding forays into black

nighttime water could not be undertaken, and the interactions between dozens or hundreds of schoolmates were only hinted at in the colony ashore. The outward face of dolphins, the ways they deal with their larger world, necessarily exists only as a hint in captive animals. But how else could one learn?

The first barrier for cetacean naturalists was simply to conceive that the problem was solvable. They had to learn to accept the possibility that one *could* learn about these fleet, remote animals whose world was separated from ours by that most difficult of barriers, the sea surface.

The second step was to learn to be content with tiny bits of understanding at any one time. Many little truths would have to be learned by many people and assembled into a coherent picture at some later date.

So, by the late 1960s, a few Western naturalists had hitched up their field pants and begun to seek out the best means and the best places to observe wild dolphins. They chose sea cliffs, they developed little radios that could be affixed to dolphin fins, and they began to watch dolphins underwater.

Probably the first concerted attempt was that of the South African team of Graham Saayman and C. K. Tayler. Saayman, a primate biologist, knew that one way to study social behavior was to start recording patterns, whatever one can see; in time, from the arid precincts of one's recorded measurements and numbers, an understanding might emerge.

Tayler and Saayman found a place on the Natal coast where two species of dolphins came close to shore. There they could watch Indian Ocean bottlenose dolphins crowd fish schools against the beach; and they could perch on a rocky promontory called Robbe Berg and look down on groups of humpback dolphins maneuvering in the surge along the rocks. Their work told us a few things about daily movements and about ecological separation between the species, and it revealed tantalizing hints of schoolwide cooperative fishing methods by the bottlenose dolphin.

Others began to test new methods. William Evans, working in California waters, was the cetologist most responsible for developing the dolphin radio tag that now allows us to follow dolphins at sea. Several vexing problems were involved. How could a radio be affixed to a dolphin so that it would stay in place? Could a package be designed which would withstand the pressures of deep dives? How could faint radio signals be made to penetrate the vapor barrier that hovers over the sea surface? Were there power sources available which would allow the tag to last a reasonable length of time and broadcast for a workable distance? Finally, could all this be made into a package that a dolphin could carry without harm to itself or disturbance to its place in the school? Evans located a fledgling electronics company, Ocean Applied Research, and together they built the first usable radio packages, which were affixed to the dorsal fins of dolphins.

By that time, William Cochran and others had designed tiny circuits for tracking the movements of migrating birds. Cochran glued these instruments to the feathers of his birds, released them, and then followed them with a big receiving antenna mounted in the back of his pickup, sometimes cruising the streets of Chicago with this rig, trying to stay within radio range of his subjects. Cochran's instruments influenced the design of the aquatic versions produced by Evans and his associates.

Evans devised modulated signals that told not only the animal's position but also the depth of its most recent dive. He was ultimately able to track dolphins for many days. Since his early studies, radio packages have become smaller and more powerful. Ranges of 30 to 40 miles have been achieved on the surface and even vertically to a receiving satellite circling above. Workers such as Bruce Mate and William Watkins have recently followed whales for very long distances, through a significant fraction of their oceanic migrations.

During the 1960s, Karen Pryor and I worked in the same institution on the island of Oahu, she as head dolphin trainer at Sea Life Park oceanarium and I as scientific director of the affiliated Oceanic Institute. Not infrequently, we collaborated on studies where trained animals were used, such as our studies, with Thomas Lang, of dolphin swimming performance in the open sea. At that time, my mind began to turn to the problem of learning about the lives of wild dolphins at sea. Hawaii was certainly a perfect place for such work. Several species of dolphins and whales occurred nearby. Sometimes they were encountered in calm, clear waters near shore. For example, off the lee of the Waianae Coast of Oahu, I knew of a school of spotted dolphins that was repeatedly seen under the lee hook of Kaena Point. Ashore of that point lay the steep, fluted slopes of high mountains. It occurred to me that there I could establish two listening stations hooked together with a long wire laid along that slope. With Evans's new radios I might be able to triangulate on animals swimming offshore, just as a sailor would make a navigational fix on points of land from offshore. But the Kaena animals proved elusive, and two other promising developments turned my mind elsewhere. The first was the location of a "resident school" of dolphins in Kealake'akua Bay, on the distant island of Hawaii, and the second was the successful launching of my first porpoise-watching vehicle, the mobile observation chamber (MOC), or more colloquially, the semisubmersible seasick machine (SSSM).

The Kealake'akua Bay topography was perfect for dolphin work. A magnificent, nearly vertical 500-foot cliff loomed over the almost-always calm and clear semicircular bay. A group of dolphins rested in the bay nearly every day. One could look down almost on top of those resting dolphins, and their behavior could often be seen in toto as they moved offshore, until they faded from view into the gray disk of the sea. My colleague Tom Dohl and I established a camp near the brink of the cliff, cutting trails through the dense brush to vantage points where telescopes and cameras could be set up. And we brought along the new viewing vehicle.

Then we set about trying to learn to recognize individual dolphins. We drew pictures of them, including scars and any distinctive markings we noted. By this means, we began to recognize some individuals over and over. The "school" proved to be a transient and ever-changing group. A broad outline of the spinner dolphin's daily life emerged from this work, and we had some glimpses of their behavior underwater, though we found that the SSSM was so slow, and so frightening to be in if the sea got at all rough, that we had to restrict her operations to within the arc of the bay.

William Evans took part in one of our early voyages in the SSSM and was so impressed with the potential of underwater viewing that he returned to the navy laboratory where he worked and designed a larger and much grander version, which placed observers within an entirely transparent plastic viewing sphere, a vessel he called the *See Sea*.

After a proud prototypical career, our SSSM was finally relegated to scrap metal. Only some years later, for studies of dolphin mortality in the yellowfin tuna fishery, did I use another such vehicle, the *Maka Ala*, or Watchful One. This craft, built from a trihull skiff, required the observer to lie down on a mattress with his or her head in a transparent plastic ship's bow. Though the *Maka Ala* was maneuverable and fast compared to the SSSM, she was a less-than-optimum platform for long-term observation. Aside from the permanent crick in the neck induced in users, her major fault was a sloping bow that prevented the observer from looking upward toward the sea surface. Only through small side panels could one see the sea surface above. This nearly eliminated the chance for an observer to identify the sex of animals that swam close to the bow, an all-important activity if one wants to unravel the social structure of a dolphin school.

A new viewing vehicle for which we have high hopes is now being built at my laboratory in Santa Cruz, California. My Swedish student Jan Östman has dubbed her the *Smygg tittar'n,* or Tiptoeing Peeping Tom. The *Smygg tittar'n* features a retractable viewing cylinder. The vessel will travel with the cylinder in the "up" position. When a subject of interest is sighted, the cylinder will be lowered, and an observer will enter and don headphones that allow communication between driver and observer. Animals will be visible in all directions, including upward toward the surface. We expect to achieve some degree of directional hearing through two hydrophones and in this way, to be able to connect dolphin sounds to their proper behavioral contexts.

The value of these viewing vessels is, first, to introduce the observer to the world of the dolphin. We naturalists need to see and understand the shafts of downwelling light in the open sea, the flicker of sunlight magnified through thousands of lenses of curved water on the sea surface, that are so constantly part of what a dolphin sees. We need to understand the underside of the surface as dolphins see it, a wavy silvery sheet extending off into the blue distance, with, directly overhead, a permanent disk of transparent water, Snell's Window, through which the dolphin sees into the air above.

We need to see other dolphins as a dolphin sees them. Traveling dolphins in a school look primarily sideways at each other for the signs and signals of their visual communication; we understand them only if we look sideways too, not down from above. Underwater, a human observer can clearly see the flashing white bellies of mating dolphins, long before the dolphins themselves come into view through the blue. That, of course, is what dolphins see too. One can see postural signals: anger shown by flashing eye whites; the wholly diagnostic jerky swimming of dolphin calves; the open spaces within schools in which babies play; mothers nursing, playing with, and disciplining their young; the serious troops of males that usually interpose themselves between the observer and the remainder of the school. And one can watch dolphins caress in what we have come to believe is a ritual of reaffirmation between animals. In short, looking from beneath the surface, one penetrates the animal's world and sees the texture and intimate structure of dolphin life.

The data from these viewing vehicles come as a collection of vignettes, each usually only seconds long. But with patience, and frame-by-frame analysis of film and videotape, much comes into focus and much can be quantified.

While I was testing my first viewing vehicle, Roger Payne began plans to watch wild right whales from a clifftop at the Golfo de San Jose on the Patagonian coast of Argentina. He built a station there, right on the brink of the cliff, and began to devise ways of extracting those vital numbers from his observations. One way, he decided, was to track the animals with a surveyor's theodolite. He set two of his student team to work with this instrument—Bernd Würsig and Melany Würsig. As a project of their own, they turned the theodolite technique to the task of describing the movements of the dusky dolphins and southern bottlenose dolphins that shared the bay with the whales. Bernd, a skilled photographer, took photographs of every school he encountered. From this trove of images, he and Melany began to assemble a dossier of animals that they could recognize. This record was a far better way to identify animals than my drawings. Many dolphins, it turned out, had subtle but definite scars and nicks on their fins which, taken together, allowed positive identification where no drawing would suffice. The method of *scars and marks analysis* has since revealed many things about how dolphins live at many places in the world oceans. It lets scientists follow individual animals over time and come to know their associates; it gives insights into such societal features as courtship, movement patterns, and site fidelity.

The Würsigs contributed another important change to our way of doing things. Once the data gathering from the cliffs and boats was done, they began a statistical search through their numbers for relationships the eye could not discern. They were able to *quantify* many aspects of the lives of the dolphins they watched, events that previously had been recorded only in words. Their papers are peppered with parentheses containing probability values and other more esoteric statistics. Though these notations may clutter the story line, they provide the most precious of scientific treasures—little truths on which future understandings of the dolphins may be anchored.

I will mention two further studies, both begun in the early 1970s, that have lifted their subjects, the killer whale and the bottlenose dolphin, from being among the least known to among the best known of large mammals. Michael Bigg, a fisheries biologist at the Pacific Biological Station on Vancouver Island, began to record the killer whales of Washington and British Columbia by the shapes of their fins, their patterns, and their scars and marks. Now every individual in these waters can be recognized and allocated to a given pod and dialect group. When a new baby is born or an animal disappears, researchers not only know that this has happened but which individual it is, its family, and its lineage, sometimes to three generations. Bigg's study is now being extended by other biologists working in Alaska, Canada, and the northwestern United States.

At about the same time Bigg was beginning his work, a team of three biologists, Blair Irvine, Randall Wells, and Michael Scott, began to assemble data from the bottlenose dolphins living in the shallow bays of the west coast of Florida. Year after year, these biologists have returned to this group of about one hundred animals to ask new and deeper questions of the dolphins. Many animals have been captured briefly and tagged; blood samples are taken which illuminate genetic relationships of nearly all animals in this population, as well as where they go, how old they are, when they give birth, what their annual hormonal cycles are, what it is like to grow up in a dolphin school, and even how thick their blubber is at any time of year (measured by ultrasound), which allows us to assess how well they are eating.

This study is backed up by other investigations of bottlenose dolphins. The Russians continue to study dolphins in the Black Sea. Susan Shane, working both in Texas and Florida, has defined the behavioral flexibility of the species by showing that the intimate patterns of daily dolphin life vary from place to place. First Elizabeth Gawain and then Richard Conner and Rachel Smolker have begun to learn about a group of Australian bottlenose dolphins that have become habituated to people at Monkey Mia, Shark Bay, north of Perth, Australia. There they can see and hear the most intimate interactions between the dolphins. Because they can wade among those dolphins that come closest to shore and because the remainder seem remarkably tame, many details of dolphin life can be seen and studied, firsthand. This work is commanding the attention of primatologists and seems to hold great promise.

The chapters that follow are samples of contemporary work on wild dolphins, ranging from the studies of young workers reporting on their first serious piece of research to established workers, those who have laid the groundwork, defining where we find ourselves now.

*A group of bottlenose dolphins (*Tursiops truncatus*) in transit or "on the march." Natural marks, scars, and dorsal fin nicks provide ways of identifying individual animals. (Photo by Bernd Würsig, Argentina.)*

CHAPTER ONE

HERD STRUCTURE, HUNTING, AND PLAY

BOTTLENOSE DOLPHINS IN THE BLACK SEA

Edited by V. M. Bel'kovich

Editor's note: This suite of three papers has been adapted from *Behavior and Bioacoustics of Dolphins,* edited by V. M. Bel'kovich (Moscow: USSR Academy of Sciences, 1978), and is presented here with the permission of Dr. Bel'kovich. The opening information on research methods, which pertains to all three papers, has been summarized from the introduction to the Russian publication. Co-authors of each section of this chapter are listed in the text.

The translations are by Vlademir Gurevich. In editing these translations for readability, I compared them to two other unpublished translations. I also consulted a professional literary Russian-English translator and arranged for V. M. Bel'kovich to peruse the final results. However, I am solely responsible for any distortions, omissions, or inaccuracies resulting from this editorial process.

—*K.P.*

DOLPHIN HERD STRUCTURE

V. M. Bel'kovich, A. V. Agafonov, O. V. Yefremenkova,
L. B. Kozarovitsky, and S. P. Kharitonov

INTRODUCTION

At present, no precise data exist on the population structure of the Black Sea bottlenose dolphin (*Tursiops truncatus*), their migratory patterns, the size of the areas they inhabit, the fluctuations in these areas, the numbers of animals in each population, or other similar questions. One approach is to conduct investigations in a specific area of the ocean which may serve as a model toward finding acceptable answers to these questions.

A portion of the coastline of western Crimea where human interference would be minimal was selected as the biological observation area ("biopolygon"). The investigations were conducted in the spring, summer, and fall over a three-year period.

Observations were made utilizing the following methods:

a. Stationary observation posts onshore (OP). These were located in the central part of the observation area and permitted observations along 10 kilometers (km) of coast and up to 3 km offshore. Temporary mobile observation posts were used to bring a total of 80 km of coastline under observation.
b. Dolphin movements were accompanied from the shore for distances up to 25 km on horseback, on bicycles, or by car.
c. Observations were made by helicopter of about 250 km of coastal water and as far as 20 km from shore.
d. Dolphins were observed and tracked at sea by motorboat and catamaran along approximately 40 km of coast and as far as 10 km offshore.
e. Bioacoustic, film and still photography, and telemetric data were obtained, as well as hydrologic and meteorologic data.

THE BOTTLENOSE DOLPHIN

Based on our own observations in the eastern Crimea as well as along the Georgian shore and on questionnaire data obtained from fishermen, we accepted as a working hypothesis that the Black Sea bottlenose dolphins are relatively sedentary, settled animals and that their populations consist of local subpopulations.

Aggregations inhabiting the same coastal area were considered to be a local subpopulation. To designate the structural subdivisions of such a subpopulation of bottlenose dolphins, we used the terms "herd," "group," and "school." A herd is a natural congregation of dolphins that uses a certain home range, has prolonged independent existence, exhibits interrelated behavior among individuals, and apparently shares a kin relationship. "Group" is used to designate smaller units in this herd; groups are constituent parts of the herd and cannot exist without the herd over a long period. A "school" is our term for an unstable association of several herds.

The structure of a population, as well as of any part of it, can also be characterized by sex, age, and spatial and behavioral differentiation. When observing bottlenose dolphins, we were not able to determine the sexual composition of the herd, since this species does not have sexual dimorphism as does, for example, the killer whale (*Orcinus orca*). However, if an animal was seen with a calf, we assumed, with a high degree of probability, that it was likely to be a female, because observations in oceanariums show that a calf spends most of its time with its mother.

In most observations, we designated the age composition of the herd simply as adults and calves. Calves younger than one year are differentiated easily by length, being approximately one-third or one-half the body length of an adult animal. As a rule, when in motion, calves stay beside and slightly behind the female, surfacing simultaneously with her. Sometimes, if the animals were not too far away, we were also able to discriminate juveniles or adolescents by length, their length being approximately three-fourths that of adult animals.

Spatial and Behavioral Structure

[When we saw dolphins in the open sea, most often we saw a group or several groups of animals.] The animals comprising one group usually stayed closer to each other than to animals from other groups. Each group was also characterized by a harmony of movement, traveling in synchrony with each other more than with the herd in general.

In bottlenose dolphins, we are dealing with a herd structure that is typical for most cetaceans (Yablokov et al. 1972). Groups usually consisted of two, three, four, or five animals, as has been noted for some other toothed whales such as the white whale (Bel'kovich 1960). Individual groups within the herd were easy to differentiate while the herd was moving. We also observed various types of herd formations that we named "team" (*tsug* or sled dog team: Ed.), "echelon," "front," "double front," "tight group," and "line" (fig. 1.1).

When dolphins moved in the front formation, they distributed themselves evenly, and group structures seemed to disappear. It was also difficult to observe group structure during fear reactions; in these situations, animals gathered together underwater and emerged in a densely packed group far away from the danger. Group structure disappeared also in diffuse searching but reappeared immediately during orientation reactions, as has been observed for white whales (Bel'kovich 1960). The group structure in bottlenose dolphins is probably based on family relationships (a female and her calves), although there are some observations of male groups in Indian Ocean bottlenose dolphins (Tayler and Saayman 1972).

Two types of spatial differentiation of bottlenose dolphin herds were defined, both directly related to orientation and navigation. First, we observed dolphins acting as scouts, a group of two or three animals searching the inshore area at a distance from the main herd (for more

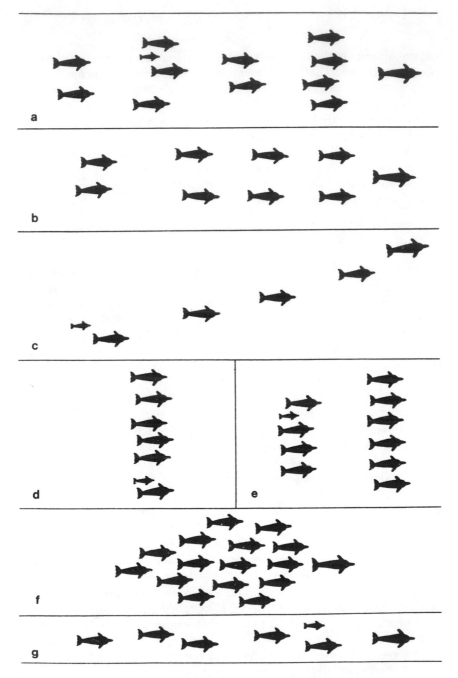

Fig. 1.1.   Bottlenose dolphin herd formations. a. Formation; b. Team;
c. Echelon; d. Front; e. Double front; f. Tight group; g. Line.

details, see "Searching and Hunting Behavior in the Bottlenose Dolphin in the Black Sea," below). Second, we observed leaders of the herd, or dominant dolphins. From the shore, we were repeatedly able to observe that a herd was usually headed by one or two large animals. Similar situations were observed from the air for bottlenose dolphins and previously for white whales (Bel'kovich 1960). The dominant animal apparently determines the degree of hazard or investigates anything "new," as is typical for captive bottlenose dolphins in oceanariums. For instance, in one oceanarium, two dominant females separated from the rest of the group and moved toward a risk situation (Bel'kovich et al. 1969). In the sea, our catamaran was studied by a very large dolphin that swam 300 to 400 meters (m) away from the herd and approached within 15 to 20 m of the catamaran.

The structure of a bottlenose dolphin herd is variable. For example, one typical group, consisting of a female, her calf, and another large animal (nicknamed "Hippo"), was observed regularly in the central part of our study area, in a herd consisting of twelve animals. For several days, however, this group foraged in the eastern part of the district, close to shore, though the rest of the herd was absent.

Structural Dynamics of Groups and Herds

In tracing the composition of groups encountered over a period of three years (winter months excepted), we discovered a certain regularity. During Year I, groups of three dolphins with one calf and of four dolphins with one calf were encountered. During Year II, schools passing through the observation area most often consisted of groups of three or five animals. Groups of three animals sometimes included a calf or sometimes consisted of three adults. A group of five animals consisted of two calves and three adults. Since we observed such groupings throughout whole field observation seasons, we assumed their stability continued for several months at least.

Group structure in Year III was characterized predominantly by groups of two or three animals; the group of five animals (three adults and two calves) was not seen (fig. 1.2). Changes in the number of animals in a group (fig. 1.3) might be related to estrus, pregnancy, birth, a calf's mortality, the reaching of sexual maturity by juvenile dolphins, cooperation during hunting, and so on.

The central questions, to us, are the constancy of the basic composition of groups, the stability of the number of groups in a herd, and the stability of the herd as an element in the local population.

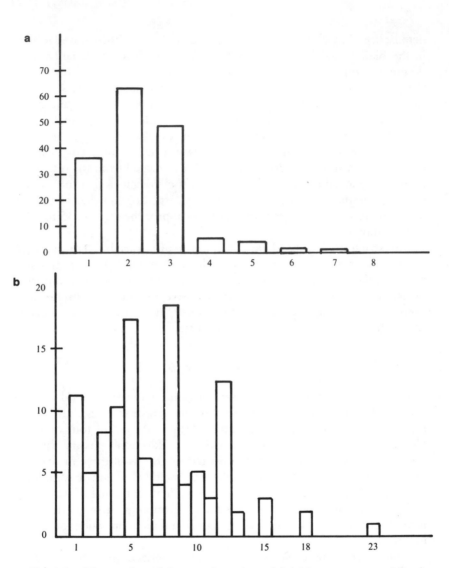

Fig. 1.2. The number of times various sizes of dolphin groups were sighted in year III: horizontal axis = number of dolphins in group; vertical axis = number of sightings of groups of a given size. a. Observation Post #1; b. Observation Posts #3 and #4.

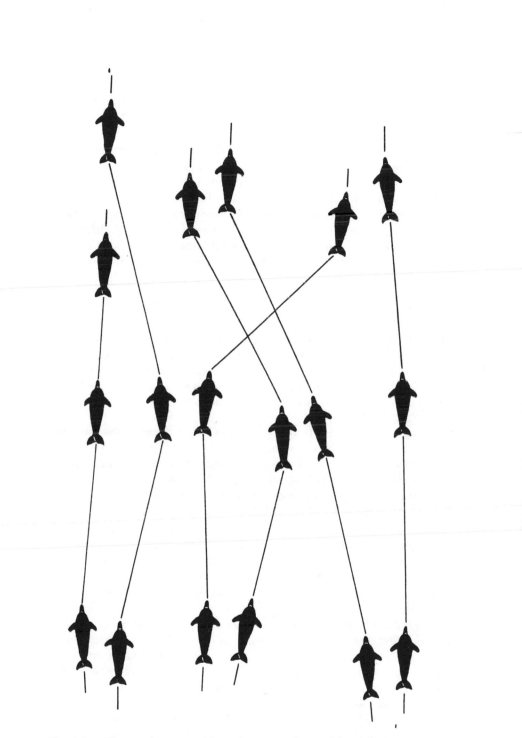

Fig. 1.3.   Changes in composition of a group observed from the catamaran.

Specific Herds Observed across Three Years

To identify herds and groups of bottlenose dolphins swimming through our observation region, we paid special attention to the following: (1) the total number in the herd, (2) the number of groups and their arrangement in the herd, (3) the number of young and their distribution within groups, (4) coordination of the groups while hunting, (5) the behavior of individual animals, and (6) the presence of animals with natural markings.

We should note that occasionally we observed several herds uniting into a school; at these times, we could see twenty to thirty animals at once, although the size of normal herds, across three years of observation, did not exceed thirteen to fifteen animals. These schools were observed rarely and usually formed and broke up during a single observation period.

Observations during Year I   During this period, groups of two to three dolphins predominated overwhelmingly. Sightings of large groups decreased sharply when the occurrence of small groups increased. Occasionally, a herd would be observed far out at sea while a group of scouts passed along shore; later, the scouts would join the herd. Since these larger herds of bottlenose dolphins seldom approached the shore during this year, it is impossible to judge their numbers from the data.

Year II   During Year II, the herds came much closer to the shore than in Year I. Usually, groups and herds of from three to twelve to fourteen animals were observed. The average number of animals per herd increased toward the end of the season (p 0.05). Data from Year II make it possible to assume the presence of two different herds of bottlenose dolphins in the study area. The first herd averaged approximately twelve animals in September and October. Only two calves were observed in this herd. One group of four animals—one calf and three adults—were together consistently. While in motion, the animals in this herd stayed more closely together than the animals in the second herd. On September 17 and 19, an animal with a natural mark on the dorsal fin was observed in this herd (fig. 1.4a, b).

The second herd consisted of thirteen animals by September–October and included three calves, occasionally one juvenile, and a rather distinct group consisting of three adults and two calves. The two groups in this second herd were clearly separated, and it was relatively easy to keep track of them. A dolphin with two white stripes on the left side of the dorsal fin and two white spots on the right side was observed in this herd (fig. 1.4b). This animal, which did not have a calf, was first observed on May 27 in a herd consisting of six animals. It was seen again on August 2

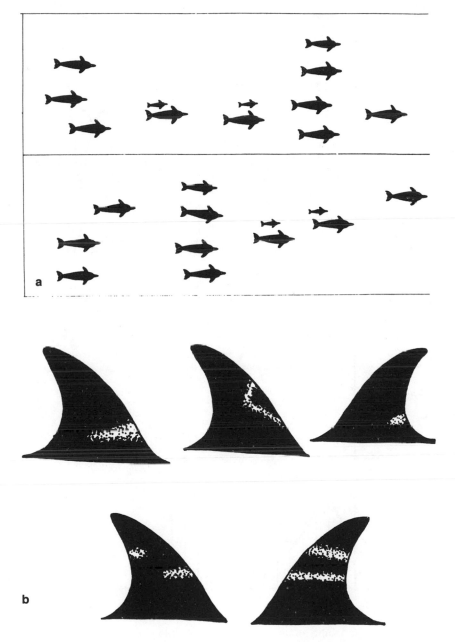

Fig. 1.4.   a. Formations of a herd observed on September 17 and 19, Year II; b. Natural markings on dorsal fins.

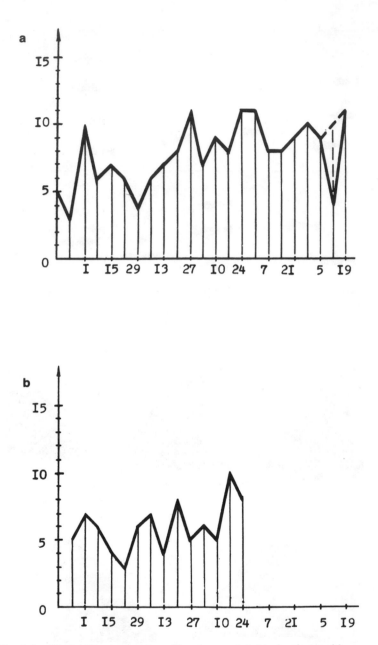

Fig. 1.5. Variability in the size of bottlenose dolphin herds (weekly averages). Vertical axis: numbers of dolphins in group; horizontal axis: dates at two-week intervals across observation periods. a. year II; b. Year III.

in a herd of eight to ten animals (with two calves) and on September 14
and October 28 in a herd of approximately ten individuals (three calves).
On October 31 and November 2, this dolphin was seen in a herd of nine
animals with one calf. A second marked dolphin was recorded on October 17 and 28 in a herd of thirteen animals (three calves) and on October 31 in a herd of nine individuals. In the two latter cases, the two
marked dolphins were together.

The consolidation of animals within the herds toward fall may be correlated with reduced distribution of groups (fig. 1.5). Apparently, at the
beginning of summer, the herds were more dispersed, and not all groups
could be seen by observers. A second possibility is that the herd gradually
acquired new groups, which had previously existed independently or as a
part of another herd. The data from Year II do not allow a choice between
the two hypotheses. The data do show, however, that herds were more
dispersed at the beginning of the field season. Such variability in the numerical composition of herds suggests that some groups are incorporated
into the herds at some times and break away at others.

Year III   In this observation period, the number of individual animals
in groups or herds was, typically, one, two, four, five, eight, or twelve.
Comparing groups and herds as to age composition, behavioral peculiarities, and presence of marked animals, we determined that several specific herds repeatedly passed through the observation area.

In the western part of the central sector, a herd of five animals was
observed about thirty times, across the entire season, both from the shore
and from the catamaran. This herd had one calf (not always noticed
when the group was observed from a distance) and one dolphin that was
much larger than the rest of the group. This large animal was the first to
approach the catamaran and often swam in front of the others, particularly when the herd was fleeing. The dolphin with a white spot on the
dorsal fin was observed twice in this group. Animals in this herd tended
to form two stable groups, one consisting of three animals including a calf,
and the other of the large dolphin paired with another animal (fig. 1.6).
Group structure in this herd tended to have a constant composition.

In the central part of our observation district, we could distinguish
three herds. Herd 1 was observed between July 6 and 16. On August 10,
four different observers obtained identical descriptions of this herd. The
herd consisted of ten to twelve animals including two calves. A typical
"spiral" movement of the herd during hunting was observed. Another detailed observation of this herd was made on July 14 when it passed
through in "march" formation or quiet travel without hunting (fig. 1.7).

On August 10, three pairs were identified in this herd. Also, in front of
the herd was a large dolphin; the distance between this individual and the

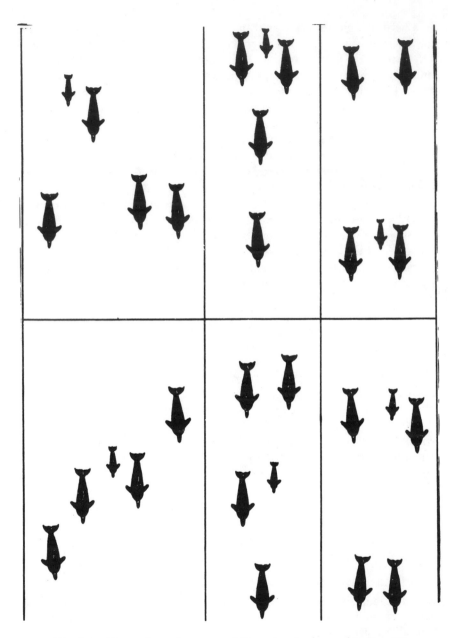

Fig. 1.6. Formations of a group of five animals observed in Year III.

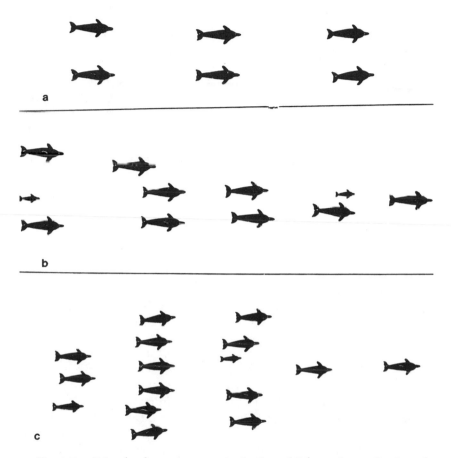

Fig. 1.7. Other herd arrangements in the "march" formation. a. Portion of Herd I; b. Herd I; c. Herd II.

first pair was constant. At the end of the herd was a group of three dolphins, consisting of two adults with a calf between them. Usually, calves stay with adult animals during a hunt, and their positions in the herd are fixed; in this herd, the two calves were absolutely independent when the rest of the animals were busy hunting. On August 10, after a normal hunt, one of the calves was seen swimming about 20 m in front of the herd, slapping its tail and turning on its back, for about ten minutes.

On August 12, a herd of six to seven dolphins divided in the following manner while hunting: first, there was a group of two adults, then a second group of two to three adults, and then a third group of two calves.

Fig. 1.8.   Abnormalities of animals in Herd II, central district.

The first and second group were separated by 200 to 250 m while hunting, and the third group, the two calves, was always in between the adult groups. This distribution of animals lasted about fifty minutes. We considered the calves' behavior to be play: it consisted of continuous jumping and flashing of heads and tails; the intervals between respirations were never longer than five seconds. Once every five to seven minutes, an adult dolphin would surface next to the calves and then disappear. On the basis of the characteristic independence of these two calves, we feel that these six to seven animals should be considered part of Herd I. Probably, the dolphins previously observed on June 12 and 13 were also a part of this herd, since the school numbered about ten animals, two of them calves, and since tandem formations were characteristic of the herd, and a group of three dolphins, two adults and a calf, brought up the rear.

Herd II was seen on August 17 and 18, August 20 and 21, and September 8, 25, and 30. The closest approach to shore by this herd was on August 20, when fifteen to seventeen animals were seen. The "march" formation of the herd is shown in figure 7c. Three groups, with a distance of 30 to 50 m between them, were clearly determined. One calf and one juvenile were seen. In front of the herd a large dolphin, separated from the one or two others that followed it, was constantly seen. Two animals in this herd had recognizable body abnormalities (fig. 1.8). The number of animals in this herd varied; for example, on September 8, a marked dolphin, which had been noted in previous years, was seen in the herd where the group of five animals was observed. Later, this group was absent from the herd with the marked dolphin. One may suppose, therefore, that groups of bottlenose dolphins retain a constant composition only for a limited time, as has been observed in other species of dolphins (Bel'kovich and Yablokov 1965). Obviously, new births and approaching sexual maturity in dolphins bring about changes in old groups and lead to formation of new groups.

Toward the end of August and in September, three groups were clearly

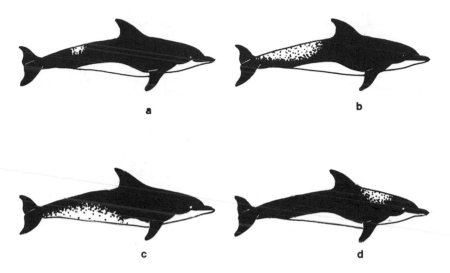

Fig. 1.9. Naturally marked animals in Herd III, central district.

seen in this herd: one of two to three dolphins and two others of approximately six each. On September 28–30, this herd, with a marked dolphin and one calf, again consisted of eight to ten animals. During this period, there were also brief sightings of groups of five to six, nine to ten, and one and two dolphins. Presumably, all these dolphins belonged to Herd II and constituted its groups. For a period of a week before August 17, when Herd II first appeared, there were no dolphins in the observation area.

On August 24, a new herd appeared which differed sharply from Herd II. Herd III was observed on August 24–26 and September 29; it consisted of eight to ten animals. During each of the three encounters, the animals were engaged in intensive hunting, hence the march formation was not seen in this school. During hunting, the herd dispersed over a huge area. The number of animals in this herd increased gradually across the three encounters. Conspicuous in this herd was a group of four animals (with one juvenile) divided into two pairs, an animal with a sickle-like dorsal fin being in one of the pairs. A conspicuous characteristic of the animals in this group was the presence of light-colored portions of skin. The dolphins may have had a skin disease, or such pigmentation might be inherited and the animals related. Four marked animals from this herd are shown in figure 1.9a–d. The marked animals could be seen clearly each day. Since these marked animals were not seen again from this observation post, it may be assumed that this school appeared in the central region in Year III only during three days in August. However, the

Table 1.1.   Year III Observations of Resident Dolphins

Number of Animals in Group	Number of Observations		
	TOTAL SIGHTINGS	GROUPS INCLUDING A JUVENILE	GROUPS INCLUDING JUVENILE "KIDDO"
1	16	6	3
2	18	9	3
3	16	13	8
4	12	12	7
5	2	2	2
6–7	1	4	—
8	1	1	—
Total	66	47	23

group of four animals including a juvenile and the animal with the sickle-shaped dorsal fin was very similar to a group of dolphins frequently seen in the eastern part of the district. The dolphin from Herd III shown in figure 1.9b was also observed with another pair of adult dolphins in the eastern region of the observation district.

One of the observations of this herd, on August 25, clearly showed that animals in a given herd may disperse over a large area and cannot always be seen all at once. At 1205 hours, a herd of eight animals was noted in the western part of the observation area from the catamaran. After twenty minutes, the herd passed by a shore observation post, OP3, where eight dolphins were observed. At OP4, only five animals could be seen, and at OP1, at 1303 hours, first two and then in the distance one more dolphin were sighted. By 1350, this herd was observed in the eastern part of the observation area moving in the opposite direction; and after 1400 hours, all eight to nine animals were again at OP1 in the central part of the zone.

Observations in the eastern part of the district in the third year were made June 4 to August 9; bottlenose dolphins were seen sixty-six times. (See table 1.1.) One group of four animals appeared to inhabit the area permanently. Studied in greatest detail was a dolphin of 1.5 to 1.7 m in length, apparently of 1.5 to 2 years of age (nicknamed Kiddo). Fifteen observers saw this animal and gave similar descriptions. The animal was light gray; it had a light spot with fuzzy edges behind the rostrum on the top of the melon and a white scar of about 15 cm in length from the blow-hole all the way back to the dorsal fin and down the left side. On the right

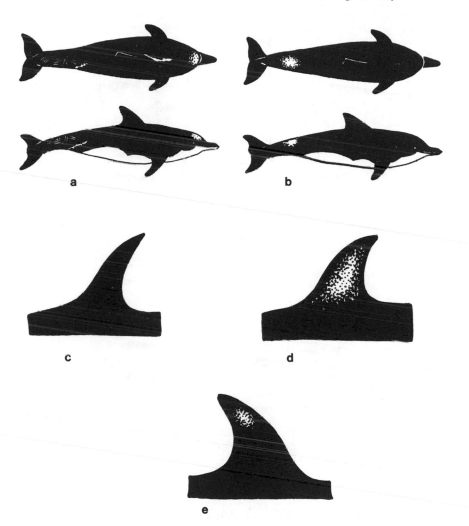

Fig. 1.10. Naturally marked animals in the western district. a. Kiddo;
c. Sickle.

side of the body behind the dorsal fin, another light gray scar was seen.
Behind the dorsal fin on the right was another light gray streaked scar
suggesting stitches, and the top of the caudal peduncle was striated with
small scars, suggesting dolphin tooth marks (fig. 1.10a). Kiddo's partner
was a large, dark bottlenose dolphin with a high, narrow, sicklelike dorsal fin, designated Sickle (fig. 1.10c). Kiddo and Sickle usually stayed together but not as stably as the second pair in this group, a pair of adult

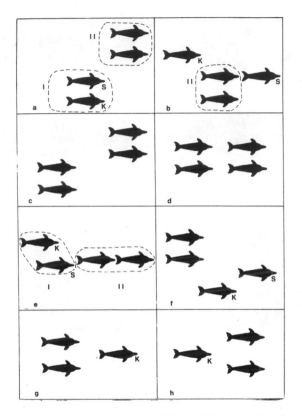

Fig. 1.11.   Formations of a recognized herd of four dolphins in the western district. I: Group consisting of Kiddo (K) and Sickle (S); II: Group consisting of a marked pair. Details in text.

animals approximately equal in size. One of them had a round, light spot on the caudal fin and a dorsal fin that was whitish on the left side, as if scraped (fig. 1.10b, d). This mark was seen four times by five different observers.

These dolphins appeared at various times during daylight, from 0500 until 2030 hours. Usually, the animals were observed in the morning hours (from 0500 until 0900, 27 times). Only one-tenth of all the records were made after 1700 hours.

Figure 1.11 shows formations used by this group of four animals during hunting or while reconnoitering the bay searching for fish. While searching for food, the dolphins divided into two pairs (fig. 1.11a). Often, Kiddo or Sickle separated from the group, but not once was either of the second pair of dolphins observed to detach itself (apart from instances of intensive, diffuse hunting when the formation was completely

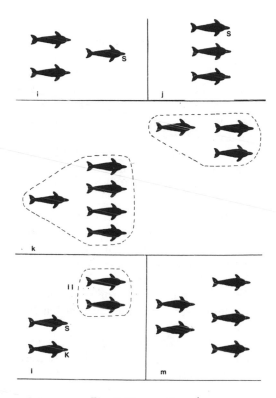

Fig. 1.11   continued

disrupted). If one animal had left, the three remaining animals arranged themselves as shown in figure 1.11g-j. In many sightings from shore, Kiddo was first observed; it would then disappear, only to reappear with Sickle, and then, after a little while, the second pair would appear from deeper water, as if led in by Kiddo.

We were interested in the interaction (if any) between this group of four dolphins and other dolphins. On four occasions, we observed more than four dolphins in the region simultaneously. For instance, on July 10, a herd consisting of eight animals (three and five) was observed frequently (fig. 1.11k). This school did not come closer than 400 m to the shore and probably was not connected with the observed group. On July 16, a fifth dolphin was observed twice with the local group; it joined the second pair of animals (fig. 1.11m). Once, on July 20, the group of four emerged from the bay and moved along the coast to the central part of the observation area, where they were joined by another two to three dolphins. The group of four, however, was clearly visible in the herd of six or seven that had formed.

We did not observe these four animals during the operation of OP7 in the central part of the observation area. Later, however, once on August 25 and twice on August 26, a school of eight to ten dolphins was observed at OP1, all of which were conspicuous for light markings. One of a pair had a round white spot on the caudal peduncle and apparently was the marked dolphin from the second pair of the four observed in the eastern part of the area (fig. 1.10b). Another had a completely white caudal peduncle. This animal was observed from the motorboat on August 4, in a group of three to four animals moving toward OP1. It can be assumed, therefore, that the group of four dolphins observed continually in the eastern area was part of a larger herd of dolphins and maintained contact with them.

To complete the picture, it should be added that a single animal was sighted in the central part of the observation area six times between August 21 and September 14. It often stayed in one area and moved slowly about. Once it swam in a small circle for several hours. The behavior of this single bottlenose dolphin differed markedly from the norm and did not resemble the behavior of a scout. [*Editor's note:* In captivity, this circling behavior is sometimes seen in a dolphin that has lost a close companion.—K.P.]

Sometimes we were able to observe herds composed of much larger numbers than mentioned in this study; these probably were not herds but schools. The aggregation of fifteen animals observed from August 17 until September 8 might be considered such a school. On August 14, a school of approximately twenty-three animals was sighted, consisting of two aggregations, one of fifteen animals and one of eight. They moved at a distance of several dozen meters apart.

HARBOR PORPOISES AND COMMON DOLPHINS

Harbor porpoises (*Phocoena phocoena*) normally appeared in the observation area either as solitary animals (36% of observations) or in a small group of two (32%), three (14%), or four (7%) animals; sometimes, groups of six to eight animals (7%) were seen, and in May of Year II, a school of twenty to twenty-five harbor porpoises was observed. Adult-calf pairs were observed in 10 percent of the cases.

Solitary animals were seldom seen. We frequently observed the incorporation of small groups into big ones, usually when the animals were hunting for food. These larger groups usually broke up quickly. For example, on July 17, a school of eight porpoises repeatedly broke up into groups of two or three and reformed again; on July 25, a group of five would break up into two groups and then recombine.

During Year II, several harbor porpoises with natural marks were observed; unfortunately, we never saw these animals again. But in Year III, we saw a marked harbor porpoise six times during two weeks—a juvenile with a jagged torn dorsal fin, always with a large animal (presumably its mother). Sometimes a third, even larger animal was also noted with these two. It behaved quite independently, swimming away, sometimes out of sight, and then returning and traveling with them. This was probably always the same individual, but it was impossible to say as it had no identifying markings. On August 5, mating was noted in this group of harbor porpoises.

In addition, a harbor porpoise with a dorsal fin similar to that of a bottlenose dolphin was seen three times. These data allow us to assume that a local population of harbor porpoises inhabits the southern part of the peninsula.

Herds of common dolphin (*Delphinus delphis*) consisting of from several individuals up to several hundred were encountered frequently. During migratory movements, groups of two to four animals were clearly distinguishable and were often observed leaping in sequence, one after the other. These dolphins never came close to the shore.

CONCLUSION

Analysis of the data on numerical composition of bottlenose dolphin herds, their age structure, the composition of groups, their behavioral peculiarities, and observations on marked animals showed that almost always the same herds or parts of herds swam through the observation district. Observations during Year III are clearer than those of Year II and show stability and permanence. The main structural pattern of this bottlenose dolphin population consists of groups of animals incorporated into herds; a series of herds form the local population. A herd of bottlenose dolphins is characterized by relative stability and long-term independent existence.

The group structure of dolphin herds is a very sensitive mechanism for providing optimal conditions for spatial and temporal use of the environment as well as adequate hunting methods under changing environmental conditions. Nevertheless, we can assume the presence of a constant "central" group (one or several) that makes up the core of the herd. Other groups may possibly attach themselves to this core. In addition, individual groups may begin to exist independently and form the beginning of a new herd. These interesting problems in population dynamics may be illuminated in further investigation.

The observations by Tayler and Saayman of bottlenose dolphins in the

Indian Ocean (Tayler and Saayman 1972) were primarily ecological in nature but revealed that the dolphins swam in schools of several hundred animals and that these schools consisted of distinctly separate herds of twenty to fifty animals. Groups of two to five animals each were noted in the herds, as with the Black Sea dolphins. Also noted in the Indian Ocean dolphins were animals with natural markings which reappeared in the observation area over a period of several years.

Data on harbor porpoises and common dolphins are unfortunately very scanty. However, observations on harbor porpoises provided interesting material on herd structure and demonstrate the possibility that some subpopulations of this species have a local distribution.

SEARCHING AND HUNTING BEHAVIOR IN THE BOTTLENOSE DOLPHIN (*Tursiops truncatus*) IN THE BLACK SEA

V. M. Bel'kovich, E. E. Ivanova, O. V. Yefremenkova, L. B. Kozarovitsky, and S. P. Kharitonov

INTRODUCTION

Our knowledge of dolphin feeding behavior obtained in oceanariums has little resemblance to their hunting behavior under natural conditions. A few remarks on hunting in various delphinids (porpoises, pilot whales, common dolphins) are given by W. E. Evans and J. Bastian (1969). A. G. Tomilin (1957) gave an interesting and quite detailed description of Black Sea bottlenose dolphins surrounding a school of fish, bringing it to a halt, and consuming it. He also describes a group of bottlenose dolphins that had driven fish against the shore. D. A. Morozov (1970) described hunting by bottlenose dolphins in the Crimea. A detailed description of bottlenose dolphins hunting in the Indian Ocean is given by Tayler and Saayman (1972), who observed two types of hunting, by groups scattered along the shoreline and by herds driving the fish into one clump and seizing them from the sides and from below.

BOTTLENOSE DOLPHINS

During our expeditions, we observed and recorded several hundred episodes of hunting. The dolphins demonstrated a great complexity and variety in their feeding behavior; but, like other animals, they also obtained their food with some standardized procedures. Observations of dolphin hunting situations accumulated over three years made it possible to single out standard techniques in the fantastic interweaving of many elements of hunting behavior.

During Year I, small groups of dolphin "scouts" were observed swimming close to shore. Herds seldom appeared in the inshore observation area, usually appeared suddenly, and then only if the scouts discovered a substantial amount of food (Bel'kovich et al. 1975).

In Year II, the situation changed drastically. Groups of dolphin scouts were rarely seen; instead, the whole herd spent a lot of time in the observation district. Moreover, their collective hunting was performed entirely differently. Instead of groups of scouts, we frequently observed individual dolphins swimming about 100 to 700 m ahead of the herds. These were not solitary dolphins but what we called "second year–type scouts."

Finally, in Year III, the dolphins' behavior differed again. They usually swam 300 to 800 m from shore. Apparently, they were trying to avoid a stationary net that had been set by fishermen within the observation district during June and July. The entire herd of bottlenose dolphins hunted as a unit; however, in the second half of that summer, groups of dolphin scouts were seen as far as 2 to 4 km from the main herd, a behavior that differed considerably from the preceding years.

Detailed analysis allowed us to construct a picture of the basic types of hunting behavior in bottlenose dolphins. These behavior patterns were often combined with one another and, interestingly, might be alternated even during a single hunt. This clearly demonstrates the plasticity of the exploratory reactions of the dolphins, which permits them to react effectively to changes in such factors as the species, behavior, and number of fish, the weather, the presence of boats, the behavior of people, and many other variables, as well as the number of dolphins participating in the hunt and the coordination of their own actions.

Hunting takes place in three phases: the search, detection, and the catch. Since detection of the fish is very difficult to separate from the catch, we will consider hunting in two main phases, the search and the catch.

Searching for Fish

Whole-Herd Searching   Searching by a whole herd of dolphins, as indicated by changes in the direction of movement, was observed on the occasions listed in table 1.2. Sometimes in whole-herd searching, the herd moved in front formation, side by side in a straight line (table 1.3). Most often, the search was conducted along a complex trajectory (table 1.4).

Sometimes, the herd made complex, purposeful patterns in the sea,

Table 1.2. Observations of Whole-Herd Searching

Date	Time	Number of Dolphins
Year II		
19.6	06.23–06.54	5
20.6	11.05–11.38	9
20.6	13.40–14.45	9
27.6	05.30–08.30	10
30.6	12.10–12.35	8
06.7	07.50–08.20	6
08.7	12.01–12.17	7
16.7	07.12–09.16	8–9
16.7	11.45–12.04	7
26.7	11.00–11.45	7–10
02.8	17.53–18.17	8
04.8	10.05–10.27	4
17.9	08.20–10.49	12
17.10	10.30–11.00	8
Year III		
06.6	14.55–15.22	10–12
09.6	16.45–17.27	8–9
19.6	08.39–09.20	5
12.6	18.20–19.05	10–12
13.6	05.38–09.22	7
13.6	15.04–15.35	9
09.7	13.42–15.40	6
31.7	13.05–13.58	8
24.8	16.12–16.40	5–7
25.8	14.00	8–10
28.8	11.41–12.21	10–12
28.8	17.24–17.51	8

Table 1.3. *Whole-Herd Searches in "March" Formation*

Date	Time	Number of Bottlenose Dolphins
Year II		
31.7	12.44–19.09	18–22
06.8	12.26–13.12	22
Year III		
22.6	12.47–12.57	8–9
18.8	08.11–08.45	6
25.8	14.00	8–10

Table 1.4. *Whole-Herd Searches in Complex Patterns*

Date	Time	Number of Dolphins	Nature of Search
Year II			
22.6	12.47–12.53	8–9	In a spiral
17.6	11.55–12.35	6	By circling
19.7	10.11–10.53	32	By squares
21.7	12.22–13.15	11	By squares
04.8	08.51–09.10	6	Stepwise
19.8	11.50–13.15	13	Stepwise
17.8	10.53–11.22	8	Stepwise

moving synchronously in a spiral or circle with all members acting in unison; we presumed that they were trying to surround fish deliberately. The level of cooperation by the dolphins was also striking when changing direction in the front formation. All the dolphins would come to the surface and arrange themselves in a row; not until that was accomplished would they simultaneously disappear under water. The dolphins that then reappeared would bob like cork floats here and there in the waves; in 15 to 30 seconds, a straight row of dolphin backs and fins would be visible, and in the next moment, a synchronized dive was again executed which lasted 30 to 70 seconds. This type of behavior might occur not only during the

Table 1.5.   Searches in Small Groups

Date	Time	Number of Dolphins	Number of Groups
Year II			
20.6	11.05–11.33	9	2–3
06.7	11.18–11.37	14	4
16.7	14.35–14.45	7–9	2
25.7	09.00–09.52	14	2
26.7	11.00–11.45	7–10	2
31.7	12.14–13.09	18–22	5–6
01.8	10.34–11.30	13.	2
06.8	12.26–13.35	22	5–6
19.8	11.50–13.16	12–13	4
19.9	10.20–11.12	10	3
03.10	10.40–11.05	10	3
13.10	10.38–10.56	11	2
28.10	13.25–14.15	13	2
Year III			
05.6	15.10–16.39	3	2
08.6	11.07	7	2
12.6	14.03–16.02	8	2
13.6	05.38–09.22	7	2
26.6	12.09–12.58	4	2
19.7	15.55–16.30	7	2
10.7	05.30–08.40	11	3
10.7	13.10	11	3
26.7	11.41–12.21	10–12	3–4
08.9	13.55–15.04	20	5–6

search phase of hunting but also during the catching phases, whenever dolphins were foraging on small pelagic fish such as mackerels and sprats.

Group Searching   In addition to whole-herd searches, searching could be conducted by individual groups. This type of searching allowed the dolphins to spread out and cover much wider areas; the configuration

Table 1.6.   Diffused Searching

Date	Time	Number of Dolphins
Year II		
06.8	12.26–13.35	22
Year III		
08.7	11.07	7
10.7	05.30–08.40	11
18.7	12.40	4
29.7	09.32–10.02	12–15

of the herd changed continually due to the movement of the groups. This type of searching was observed on the occasions listed in table 1.5. The groups "combed" the water by using diagonal and front structures to search for fish (and also sometimes during the catching phase). We also observed a diffused search pattern in which the dolphins scattered through a specific water area moving in different directions (table 1.6).

Hunting was usually a very dynamic process; techniques might change from one moment to another. For instance, a herd might enter the bay in front formation and then break down into small groups and spread out over a wide area of water. When one of the dolphins located a fish and moved toward it (detection phase), usually the rest of the herd approached that individual immediately and general feeding began.

Dolphin Scouts   The search for fish by dolphin scouts is a specific type of group search. Owing to its distinctive characteristics, we classified this type of search separately. Hunting with scouts usually proceeded as follows: a group of 2 to 4 dolphins would travel 20 to 300 m offshore, following the contours of the coast. That the dolphins were searching for fish was indicated by detailed investigation of the shoreline, sharp changes of speed, and changing of respiratory rhythm from 15-second intervals up to 6 minutes (while diving). Often the rest of the herd swam on a parallel course up to several kilometers farther out to sea. When scouts detected fish, the rest of the herd immediately joined them and hunting began.

Occasionally, scouts detached themselves from a herd passing by to investigate fishing nets in the bay in the observation area and then rejoined the herd. In Year I, the researchers frequently accompanied and tracked dolphin scouts along the shore on bicycles or horseback (Bel'kovich et al.

Table 1.7.   Searching by Scouts

Date	Time	Number of Dolphins in School	Number of Scouts
Year II			
15.6	20.41–20.50	9	3
27.6	08.20–12.30	10	1
24.7	16.20–18.40	10	1
25.7	05.02–05.16	22	2
31.7	05.51–06.50	12–13	1
24.8	14.52–15.18	12–14	3
07.10	13.03–13.43	12	2
11.10	10.46–11.20	11–12	2
Year III			
05.6	15.10–16.39	—	3
11.6	12.33–13.07	8–9	1
11.6	09.10–10.44	—	3
12.6	18.20–19.03	10–12	—
13.6	05.38–09.22	9–10	—
04.8	13.43–14.23	10	—
26.8	11.41–12.21	10–12	—
31.8	08.55–09.11	—	—

1975). Scouting was observed numerous times during Year II and III, but the types of reconnaissance differed (table 1.7).

Usually, the scouts investigated the shoreline and, obviously, foraged on the fish they found. As a rule, the dolphin closest to shore was most successful. The prey was usually gray mullet (*Mugilus sp.*), which the dolphin would drive into the shallows and force to the surface, or against the beach, without allowing the fish to escape sideways or downward. We called this type of hunting "attack against the shore" (fig. 1.12). The role of dolphin scouts is not equal; the one closest to shore works hardest and also has the most opportunity to fill up.

The reason for the change of functional roles in dolphin scouts across three years has not been determined. Also, we wondered if one group of scouts is sometimes substituted for another. Based on many hours of ob-

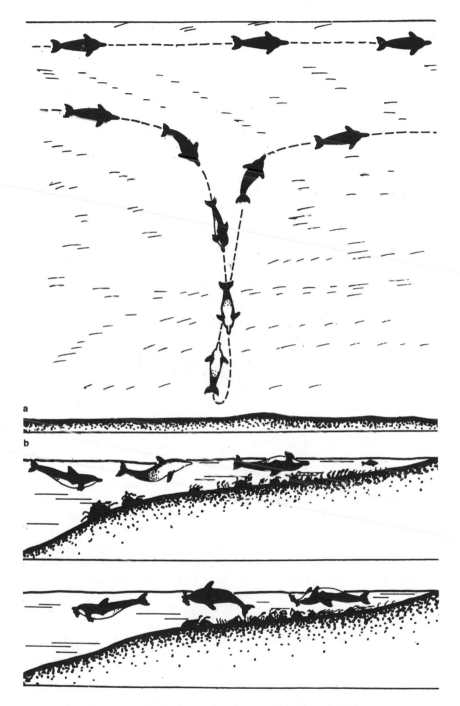

Fig. 1.12. Attack against the shore. a. Top view; b. Side view.

servations, during which several animals with natural marks were observed, we assumed that different dolphins played the scout role (Bel'kovich et al. 1975) and therefore that strict specialization within dolphin herds is not the case.

Several observations allowed us to assume that dolphin herds received information from the scouts concerning the number of fish found (Bel'kovich et al. 1975, Bel'kovich and Dubrovsky 1976). For example, only part of the herd might detach itself to join the scouts. Similar behavior has been described for the Indian Ocean bottlenose dolphin (Tayler and Saayman 1972). One of the dolphins detected a school of sardines, and immediately the whole herd of about 200 animals approached the place and started to hunt. The authors assumed that dolphins were able to detect the location of a school of fish by listening to the echolocational signals of the scouts.

We assume that several observed formations, such as the herd swimming in squares, spirals, or a stepwise pattern, are related to searching for fish. Since these formations were not monitored acoustically, however, we cannot unequivocally assign them only to searching. In several cases, these techniques were very similar to those used for catching small pelagic fish such as jacks and sprat. But since we could not observe or monitor the actual feeding, either acoustically or visually, we will regard these formations as search behavior for the time being.

Catching Fish

Let us now consider the final phase of hunting behavior, which usually results in catching fish. We should keep in mind that what we see in oceanariums, where dolphins eat dead fish, is quite unlike what occurs in the wild. The main difference is that to catch a fish, a wild dolphin has to move actively and to interact very closely with the other members of the herd or group. Analysis of our many observations of a wide variety of hunting behavior suggests that the success of any catch depends on restricting the activity and maneuverability of the prey. This is achieved by several methods, which we will consider in more detail.

The "Carousel" and "Kettle" Techniques While hunting for such pelagic species of fish as jacks, the herd usually surrounded the school, forcing the fish to travel in a circle that was gradually tightened by the dolphins swimming in larger circles around the moving mass of fish. This technique, which we called the "horizontal carousel," was used often (fig. 1.13a; table 1.8). The number of dolphins involved in any specific episode of hunting by means of the carousel formation probably indicated the abundance of fish (so far, we have not been able to ascertain

Fig. 1.13. The "carousel" hunting formation: a. Horizontal carousel;
b. Vertical carousel.

Table 1.8.   Horizontal "Carousels"

Date	Time	Number of Dolphins
Year II		
09.6	16.49	8–9
29.6	09.12	17
24.7	16.20	9–10
31.7	05.51	12–13
01.8	19.34	13
17.8	10.57	8
18.8	07.45	8–10
19.8	10.31	8
06.10	16.20	10
07.10	13.03	12
10.10	11.23	10
Year III		
06.6	14.55	10–12
12.6	18.20	10–12
10.7	05.30	11
10.7	18.51	8
04.8	13.41	10
17.8	05.30	6
24.8	09.42	5
25.9	08.31	13–15

which dolphins took part in carousels when not all were present; it could be the hungriest, the most dominant, or simply those that found the fish first). Sometimes several carousels took place one right after the other; possibly some fish escaped the first encirclement, or possibly the fish sometimes occur in several small schools.

Two other fish-catching formations were dubbed the "vertical carousel" (fig. 1.13b) and the "kettle" (fig. 1.14). These formations were often observed after a horizontal carousel, apparently in the presence of significant congregations of fish. The animals tightened their circles and then started diving either under the mass (vertical carousel) or from several directions into the mass of fish (like stirring a boiling kettle). These tech-

Fig. 1.14. The "kettle" hunting formation.

niques, which sometimes involved only a very small part of the herd, were observed frequently (table 1.9).

The kettle formation and both types of carousels were characterized by swift swimming, noisy surfacing, and brief appearances at the surface. Small fish were impossible to see in the dolphins' mouths, but big mullet were often seen. Hunting by means of the carousel formation was first described by Morozov (1970) in Crimean waters. Similar hunting techniques were observed in the Indian Ocean bottlenose dolphin by Tayler and Saayman (1972). Tayler and Saayman also observed a herd of dolphins that had trapped a school of sardines in a bay; some acted as guards, preventing the fish from leaving the bay, while the other dolphins fed on the school.

Carousel techniques can be conducted by the herd as a whole or by its parts (groups). A school of fish could be surrounded by several means, including (1) the dolphins curved around a school of fish from one side and then closed the circle by means of a carousel (on Oct. 6 and 8, Year II: fig. 1.15); and (2) dolphins surrounded a school simultaneously from both sides in a fork formation and then formed a closed carousel (e.g., Oct. 7, Year II: fig. 1.16).

Table 1.9.   Vertical "Carousels" and "Kettles"

Date	Time	Number of Dolphins	Type of Activity
		CAROUSEL	
Year II			
06.8	12.6	22	Vertical carousel
18.8	06.35	8–10	Three carousels by 2, 3, and 4 dolphins
24.8	14.52	12–14	Three carousels by 2, 3, and 5 dolphins
19.9	10.35	10	Carousel twice
Year III			
05.6	10.26	2	Carousel twice
05.6	15.10	3	Carousel twice
07.6	07.19	8	Carousel
09.6	16.45	8–9	Carousel
10.7	05.30	11	Vertical carousel of 5–6 dolphins
10.7	16.15	8–10	Vertical carousel by 6–8 dolphins
16.7	09.30	3	Vertical carousel
28.7	11.12	2	Vertical carousel
		KETTLE	
Year II			
19.6	05.06	5	Kettle
02.8	05.38	7	Kettle
19.8	13.05	15	Kettle of 8 dolphins
23.9	11.56	10	Kettle of 4 dolphins
06.10	16.25	10	Kettle of 3 dolphins
01.10	13.03	12	Two kettles and then three more
13.10	10.38	11	Kettle
17.10	10.30	8	Kettle
17.10	14.56	8	Kettle
Year III			
06.6	15.05	10–12	Kettle of 5–6 dolphins
09.6	16.45	8–9	Several kettles
12.6	14.03	8	Two kettles of 4–5 dolphins
10.7	13.26	12	Kettle of 4–5 dolphins
04.8	13.43	10	Kettle
20.8	16.05	10–12	Two groups, two kettles
26.8	11.41	10–12	Kettle of 6 followed by kettle of 3 dolphins
27.9	07.45	8–10	Kettle of 5–6 dolphins

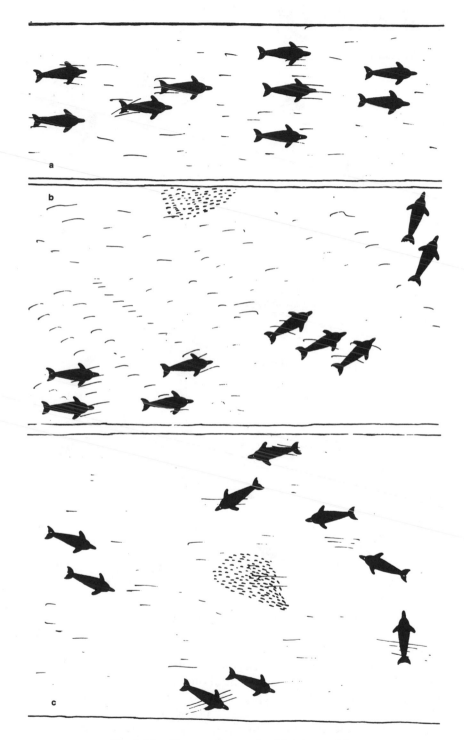

Fig. 1.15. Surrounding a school from one side.

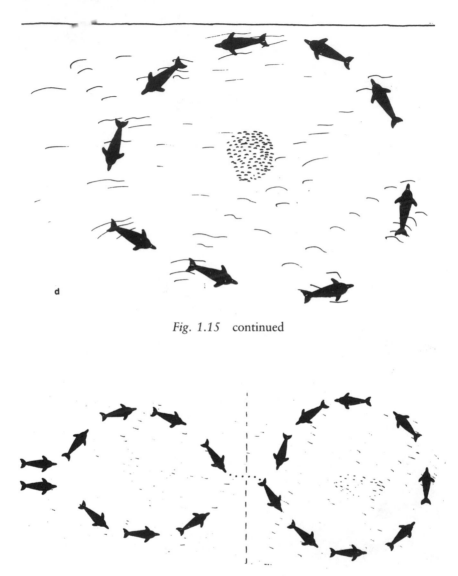

Fig. 1.15 continued

Fig. 1.16. Surrounding a school by bracketing it.

The size of the dolphins' carousel once reached up to 150 m in diameter (June 12, Year II) but also could be about 50 to 70 m across (June 6, Year III) or even 25 to 50 m (Aug. 17, Year III). The size of the circle was normally determined by the initial size of the school of fish. As mentioned, not all dolphins in the herd necessarily participated in a carousel

Table 1.10.   Carousels in Which Only a Portion of the Herd Participated

	Number of Dolphins	
	---	---
Date	TOTAL	PARTICIPATING IN THE CAROUSEL
Year II		
01.8.75	13	5
17.8.75	8	4
Year III		
18.8.76	8–10	4–6
Year II		
19.8.75	8	3–4, 7–8
07.10.75	12	8

(table 1.10). The kettle formation and both types of carousel could form closed circles, usually counterclockwise, or partly open circles.

The "Wall" Method   Besides surrounding or carousel techniques, the dolphins also used an enclosure-type formation to drive schools of fish toward the shore, toward fishermen's nets placed in the water, or toward a "wall" of one or several dolphins. One of these methods, driving the fish against the shore, was used by bottlenose dolphins rather frequently (fig. 1.17c; table 1.11).

During a drive, the whole herd might participate, either in a front formation or in separate groups. Possibly the choice of method was determined by the species and abundance of fish. During the terminal phase of hunting, dolphins moved very swiftly and seized fish in the foam, very close to the shore. For example, a hunt on July 24 (Year II) was described as follows: "The dolphins fly completely out of the water; they fly more than they swim. The fish, a school of large gray mullet, also shoot out of the water for a distance of 4 to 7 m. Often the dolphins and fish fly simultaneously; the dolphins appear to be rained on by falling fish."

As we have indicated, the dolphins themselves could serve as obstacles to the fish. Table 1.12 shows data on occasions when dolphins were observed using each other as a wall against which fish could be driven. We also observed wall-type hunting wherein the dolphins chased fish to shore in two almost parallel columns; here, apparently, both the shore and the parallel columns of dolphins functioned as walls (fig. 1.17a). Dolphins played functionally equal roles, regardless of the number of dolphins

Fig. 1.17. "Wall" formation used in driving fish. a, b. Between two groups of dolphins; c. Against the shore.

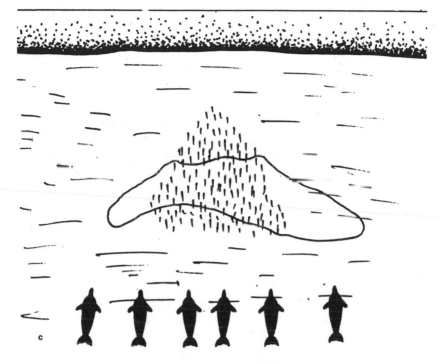

Fig. 1.17 (continued).

Table 1.11. Driving Fish in "Front" Formations

Date	Time	Number of Dolphins	Type of Activity
Year II			
24.7	16.20	9–10	Front
30.7	12.44	9	Front
01.8	10.34	13	Double front
19.8	12.57	15	Front
17.9	09.29	12	Three groups of 3-4-2
19.9	10.20	8	Two frontal attacks
28.10	13.25	13	Corridor
Year III			
11.7	10.34	11	Three-dolphin attack
28.7	11.12	2	Series of attacks
04.8	13.43	10	Frontal attack by entire school

Table 1.12. Driving Fish Against a "Wall" of Other Dolphins

Date	Time	Number of Dolphins	Type of Drive
Year II			
20.6	11.05	9	Two groups
27.6	17.44	9	Two groups
24.7	16.20	9–10	Two dolphins drive fish toward shore
25.7	09.00	14	Two groups (4 and 4) and two groups (3 and 3)
30.7	12.44	9	Front against front
17.8	10.53	8	Triangle
19.8	10.31	8	Triangle
19.8	11.50	13	Two fronts
19.8	12.52	15	Two groups
17.10	10.30	8	Groups of 2 and 6 dolphins
Year III			
07.6	14.20	7	
09.6	16.45	8–9	Two dolphins drive fish toward school of 7 dolphins
13.6	05.38	9–10	"Wall to wall"

Table 1.13. Incidents in Which a Single Dolphin Acted as the "Wall"

Date	Time	Number of Dolphins
Year III		
05.6	15.10	3
06.6	11.07	7–8
11.6	09.10	3
12.6	05.19	5
12.6	14.03	8
13.6	15.08	9
26.6	12.09	4
09.7	13.42	6
10.7	05.30	11
18.7	06.07	6
31.7	05.51	12–13
04.8	13.43	10

Table 1.14.   *Driving Fish Against a Stationary Fishing Net*

Date	Time	Number of Dolphins	Type of Activity
Year II			
27.6	12.49	9	Two groups
Year III			
08.6	11.07	4–5	By means of arc and front
11.6	09.10	3	Singly and by group
11.6	12.33	3	Singly and by group
12.6	14.03	8	By front, twice

hunting. For example (July 24, Year II), a pair of dolphins that failed to drive a school of fish onto the shore drove them instead back to the herd. The entire herd then pinned down the fish in the front formation. A herd could also divide into two approximately equal groups that would then drive the fish against each other (June 27 and July 30, Year II; June 13, Year III.) Even one dolphin could function as a wall, since its presence would slow down the movement of the fish school and prevent large solitary fish from getting away. These techniques were frequently described during the second and third years of observation (see table 1.13).

The wing of a stationary fishermen's net was also utilized as an obstacle against which fish could be driven. In this case, the dolphin groups often drove the fish along one wing of the net toward the shore. Sometimes some animals jumped over the net (May 26, Year II: table 1.14).

Chasing   Sometimes solitary animals or a group of animals chased fish and, without restricting the mobility of the fish school, simply overtook them. Chasing could occur in front formation or in diffused formations, probably when foraging on small fish; both formations might occur simultaneously, with part of the herd pursuing fish in unison and the rest swimming in a diffused pattern. Observations of chasing of fish in frontal formation are presented in table 1.15. Pursuit in diffused formation was also seen several times (table 1.16).

Hunting in frontal, spiral, or circular formations was characterized by animals submerging for periods more prolonged (3–4 min.) than under normal swimming conditions. This was probably related to hunting for bottom fish. Similar types of foraging on the sea bottom (benthic foraging) were seen for other groups of bottlenose dolphins (table 1.17).

Table 1.15. Chases in Frontal Formations

Date	Time	Number of Dolphins	Type of Pursuit
Year II			
27.6	05.30	10	Front consisting of a group of 3 dolphins
05.7	14.30	7	Single file
31.7	05.51	12–13	Front consisting of a group of 5–6 head toward the shore and then toward the sea
02.8	05.38	7	Front
18.8	06.35	8–10	Sprints in synchronism by 2–3 dolphins
18.8	07.45	10–12	Front by entire school toward the sea
19.9	10.20	8	Front

Table 1.16. Chases in Diffused Formations

Date	Time	Number of Dolphins
Year II		
27.6	05.30	10
05.7	14.30	7
02.8	05.38	7
Year III		
26.8	11.41	10–12
17.8	14.27	2–3
08.9	13.55	20

Table 1.17. Benthic Foraging

Date	Time	Number of Dolphins	Duration of Dive, Minutes
Year III			
07.6	14.20	7	1
12.6	07.31	6	1–2
29.7	09.32	12–15	1–3
18.7	06.00	5–7	1.5–2

SOLITARY HUNTING

We were primarily interested in the hunting techniques used by herds or groups of dolphins; however, we were also able to observe hunting by single animals, sometimes acting alone even when other dolphins were not far away. Individual hunters tried to drive the fish either toward shore or toward the wing of a stationary fishing net, to bring the fish to the surface. These types of hunting represented the wall technique but as used by solitary animals. Data on single dolphins using the attack against the shore technique, or using a net as a wall, are given in table 1.18. Table 1.19 shows instances of direct chasing by single dolphins.

The most commonly observed method of hunting by solitary dolphins was the attack against the shore. The dolphin would suddenly change direction and rush toward the shore, close to the surface with its dorsal fin

Table 1.18.   Single Dolphins "Attacking Against the Shore" or the Net

Date	Time	Number of Dolphins	Number of Attacks
Year II			
17.6	10.00	4	1
21.6	10.08	7	1
25.6	08.03	8	2
25.7	18.34	1	1
26.7	19.00		1
30.7	12.44	9 (two groups)	Several by both groups
31.7	05.51	12–13	Double attacks
31.7	12.14	18–22	Several dolphins
03.10	10.40	10	1
Year III			
04.6	18.29		2
	18.33		
10.6	16.00	4	1
11.7	10.34	7–8	1
04.8	13.43	10	2
20.8	05.45	2	2

Table 1.19.   Single Dolphins Chasing Fish Down

Date	Time	Number of Dolphins
Year II		
19.6	16.55	8–9
Year III		
12.6	07.31	1
04.8	13.43	4
24.8	16.08	5–7
26.8	11.41	10–12

showing above the water. At a distance of several meters from shore, the dolphin usually turned upside down and grasped a fish (normally a mullet), then immediately turned and swam back out to sea, sometimes still upside down, tossing the fish into the air once or twice (which facilitated swallowing it head first).

How can we explain the upside down technique so frequently used? One hypothesis is that under shallow water conditions, with good visibility, the dolphin does not use its echolocational apparatus. Fish are detected using acoustical channels, and then during the chase and catch phases, the dolphin uses its sight; since a dolphin's vision field points down, and the fish stays up at the surface, the animal turns upside down to see the prey better. Another speculation is that the dolphin is trying to protect its pectoral fins from damage. A third is that it is more convenient to catch a fish that is swimming at the level of one's forehead, while upside down. Obviously, a combination of all these reasons may be in effect simultaneously; sometimes even in deep water, dolphins caught fish by bearing down on them in upside down or sideways positions.

Jumping during Hunting

Jumping activity (fig. 1.18) was frequently observed. But out of five different types of jumps recognized by us, only two were seen during hunting.

Low Horizontal Jump [*Editor's note:* U.S. terminology: "porpoising."—K.P.].   The dolphin jumps across the surface 30 to 50 cm out of water and reenters without a splash (fig. 1.18). This type of jumping is

Fig. 1.18. Bottlenose dolphin leaps. a. Candle; b. Horizontal leap; c. 180°
turn.

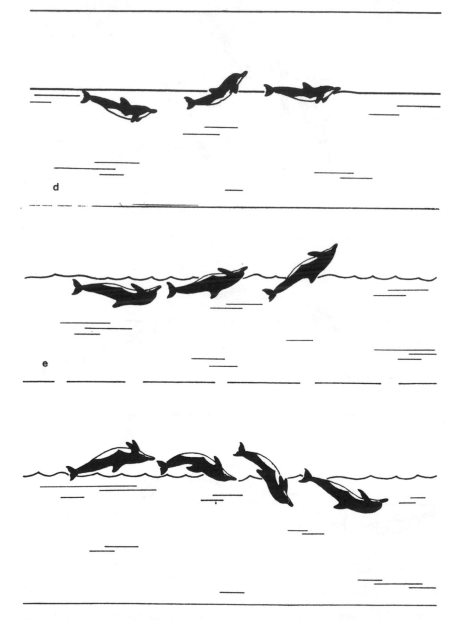

Fig. 1.18 *(continued)*. d. "Log" jump; e. Back jump.

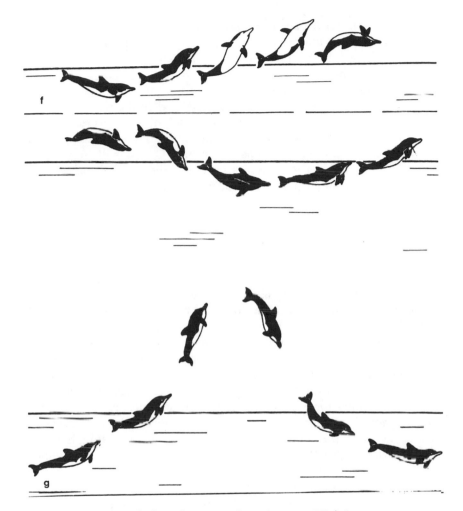

Fig. 1.18   (continued). f. Back jump with 360° turn; g. High jump.

typical in common dolphins (*Delphinus delphis*) while moving at moderate speed and in bottlenose dolphins when moving fast. These jumps were performed either by individual animals or by the herd as a whole, apparently during the chasing of fish, and were typically observed while the animals were moving in line, with the whole herd jumping synchronously.

Jump as a "Log" [*Editor's note:* "Breaching."—K.P.].   In this jump, the dolphin usually comes out of water obliquely and returns back to water on its side. Very often, only half or two-thirds of the body length is out of water; the tail usually remains underwater (fig. 1.18d). This type of jump predominated during hunting; apparently, the animals used the body splash to drive fish. Another type of aerial behavior, called "candles," consisted of vertical surfacing with the head up, followed by immediate vertical submerging (fig. 1.18a). This was frequently observed after successful hunting. In the opinion of V. M. Bel'kovich, the candle behavior may help to settle the consumed fish. Some dolphins performed as many as 5 to 8 candles in a row; once 13 candles were counted in a single individual, each rise lasting as long as one second. Occasionally following a hunt, an unusual behavior was seen in which the rear half of the body was thrust into the air, tail upward, which might also assist to settle the fish.

HARBOR PORPOISES

The differentiation of hunting and migratory patterns in the harbor porpoise (*Phocoena phocoena*) is much more difficult than for bottlenose dolphins, primarily because of the long periods of diving and breath holding (up to 6 min.), during which the animals may travel far away from the initial dive site. Under conditions of good visibility and a quiet sea state, differentiation is possible; then hunting behavior is easy to distinguish from migratory moving. The harbor porpoise's usual traveling pattern consists of a series of short dives (5–10 sec.) after which the animals remain under water for one to six minutes; the sequence is then repeated. During typical migratory moving, harbor porpoises swim in a straight course, normally traveling about 10 km/hour. During hunting, however, the animals usually move much more slowly (2–3 km/hr.). A good indication of searching-hunting behavior was slow movement through an area of several hundred meters in diameter. Sometimes, but not always, such movement attracted seabirds such as cormorants and gulls. The most frequent swimming pattern during such periods of hunting was the shuttle, in which animals moved back and forth in loops of varying length (fig. 1.19). The animals often swam across the current, diving a few dozen meters in one direction, then turning back. Individuals in groups of two or three usually moved synchronously several meters apart. Solitary animals during hunting frequently traveled in loops, or circles (fig. 1.19c, d).

Sometimes, more complex types of hunting interactions were observed within small groups of animals (three to five individuals). For example, on September 25, a single porpoise in a group of six separated itself and moved in shuttle fashion toward the other five, possibly chasing fish toward the others in the wall maneuver. Carousel and kettle formations

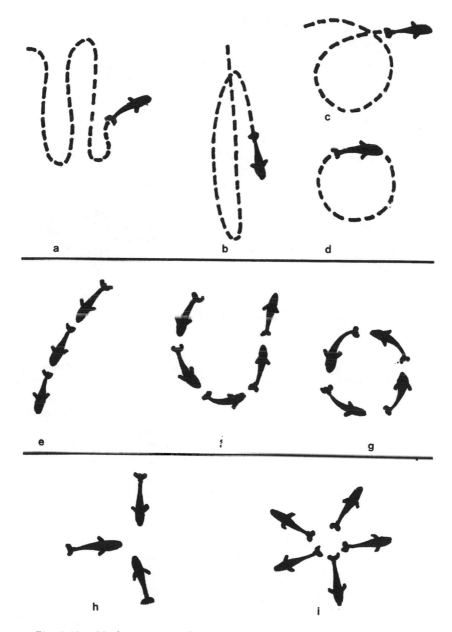

Fig. 1.19. Harbor porpoise foraging techniques. a. S-curves; b. Shuttle;
c. Loop; d. Circle; e, f. Forming carousel; g. Kettle; h, i. Flower.

Fig. 1.20.   Sexual behavior in harbor porpoises.

were also observed repeatedly (fig. 1.19e, g). On October 5, Year III, we observed a wall formation: three animals lined up in a row and swam rapidly toward the shore.

In the hunting area, harbor porpoises were often out of sight for periods of from ten minutes up to several hours; their appearances and disappearances were very hard to document. We did observe that hunting behavior in harbor porpoises alternates with migratory behavior. A group of the animals might move for a considerably long time in one direction and then suddenly stop this "purposeful" motion and start to hunt. Soon after, animals would gather again and continue purposeful movement.

Individual behavioral patterns in harbor porpoises were not as abundant as in bottlenose dolphins. Two types of jumps were observed: a low, horizontal jump and the log-style jump. We frequently saw tails in the air or slapped on the water surface. Animals in groups sometimes leaped in sequence or toward each other. On August 5, detailed observations were made of sexual behavior in a group of three harbor porpoises, one of which was the "marked" calf mentioned previously. Two adults repeatedly leaped out of the water, jumping over each other, and were seen swimming belly to belly for several seconds (fig. 1.20). This sexual behavior was observed for more than five minutes. During this time, the calf swam either behind or ahead of the adults.

COMMON DOLPHINS

Migrational movements of common dolphins are characterized by great impetuousness. The animals usually move very swiftly and often jump out of water. Division into small groups (one to four animals) is very typical. In each group, jumping occurs in sequence, one after another. The hunting behavior of common dolphins was observed while accompanying them aboard a catamaran. In Year II, behavioral patterns similar to the carousel and kettle feeding formations of bottlenose dolphins were observed. Some observations allowed us to assume the presence of scouting dolphins. Jumps in common dolphins are similar to that of bottlenose dolphins, but during jumps, common dolphins curve their bodies less than bottlenose dolphins. The jumps of common dolphins are described in detail by G. Pilleri and J. Knuckey (1967).

CONCLUSIONS

Detailed analysis of several hundred hunting situations reveals great variability in the searching-hunting behavioral patterns of dolphins. We determined and described several previously unknown types of hunting behavior of dolphins in the wild. These behavioral actions—search, detection, and catching fish—are combined with one another or alternated during a hunt, depending on the species and abundance of fish, the climatic conditions, the presence of obstacles, and the number of animals participating in hunting. We are struck by the many methods of hunting and, more important, by the high plasticity of the dolphin's behavior, which allows them to react adequately to changing environmental conditions.

DOLPHIN PLAY BEHAVIOR IN THE OPEN SEA

V. M. Bel'kovich, E. E. Ivanova, L. B. Kozarovitsky, E. V. Novikova, and S. P. Kharitonov

One of the most frequently observed behavioral patterns of bottlenose dolphins is referred to, more or less conventionally, as play. We designate as play those situations in which dolphin activity is occurring which is not directed toward the satisfaction of hunger, migration, or any other utilitarian needs. The conventional nature of this definition should be noted; often, it is difficult to determine where "pure play" is being observed among wild animals and where hunting behavior is mixed with elements of play.

Play is one of the most complicated forms of animal behavior and is difficult to analyze. Arguments about the nature of play are long-standing. Science is lacking not only a single definition of play but even a relatively acceptable point of view on defining this behavior. A great contribution to the theory of play was made by Dutch psychologist F. Boytendeik. He noted accurately that play is usually seen in animals that utilize hunting for their survival. He considered play to be a specific orientational-investigational activity relating to the surrounding environment.

Play behavior is one of the primary forms of activity among bottlenose dolphins living in oceanariums and has been described in detail in a great number of publications (e.g., McBride and Hebb 1948; McBride and Kritzler 1951; Caldwell 1956; Brown and Norris 1956; Andersen and Dziedzic 1964; Tavolga 1966; Bel'kovich and Yablokov 1965; Bel'kovich et al. 1969; Voronin et al. 1971; Krushinskaya 1972; Tayler and Saayman 1973). According to the data of these and other authors, bottlenose dolphins might be equal to apes, or even surpass them, in their variety of play activities and, mainly, in the duration of episodes of play.

Manipulation of objects, chasing, and different types of jumping activities are all basically manifestations of play behavior of dolphins. The monotonous environmental conditions and the "easy" life in oceanariums naturally stimulate and intensify substitute activities such as searching, play, and sexual behavior. It is important to note that dolphins often transform other types of activity into play. For instance, when training dolphins to retrieve different objects, it is not always necessary to use fish as a reward. Apparently, the animals receive positive reinforcement just from playing with the objects (Bel'kovich et al. 1969; Bozhedomov et al. 1975). It should be noted, however, that not all dolphins are eager to manipulate objects.

Very few prolonged observations of dolphin behavioral patterns in their natural habitat have been conducted up to now. It is important to evaluate the true value and significance of play behavior for bottlenose dolphins under natural conditions. Dolphins display considerably less play activity in the wild than in captivity. However, we observed several types of play activity in bottlenose dolphins in the Black Sea: (1) different types of breaching and jumping; (2) animals playing with jellyfish, different species of food fish, and each other; and (3) complex turns, dancing, tail splashing, and other activity. The dolphins also frequently engaged in chasing, attacks, tactile interaction, and posturing as well as extensive play behavior among calves. Individual dolphins in the herd could sometimes be easily identified based on the amount of play and individual peculiarities of play behavior.

Since we seldom observed bouts of behavior we could define strictly as play, we agreed to count occasional episodes of play displayed during migration, searching for fish, and hunting. One of the primary manifestations of play behavior of dolphins apparently is increase of locomotory activity (chasing, different fancy jumps, twisting, sharp turns, roll-overs, and different unusual poses) that does not provide any benefits in hunting. Play and hunting activities can be separated, since all dolphins are normally involved in hunting, and only a few out of all the herd are involved in play activities. Furthermore, behavior related strictly to hunting could be identified by acoustic signals, by the sight of jumping fish, and sometimes by the characteristic behavior of birds.

Elements of play behavior were identified in sixty-six sightings in year II and seventy-eight sightings in Year III. Of course, the "purest" play behavior was seen in calves. However, we are not only going to consider play in juveniles; older bottlenose dolphins also play, in singles, in pairs, and sometimes in groups. Several types of play were determined in the inshore zone where these observations were conducted.

ACTIVE PLAY

Jumps

We classified several types of jumps as active play. Jumping was rarely seen during migrational moving and when searching for food. Most jumps took place during the hunt and immediately after it as well as during rest periods when the dolphins lingered together in one place. We observed many different types of jumps, with complete or incomplete separation from the water, including coming out into the air in vertical and horizontal candles for different heights and lengths of time; bending the body while falling on the side, belly, or back; and striking the water surface with the head, sides, belly, or tail (fig. 1.18).

These jumps could be divided into several types according to their shape and possible meaning. Simple horizontal jumps were usually performed when an animal swam at maximum speed, but a horizontal jump with a slap against water by the side, belly, back, or head apparently could express excitement. A jump with rotation around the body's axis, which we observed three times, was thought to be an individual achievement; we used it as a recognition mark. During our observations on one of the herds of bottlenose dolphins, we frequently observed a "jumping dolphin" (which we also tend to consider to be a single individual) making high archlike jumps about 1 to 3 m out of the water, with a smooth return back into water, rostrum first. Possibly this jump can be considered as "play," but these jumps may also carry some other functions, such as display, because they were occasionally observed when two schools met in the observation area.

Normally, bottlenose dolphins jump separately, one by one, but on several occasions, a group of dolphins jumped simultaneously, for example, while moving in lines or in rows. We observed several simultaneous and parallel jumps by two animals (June 10) as well as synchronous jumps in the candle form by an adult and a calf (October 6) and high archlike jumps by another mother and calf (October 13). On August 26, we observed a long jump by one dolphin across two others.

Complex Turns

Other forms of behavior we categorized as active play consisted of complex turns: rotations without separating the body from the water; unusual swimming patterns (on the side or belly up); showing the tail flukes and caudal peduncle, rhythmical waving with the tail in the air, and splashing with the tail on the water surface; turning upside down with the pectoral fins sticking out of the water; tail-walking; and holding the head above the surface, alternating with fast swimming and sharp turns. Exposure of the tail above the surface was observed frequently during quiet swimming as well, especially after a successful hunt. Standing on the tail, or tail-walking, was first noted on September 8 in a school of fifteen animals. One dolphin suddenly raised itself out of the water by more than two-thirds the length of the body, furiously working the tail to remain in an upright pose for approximately ten seconds (fig. 1.21).

Fig. 1.21.   Tail-walking.

"Chasing"

Chasing is a game in which the animal being chased often jumps out of the water and the pursuer sometimes jumps with it simultaneously. Some of the horizontal jumps described in the previous section were the result of chase games. During Year I, we listened to the sounds of young dolphins during play. It seemed that a typical sound resembling double knocks on a door served as a signal for initiating chasing, bouts of synchronized horizontal jumps, and frontal attacks.

"Frontal Attacks"

In this maneuver, two dolphins swim swiftly toward each other and, when very close, suddenly turn aside, or jump out of water horizontally in what we called a "counterjump" (fig. 1.22A,a). Usually, the dolphins passed each other in midair, but sometimes they made "candle" jumps (fig. 1.22B,b), falling on the tail back into water; and midair collisions were also observed.

"Free-Style" Play

Free-style play was observed when dolphins chased boats and hydrofoils or rode the bow wave on passing vessels.

MANIPULATION[1] OF JELLYFISH AND FOOD FISH

We often saw animals tossing jellyfish or food fish in all directions with their rostrums and tails. When these objects were absent, the dolphins sometimes simulated them by tossing a small amount of water from the mouth or by making tossing movements with the tail.

TACTILE PLAY

Tactile activity was usually associated not so much with play as with sexual behavior. This included animals rubbing bodies, jumps with touching (mainly various crisscross jumps), and rubbing another animal with the head, pectoral fins, or upper parts of the bodies.

---

[1]The term "manipulation" is used arbitrarily here with reference to dolphins.—K.P.

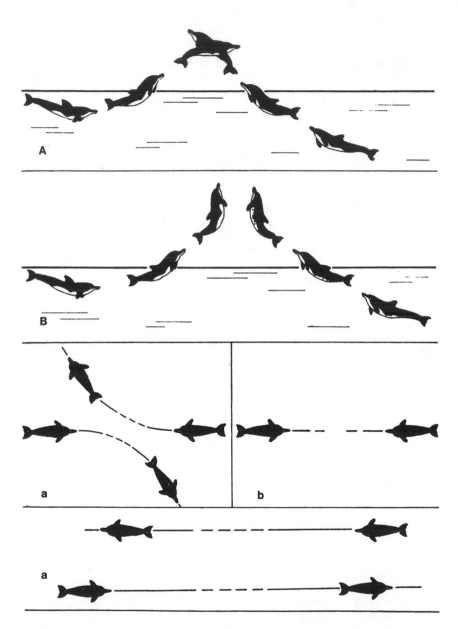

Fig. 1.22. Two versions of "frontal attack" jumps. A, B. Side views; a, b. Top views. Solid line = movement through water; broken line = movement through air.

THE PLAYERS

Having identified a variety of activities we classified as play, let us consider the frequency with which these different elements of play were observed among calves and adults and among single animals versus groups of dolphins.

Incidents of individual play in calves is given in table 1.20. When several calves were involved in simultaneous play, the types of activity were diverse (table 1.21). Four calves were seen in the herd on July 10 playing not only in pairs but together in one group. Play was manifested by jumps, chasing, and sharp turns. Usually, adults did not join juveniles in

Table 1.20.   Individual Play in Calves

Date	Time	Number of Adults	Number of Juveniles	Nature of Activity
Year III				
06.6	14.55		1	Leaps
10.6	10.16	10–12		
10.7	10.45	6–7	1	Leaps
18.8	08.11	11	2	Turns
20.8	12.15	6	1	
26.8	11.41	4	1	
02.9	07.27	10–12	2	

Table 1.21.   Group Play in Calves

Date	Time	Number of Animals	Number of Juveniles	Nature of Activity
Year III				
10.7	05.30–08.40	11	4	Leaps, turns, "chasing"
12.7	06.10–08.40	7	2	Leaps, turns
26.8	11.41–12.21	10–12	2	Leaps

Table 1.22.   Play in Single Adults

Date	Time	Number of Dolphins	Nature of Activity
Year II			
03.6	05.08–05.35	9	Jumps
22.6	12.47–13.06	8–9	Jumps
10.7	05.30–08.40	11	Jumps, spins
10.7	19.34–19.57	8	Jumps, spins, on back, tossing fish with rostrum
10.7	10.34–10.38	8	Leaps, tossing fish with tail
04.8	14.05–14.12	10	Leaps, tossing fish with tail
18.8	08.11–08.45	6	Leaps, tossing fish with tail
24.8	16.00–16.34	5–7	On back, tossing jellyfish with beak
26.8	17.24–17.51	8	Five candles, falling off to one side, tossing jellyfish with tail

their play, but twice in Year I, and again on July 10 in Year II, we saw an adult and a juvenile leaping together.

Adult dolphin play involved single animals, pairs, and groups. The most characteristic play activity of single adult dolphins was leaping (9 times). We also noticed single adults tossing jellyfish or food fishes with their tails (5 times: table 1.22). Individual specialization in various kinds of play offered possibilities for analyzing the structure of groups. If, for example, only one dolphin had the habit of tossing a fish in its jaws while swimming on its back, we can assume that we saw that same dolphin on different days (July 10 at 1934 hours and August 24 at 1600 hours); on other days (June 11 and August 4, 18, and 26) we recognized another "specialist" by the way it tossed objects with its tail.

Play between two adult dolphins included turns, spins, tactile interactions, chasing, and manipulations with beak and tail (table 1.23). Play among adult dolphins in groups included pinwheels or forward flips, jumps, tactile interactions, and surfing on waves (table 1.24).

We think it is important to study acoustic signaling in dolphins during play activity. It is known from the literature that in some species of mammals special signals apparently occur prior to play behavior, as, for example, game mimicry in rhesus monkeys and lowering the thorax and

Table 1.23.   *Play Between Pairs of Adults*

Date	Time	Number of Dolphins	Nature of Activity
Year II			
10.7	05.30–08.40	12	Jumps, tactile interaction, pinwheels
10.7	16.11–16.35	8–10	Jumps, turning, tactile interaction
10.7	19.34–19.57	8	Pinwheel
25.8	12.52	2	Five synchronized "frontal attack" jumps
26.8	17.24–17.51	8	Sequential simulation of tail tossing every 3–7 minutes
27.9	14.40	12	Jumps, tactile interaction

Table 1.24.   *Play Among Groups of Adults*

Date	Time	Number of Dolphins	Nature of Activity
Year II			
22.6	12.47–13.06	8–9	Jumps
10.7	05.30–08.40	11	Pinwheels, jumps
10.7	19.34–19.57	8	Pinwheels, jumps by four dolphins, spine forward
04.8	14.05–14.12	10	Jumps, manipulations with beak and tail
19.8	11.50–12.15	4	Tactile interaction
24.8	16.00–16.34	5–7	Surfing on wave

front legs to the ground in dogs and cats (Loizoa 1966). We have mentioned a sound signal as a possible initiation to play, followed by synchronous jumps. Finally, we conclude these preliminary observations on play by noting that the extraordinary play activity of dolphins can be explained at least in part by the fact that dolphins have practically no enemies and feel free and unfettered in the ocean.

REFERENCES

Altman, S. A. 1962. Social behavior of anthropoid primates: Analysis of recent concepts. In: Roots of behavior, ed. E. L. Bliss. New York: Hoeber.

Andersen, S., and A. Dziedzic. 1964. Behavior patterns of captive harbor porpoise (*Phocoena phocoena*). Bull. Inst. Oceanography 63 : 1316, 1–20.

Bel'kovich, V. M. 1960. Some biological observations of a beluga whale from an aircraft. Zoological Journal 39 : 9.

Bel'kovich, V. M., F. V. Andreyev, S. D. Vronskaya, and A. I. Cherdantsev. 1975. A study of bottlenosed dolphin behavior in nature. In: Marine mammals: proceedings of the sixth all-union session on the study of marine mammals. Kiev: Naukova Dumka. 1 : 24–25.

Bel'kovich, V. M., and N. A. Dubrovsky. 1976. Sensory bases of orientation of cetaceans. Leningrad: Nauka Publishers.

Bel'kovich, V. M., N. L. Krushinskaya, and V. S. Gurevich. 1969. Dolphin behavior in captivity. Nature 11 : 18–28.

Bel'kovich, V. M., and A. V. Yablokov. 1965. School structure of toothed cetaceans (*Odontoceti*). In: Marine mammals. Moscow: Science. 65–69.

Bozhedomov, V. P., L. B. Kozarovitsky, E. I. Pinkhasova, and A. A. Sabitov. 1975. Training dolphins to carry articles. In: Marine mammals: Proceedings of the sixth all-union session on the study of marine mammals. Kiev: Naukova Dumka.

Brown, D. H., and K. S. Norris. 1956. Observations of captive and wild cetaceans. Journal of Mammalogy 37 : 3, 311–326.

Caldwell, D. K. 1956. Removal of object by a dolphin. Journal of Mammalogy 37 : 3, 454–455.

Caldwell, M. C., and D. K. Caldwell. 1965. Individualized whistle contours in the bottlenosed dolphin (*Tursiops truncatus*). Nature 207 : 4995, 435–476.

Evans, W. E., and J. Bastian. 1969. Marine mammal communications: Social and ecological factors. In: The biology of marine mammals. New York and London: Academic Press. 425–476.

Kleinenberg, S. E. 1956. Mammals of the Black and Azov Seas. Moscow: Academy of Sciences of the USSR.

Krushinskaya, N. L. 1972. Behavior. In: Whales and dolphins, ed. A. V. Yablokov. Moscow: Science. 335–360.

Loizos, C. 1966. Play in mammals. Symposium of the Zoological Society of London 18 : 1–10.

McBride, A. F., and D. O. Hebb, 1948. Behavior of the captive bottlenose dolphin (*Tursiops truncatus*). Journal of Comparative Physiology and Psychology 41:111–123.

McBride, A. F. and H. Kritzler. 1951. Observations on pregnancy, parturition, and postnatal behavior in the bottlenose dolphin. Journal of Mammalogy 32:251–266.

Morozov, D. A. 1970. The dolphins are hunting. Fisheries Management 6:16–17.

Pilleri, G., and J. Knuckey. 1967. The distribution, navigation, and orientation by the sun of the common dolphin (*Delphinus delphis* L.) in the western Mediterranean. Experientia 24:4, 394–396.

Saayman, G. S., and C. K. Tayler. 1973. Social organization of inshore dolphins (*Tursiops aduncus*) in the Indian Ocean. Journal of Mammalogy 54:4, 993.

Tarasevich, M. N. 1951. Age and sex structure of schools of the common dolphin. (*Delphinus delphis*). Proceedings of the All-Union Hydro-Biological Society 3:172–178.

———. 1967*a*. The structure of cetacean groups, Report 1: Group structure of cachalot males (*Physeter catodon*). Zoological Journal 46:1, 124–131.

———. 1967*b*. The Structure of Cetacean Groups, Report 2: Group Structure of Finback Whales. *Zoological Journal* 46:3, 420–431.

Tavolga, M. C. 1966. Behavior of the bottlenose dolphin (*Tursiops truncatus*): Social interactions in a captive colony. In: Whales, dolphins, and porpoises, ed. K. S. Norris. Berkeley and Los Angeles: University of California Press. 718–730.

Tayler, C. K., and G. S. Saayman. 1972. The social organization and behavior of dolphins (*Tursiops aduncus*) and baboons (*Papiuo ursinus*): Some comparisons and assessments. Annals of the Cape Provincial Museums 9:11–49.

———. 1973. Imitative behavior by Indian Ocean bottlenose dolphins (*Tursiops aduncus*) in captivity. Behavior 44:3–4, 286–296.

Tomilin, A. G. 1957. Animals of the USSR and contiguous nations, Vol. 9: Cetaceans. Moscow: Academy of Sciences of the USSR.

Voronin, L. G., Y. D. Starodubtsev, and L. B. Kozarovitsky. 1971. Dynamics of training behavior in Black Sea bottlenosed dolphins. In: The morphology and ecology of sea mammals (dolphins). Nature 73–77.

Yablokov, A. V., V. M. Bel'kovich, and V. I. Borisov. 1972. Whales and dolphins. Moscow: Science.

Dusky dolphins herding fish, Argentina. (Photo by Bernd Würsig.)

CHAPTER TWO

DOLPHIN MOVEMENT PATTERNS

INFORMATION FROM RADIO AND THEODOLITE TRACKING STUDIES

Bernd Würsig, Frank Cipriano, and Melany Würsig

INTRODUCTION

During the last twenty years, studies of toothed whales in their natural environment have proliferated. Although knowledge of some cetacean species has increased dramatically through careful long-term study, few of the techniques in use have changed greatly in complexity or sophistication. Major exceptions to this are (1) the use of recently introduced cytological and cytogenetic methods for ascertaining sex, phylogenetic, and genetic relationships; and (2) studies of blood hormone levels for correlation with sexual behavior (e.g., Duffield 1982; Kirby and Ridgway 1984; Wells 1984; Duffield and Wells 1986). "Old standbys" still often

Many people helped to gather information on the three dolphin communities summarized here. We thank especially R. Payne and C. Walcott for help in Argentina, K. Norris and R. Wells for collaboration in Hawaii, and J. Van Berkel and M. Webber for help in New Zealand. D. Croll, T. Jefferson, S. Kruse, K. Norris, K. Pryor, W. Schevill, S. Shane, P. Tyack, W. Watkins, and R. Wells made many helpful comments. W. Broenkow, K. Lohman, and J. Wolitzky provided computer programs and expertise in the analysis and visual display of track data; S. Sanduski and M. Benson kindly typed several versions of this manuscript. This is contribution no. 9 of the Marine Mammal Research Program, Texas A&M University at Galveston.

used in field studies of social organization and movement patterns are photographic recognition of naturally marked dolphins (Würsig and Würsig 1977, Würsig and Jefferson, in press) and theodolite tracking from shore (first introduced for marine mammals by Roger Payne in 1972; described for dolphins by Würsig and Würsig, 1979). Techniques once more popular but now used less often include tagging of dolphins with visually recognizable marks (e.g., Norris and Pryor 1970; Evans et al. 1972; Irvine and Wells 1972; Irvine et al. 1982; Scott et al. 1990) and radio tracking (Evans 1974; Irvine et al. 1981, 1982; Würsig 1982; Norris et al. 1985). Attached visual tags are now rarely used, since photographic identification using natural marks is relatively benign and is possible whenever an appropriate proportion of the study population carries recognizable marks (in the interest of cost and time efficiency, a minimum of about 20% naturally marked animals in a population is usually needed to obtain some information on movement patterns, ranges, and group structure). Radio tracking provides unique and extremely useful data on movements and dive patterns even when animals cannot be closely approached and at night but has received a bad reputation because early work relied on large transmitters attached to dorsal fins of relatively small dolphins (Evans 1974; Irvine et al. 1982; Würsig 1982). Irvine et al. (1982) showed that in some cases, long-term placement of these large transmitters could be damaging to even usually slow-moving animals such as bottlenose dolphins (*Tursiops truncatus*). Fortunately, much smaller transmitters are now available. Wells and Würsig (1983; also described in Norris et al. 1985) and Read and Gaskin (1985) used transmitters about the size of a cigarette package on Hawaiian spinner and spotted dolphins (*Stenella longirostris* and *S. attenuata*) and Canadian harbor porpoises (*Phocoena phocoena*). Würsig, Cipriano, and Webber (1985) recently used tags no larger than a man's little finger to track dusky dolphins (*Lagenorhynchus obscurus*) at ranges of over 40 km off the South Island of New Zealand.

Up to now, field studies of the behavior of toothed whales have been concerned primarily with basic descriptions of particular social communities, almost all composed of animals spending at least part of their lives close to shore. Comprehensive syntheses of cetacean behavior as related to ecological conditions have been rare, probably due to the youthful nature of field investigations. Notable exceptions are the insightful summary articles of Norris and Dohl (1980*a*) and Wells et al. (1980), general books on cetacean biology and ecology by Gaskin (1982) and Evans (1987), an article by Tyack (1986), and summary articles primarily discussing bottlenose dolphins by Leatherwood and Reeves (1982) and Shane et al. (1986).

This chapter is intended to show how two techniques of study—

theodolite and radio tracking—can be used to detect and analyze general movement patterns in an area. Such information is needed for the evaluation of ecological factors that influence cetacean movement patterns observed in different habitats. Sociobiological theory and studies of free-ranging dolphins suggest that both foraging and predator avoidance tactics in turn help to shape the social organization of cetaceans (e.g., Wilson 1975; Norris and Dohl 1980*b;* Wells et al. 1980). Thus, it seems that analysis of foraging and other movement patterns may be profitable in illuminating factors important in the structuring of dolphin social systems (Norris and Dohl 1980*a*, Würsig and Würsig 1980). For the following discussion, our main sources of comparison will be the activities of dusky dolphins in shallow water off Argentina and in deeper waters off New Zealand and of Hawaiian spinner dolphins in shallow and nearby deep water. We intend to show that differences in dolphin habitats are just as important as species differences in the structuring of daily foraging strategies, and these differences probably also affect social organization (Würsig et al. 1989). This is not at all a new concept, of course, and it has been especially well explored in terrestrial systems for wolves (Mech 1970), coyotes (Bekoff and Wells 1980), antelopes (Jarman 1974), and primates (Clutton-Brock and Harvey 1977).

METHODS

It is perhaps most helpful for future research to first give a general description of the techniques used for theodolite and radio tracking. We refer the reader to the various publications from which this summary draws its information for specific methods, details of the equipment used, and information pertaining to particular locales (Würsig 1978, 1982; Würsig and Würsig 1979, 1980; Wells and Würsig 1983; Würsig et al. 1985; Norris et al. 1985).

Theodolite Tracking

Theodolite tracking[1] utilizes a surveying instrument that measures horizontal angles from some arbitrarily selected reference point ("zero") and vertical angles relative to a gravity-referenced level vector (see Davis et al. 1981, for a discussion of surveying methodology). If the instrument's height above the sea surface and the positions of the theodolite station

---

[1] A theodolite is essentially a surveyor's transit, and we use the terms interchangeably. Thedolites used for tracking cetaceans should have a stated accuracy of 20 seconds or less in both the horizontal and vertical planes.

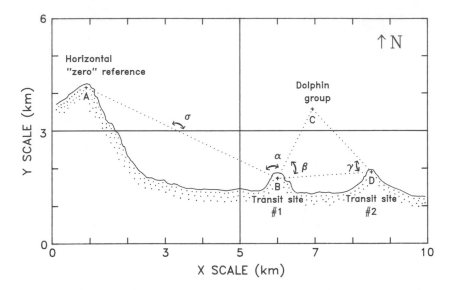

Fig. 2.1. Generalized coastline to show basic transit-tracking setup. Point A can be a lighthouse or other prominent feature found on a map. Angles σ and α as well as height and x/y coordinate of Point B must be known for calculation of x/y coordinate of Point C with the use of one transit. Alternatively, Point C can be located by using angles β and α to triangulate from known positions B and D.

and the "zero" reference point on a map of the coastline are accurately known, these angular measurements of animal or group positions can be translated into x/y coordinates on a map (fig. 2.1). Successive positions and times can be compared to calculate travel speeds of the animal or group. Position data plotted on a depth contour map can also be useful in revealing correlation of movements with features of bottom topography (Würsig and Würsig 1979, 1980).

Theodolite tracking of dolphin schools does not disturb them or alter their behavior since it is done from land. The use of this technique is limited to areas with reasonably high relief near shore and to species that regularly come within sight of land. The farther the animals are from land, the higher the theodolite site must be for reasonable precision and accuracy. As a general rule, dolphins within 5 km of shore should be tracked from a vantage point at least 20 m above the surface of the water. Below this height, the error in the measured vertical angle may become unacceptably large due to the small angle subtended by the water surface from shore to the horizon.

To track dolphins from a single theodolite station, it is necessary to know the height of the station generally to within ±10 cm, although such

Table 2.1.   Errors Associated with Incorrect Measurement of Cliff Height

Actual Cliff Height	Error in Height	Distance Error (m) TRUE DISTANCE TO POSITION ON WATER		
		500 m	2,500 m	5,000 m
15 m	100 cm high	+34	+173	+388
	10 cm high	+4	+17	+39
	10 cm low	−3	−17	−38
	100 cm low	−30	−172	−379
30 m	100 cm high	+17	185	+179
	10 cm high	+2	+8	+18
	10 cm low	−2	−9	−17
	100 cm low	−17	−85	−177
45 m	100 cm high	+12	+56	+117
	10 cm high	+2	+5	+12
	10 cm low	−1	−6	−11
	100 cm low	−11	−56	−116
100 m	100 cm high	+5	+25	+51
	10 cm high	+1	+2	+5
	10 cm low	0	−3	−5
	100 cm low	−5	−25	−51

Note: Errors are not symmetrical above and below the actual cliff height because of the differential effects of curvature of the earth at different heights and rounding of errors. (Errors of cliff height are usually due to an incorrect measurement of tidal height. Errors of positioning animals can also occur due to heat haze or swell moving the animals up and down. These noncliff height errors cannot easily be corrected, and if heat haze or swells are prodigious, theodolite tracking with one theodolite is not advised. Triangulating from two positions on shore is usually still possible.)

accuracy is not as critical from stations greater than 45 m above the sea surface or when animals are very close (less than 500 m) to shore (table 2.1). This need for an accurate height reference necessitates initial careful surveying of the site (fig. 2.2) to determine its height above the sea surface at a known tidal state (mean low water is an appropriate and commonly used reference point) as well as knowledge of tidal excursions while making observations. Tide height can be obtained from tide stakes placed in the water, from mechanical or electronic tide meters, or, as a less accurate alternative, from predictive tide tables.

Vertical angles and cliff height are used to calculate the distance to dolphins offshore. Incorrect measurement of the cliff height will result in an

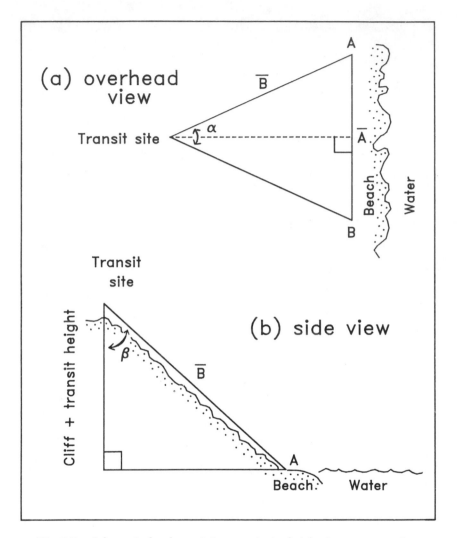

Fig. 2.2. Schematic for determining transit site height. Locate two points, A and B, at or near the waterline with the same vertical angle to the site. Measure distance A between Points A and B, and record angles α and β. From (a),

$$B = \frac{\frac{1}{2}A}{\sin(\frac{1}{2}\alpha)}.$$

From (b), cliff and transit height above water = B (cos β). When Points A and B cannot be located as shown due to topography, height of site can be calculated by setting up a second transit position a known distance from the first position and using nonright triangles for calculating B.

inaccurate calculation of the distance to a group. Inaccuracies may also result from failure to take tidal fluctuations or swell into consideration. The magnitude of such distance errors is directly proportional to the distance to a group and inversely proportional to the cliff height (table 2.2). Errors in the calculated distance traveled and speed resulting from inaccurate cliff-height measurements are also inversely proportional to cliff height. The percent error in cliff-height measurement translates roughly to percent error in distance to group, distance traveled, and speed (table 2.2). As a result, theodolite stations on higher cliffs are more "forgiving" of such errors.

For one-station tracking, it is necessary to find a point to serve as zero reference in the horizontal plane (Point A of fig. 2.1). This point should be far enough from the transit station to allow unambiguous positioning of both the station and the zero reference on a local map but not so far as to be commonly obscured by fog or heat haze. In practice, such a zero reference can be a lighthouse, small island, mountain peak, abrupt headland, radio mast, and so on. Instead of a point, magnetic or true north can be used; in this case, an external reference position is not needed. The vertical reference is fixed by careful leveling of the theodolite relative to gravity.

To convert horizontal and vertical degree readings of the theodolite to x and y position coordinates on a map, it is necessary to know the angle between the horizontal zero reference point and the baseline of an orthogonal grid system if the theodolite station and reference point do not both lie on or parallel to the baseline. In addition, the angle between the baseline and true or magnetic north should be known for calculation of the heading of moving groups or individuals. Alternatively, this can be measured directly relative to a compass rose on the map. The x/y grid may be congruent with the latitude-longitude lines on a map or may be marked out in meters (fig. 2.1).

If a high vantage point is not available near shore and theodolite tracking is nevertheless desirable, it is best to triangulate from two tracking sites and thereby eliminate the need to know precisely height above the water (fig. 2.1). The optimal baseline separation of the two stations is reached when the angle between stations and the object in the water is approximately 45°, although large variations from this optimal condition are acceptable. When two stations are used for triangulation, they can be "zeroed" on each other, and an external zero reference is not needed.

A BASIC computer program based on an iterative function designed to calculate positions of objects on a curved earth was developed in 1976 by Jan Wolitzky (now of Bell Laboratories, Murray Hill, N.J.) and was described in Würsig (1978). Many researchers have since modified this program for their own particular needs and computers, and some have

Table 2.2. Distance and Speed Errors Due to Incorrect Cliff Height

Actual Cliff Height (m)	Error in Height (%)	True Distance Offshore (m)	Distance Underestimate[a] (m)	(%)	Estimated Distance Traveled (m)	Error in Distance Traveled (m = %)	Estimated Speed (m/sec.)	Speed Underestimate (m/sec.)	(%)
15	6.7	500	30	6.0	94.0	6.0	1.88	0.12	6.0
		2,500	172	6.9	93.1	6.9	1.86	0.14	6.9
		5,000	379	7.6	92.4	7.6	1.85	0.15	7.6
30	3.3	500	17	3.4	96.6	3.4	1.93	0.07	3.4
		2,500	85	3.4	96.6	3.4	1.93	0.07	3.4
		5,000	177	3.5	96.5	3.5	1.93	0.07	3.4
45	2.2	500	11	2.2	97.8	2.2	1.96	0.04	2.2
		2,500	56	2.2	97.8	2.2	1.96	0.04	2.2
		5,000	116	2.3	97.7	2.3	1.95	0.05	2.3
100	1.0	500	5	1.0	99.0	1.0	1.98	0.02	1.0
		2,500	25	1.0	99.0	1.0	1.98	0.02	1.0
		5,000	51	1.0	99.0	1.0	1.98	0.02	1.0

[a] See table 2.1.

Note: For case where error in cliff height measurement is 100 cm too low (from table 2.1), travel speed = 2 m/sec., true distance traveled = 100 m, time between positions = 50 sec., and travel is exactly perpendicular to theodolite station (see fig. 2.3). Note that percent error in measurement of cliff height translates roughly to percent error in distance, distance traveled, and speed.

developed entirely different programs. Peter Tyack of Woods Hole Ocean-ographic Institution developed an expanded version of the Wolitzky program in PASCAL for use on Apple II microcomputers (Tyack 1982), and Kevin Lohman of Moss Landing Marine Laboratories made similar modifications for use both on Hewlett-Packard HPL-speaking computers and an HP-41CV hand-calculator. A sample computer output is shown in figure 2.3. We are currently utilizing an electronic digital theodolite (Lietz DT20E), which has an RS-232C output for serial transfer of horizontal and vertical degree readings to a computer. Although such systems make continuous, real-time data collection possible and analysis easier and faster than with a traditional mechanical analog theodolite, the same basic procedures and constraints described above apply.

Radio Tracking

Since radiotelemetry has been widely used in terrestrial wildlife studies for many years, much more published information on general methods is available than for theodolite tracking (e.g., Mech 1983, Dietz 1986). Radio transmitters have been mounted to the dorsal fins of dolphins and porpoises in several studies (Evans 1974; Gaskin et al. 1973; Leatherwood and Ljungblad 1979; Irvine et al. 1981; Würsig 1982; Norris et al. 1985; Read and Gaskin 1985; Würsig et al. 1985). Irvine et al. (1982) and Scott et al. (1990) analyzed a variety of tagging techniques used with small cetaceans. Although many of the same considerations apply, radio tracking of free-swimming cetaceans requires a different approach from that used with terrestrial animals. The choice of equipment and trade-offs among cost, practicality, and type of data desired are discussed below.

Radio tagging with relatively inexpensive and simple transmitters can provide information about individual movements, diurnal and seasonal surfacing and dive patterns, and associations between individuals. Transmission range is approximately "line of sight" and may vary from about 3 to 60 km, depending on transmitter strength, radio frequency, and the placement of receiver antennas on shore. Greatly expanded reception range can be obtained from airplane-mounted receivers but at greater cost. Depth-of-dive transmitters, which encode information on the maximum depth reached during a dive into the signal transmitted on surfacing, have been used (Evans 1971), but high cost has precluded extensive use. Depth recorders that must be recovered from animals for data transcription have been used successfully for a number of pinnipeds (Kooyman 1965; Kooyman et al. 1976; Kooyman et al. 1983), and smaller depth-of-dive recorders with a programmable sampling strategy have recently been developed (Donald Croll pers. comm. 1986, Physiological Research Laboratory, Scripps Institution of Oceanography). A tag that broadcast a

Row	Time (HHMMSS)	X (m)	Y (m)	Distance (m)	Moved (m)	Heading (deg. true)	Speed (km/hr.)	Tide (m)
1	131830	9,112	1,514	1,117	0	−1	0.0	5.07
2	131906	9,000	1,454	1,108	127	203	12.7	5.06
3	132500	8,620	2,126	1,880	772	291	7.9	5.15
4	132530	8,584	2,222	1,982	103	300	12.4	5.17
5	132845	8,782	2,681	2,326	500	344	9.2	5.33
6	132911	8,794	2,724	2,363	45	336	6.2	5.35
7	133017	8,889	2,833	2,441	145	2	7.9	5.40
8	133042	8,921	2,929	2,526	101	340	14.6	5.42
9	133115	8,930	3,024	2,617	96	326	10.5	5.45
10	133142	8,955	3,079	2,664	60	346	8.0	5.47
11	133200	8,957	3,184	2,767	105	322	21.0	5.48

Fig. 2.3. Dusky Dolphin Track Sample, November 21, 1975, Golfo San José, Argentina. Sample computer printout of 11 x/y position readings, corresponding to the first 11 positions of a dusky dolphin group in Golfo San José, Argentina, Nov. 21, 1975 (fig. 2.5). "Dist" is distance of group from transit site, "Moved" is distance between readings, "Heading" is relative to true north, and "Speed" assumes movement in a straight line between readings. "Tide" is tide above mean low water and is calculated for each reading by the computer from sample tide data entered for this date.

Pentax-type theodolite	
Height (m) from reference to zero tide:	46.18
Angle (deg) between baseline and reference:	33.00
Angle (deg) between true north and baseline:	231.00
Value of X (m) at the observation site:	9520.00
Value of Y (m) at the observation site:	475.00

signal to an earth-orbiting satellite, with practically unlimited range, was used experimentally on a spotted dolphin in Hawaii (Jennings and Gandy 1980, Jennings 1982). Unfortunately, satellite transmitters are still too large for practical use with smaller cetaceans (those less than about 4 m long), although this is likely to change within the next several years as electronic packages decrease in size (Tanaka et al. 1987, Mate 1989).

Transmitter frequencies used on dolphins have been in the 27 MHz (High Frequency, HF) and in the 148 to 160 MHz (Very High Frequency, VHF) ranges. Both have advantages and disadvantages. Longer wavelength HF signals can bend around obstructions and over the horizon but do not tend to reflect off objects and therefore give good directionality information to distant receivers. However, HF antennas generally require considerable power (several hundred milliwatts) to perform satisfactorily, and this power requirement dictates a larger transmitter package for good signal transmission. Longer wavelengths also require longer antennas, and 60 cm antennas are commonly used.

HF transmitters in a tube approximately 4 cm in diameter by 35 cm long gave consistently good signals from dusky dolphins to land stations 50 km distant (Würsig 1982), and tags as small as 1.5 × 7.5 cm have been used to track baleen whales to 30 km from surface vessels and to 300 km from aircraft (Watkins 1985, pers. comm. 1986).

VHF frequencies are relatively quiet, with little background noise (due to inherently limited range of higher frequency signals), and are used most widely for tracking terrestrial animals. Transmitters can be low-power oscillators (often with as little as 10 milliwatt output) coupled directly to efficient and relatively small antennas, as short as 10 to 20 cm. However, these higher frequencies attenuate more rapidly, often bounce off objects to give confusing or false directionality information, and do not bend with the curvature of the earth. They are therefore limited to line of sight reception, at distances from water to surface vessels or shore rarely greater than 30 km. VHF transmitters can be made extremely small, however, and this one advantage makes them desirable for close-range tracking of the smaller dolphins and porpoises.

A conventional radio receiver can determine only the general direction of the transmitter and sometimes provide a rough estimate of distance by measuring the relative signal strength at the receiver. For more accurate determination of the dolphin's position, triangulation from two sites is desirable, or the observer can home in on the signal from a boat or an airplane. Techniques for efficiently homing toward the signal are described in Mech (1983).

Two different types of radio receivers have been used in dolphin research. Automatic direction finders (ADFs) obtain direction information electronically with any of several types of antenna arrays and instantaneously display direction of the signal source on an oscilloscope (Evans 1974; Irvine et al. 1981; Würsig 1982). Present ADF systems need relatively large and strong transmitters for efficient operation. Alternatively, a simple receiver can be connected to a high-gain directional antenna (such as a Yagi-Uda array, Uda and Mushiake 1954), which is physically rotated to determine greatest signal strength in a line toward the transmitter (Norris et al. 1985, Würsig et al. 1985). This type of system is more difficult to use on cetaceans, since they often surface only briefly and the operator must rotate the antenna rapidly to ascertain direction of the signal. Weaker and relatively small transmitters can be used, however. Experience plays an important part in the successful use of directional antennas. ADF systems have usually been used with HF signals, and Yagi-Uda directionalizing has been used with VHF signals, for reasons beyond the scope of this review. For further information on antenna capabilities and constraints, see Beaty (1978).

Dolphins have been captured for radio tagging by encircling them with large nets (Irvine et al. 1981) and with breakaway hoop nets (Evans 1974) and tail grabs (Würsig 1978) for the capture of bow-riding individuals. We prefer to keep a dolphin out of water for less than 10 minutes during measurement, sex determination, and tagging. Short handling time minimizes stress and allows return of the animal to its conspecifics as soon as possible. There is a great variability in response to handling among different individuals and different species. We suspect that older individuals are generally more prone to capture shock than are younger ones and that pelagic dolphins (e.g., *Stenella* spp. and porpoises such as *Phocoenoides*) are more vulnerable than are bottlenose or dusky dolphins.

Dolphin radio tags generally have been attached through the cartilaginous connective tissue of the dorsal fin, which provides a secure mount for the transmitter and maximum exposure time for the antenna as the animal surfaces. Early tags used in a study of Argentine dusky dolphins were described in Würsig (1982). Those used more recently on New Zealand dusky dolphins and Hawaiian spinner dolphins were attached using two 3-mm-diameter bolts with corrosible nuts inserted through holes punched through the dorsal fin with an alcohol-sterilized laboratory cork borer. Two thin metal plates, padded with neoprene rubber on the inner surface where they touch the dorsal fin, were used to support the radio and attachment nuts on opposite sides of the fin (fig. 2.4). A chemically inert plastic sleeve was slipped over each stainless steel bolt before it was inserted, so as to form a secure fit of bolts through the fin and to keep movement of the transmitter from irritating the fin. Such tags can be "programmed" to fall away after times ranging from hours to weeks by altering the size of the corrosible (usually magnesium alloy) attachment nuts. We prefer to keep tags on dolphins for only a matter of

Fig. 2.4. Telonics Model RT-3 radio transmitter mounted onto dorsal fin of a dusky dolphin in New Zealand. The tag is backed by an aluminum plate padded with 3 mm neoprene rubber, and magnesium alloy bolts are programmed to corrode in about one week. An identical plate serves as washer for the bolts and nuts on the other side of the dorsal fin. (Photo by Bernd Würsig.)

days to weeks to reduce the chances of injury to the animal, and we are presently developing a tag that will be released on command by a radio signal from a nearby boat.

All figures but figure 2.4 were generated by Hewlett-Packard desktop computers, plotters, and printers. Maps were digitized on an x/y digitizing tablet connected to an HP 9816 computer.

RESULTS AND COMPARISONS

We divide the following summary of results and discussion into a consideration of movement patterns of dusky dolphins in two physically and biologically divergent habitats in the southern hemisphere and movements of Hawaiian spinner dolphins in an environment that has elements of both habitat types considered for dusky dolphins. Dusky dolphins occur off the coasts of South Africa, South America, and New Zealand; spinner dolphins occur in tropical and subtropical waters of the Pacific, Indian, and Atlantic oceans (for a general summary on distribution, see Leatherwood et al. 1983). Both species are found in a variety of habitats close to shore and at least 100 km from shore. Spinner dolphins also occur in pelagic waters (Perrin 1975), but their movements and social organization have not been well studied in the open ocean.

Dusky Dolphins, Argentina

Most nearshore Argentine waters overlie an extensive, shallow continental shelf. Dusky dolphins frequent these shelf waters and usually travel in depths less than 100 m. They generally forage for schools of southern anchovy (*Engraulis anchoita*) in small groups of about eight to ten animals, and up to about thirty such groups may be spaced from 1 to 8 km apart while searching for food. Groups appear to stay in acoustic contact with each other and are able to aggregate quickly when food is found. They then cooperate as a larger group to surround and efficiently herd fish schools toward the surface. Much social activity takes place during and after herding and feeding. Social activity is probably important for animals that hunt together, and bonding mechanisms reinforced by social activity may be necessary for the coordination of presumably sophisticated cooperation (Würsig and Würsig 1980).

Apparently because anchovy, like most schooling fishes, tend to forgo tight schooling at night and occur in more cohesive groups during the day (Marshall 1966), Argentine dusky dolphins do most searching for food and feeding during the day. As a result, small groups tend to meander, apparently searching for food, in the earlier morning hours. They often find food from about midmorning to noon and aggregate at that time. Most feeding in large groups occurs in the afternoon. The dolphins break into smaller groups again in the evening or at night, and spend the night resting at the surface with very little net movement (Würsig 1982). When food is not found in one area, they can rapidly move up to about 50 km per day to search in a different area.

Almost all that we know about the basic behavioral repertoire of this population comes from analysis of theodolite- and radio-tracking information. Tracking and photoidentification data show that dusky dolphin groups disperse and coalesce over time, so their society may be described as an open one, somewhat similar to the sometimes ephemeral nature of large groups of foraging chimpanzees (Clutton-Brock and Harvey 1977). However, tracking information shows that the diurnality mentioned above is not a strict regime; dolphins do not find anchovy to feed on every day and do not aggregate when schooling fish prey are not encountered. They may also find food at other times than midday. We have observed feeding late in the evening and indications (based on types and amount of dolphin social behaviors and the presence of large numbers of birds on the water near dolphin aggregations) of feeding during the night. We guess that this lack of rigidity in diurnal behavior reflects the natural variability in anchovy schooling behavior.

A theodolite track showing the most common diurnal progression of movements of a dusky dolphin group is given in figure 2.5. The group, composed of approximately fifteen dolphins, was first located in about 35 m water depth at midmorning (although the track shown in fig. 2.5 does not begin until early afternoon). The group moved into progressively deeper water as the day advanced and began feeding (as indicated by birds aggregating near the dolphins) over a series of underwater rises, in about 20 m depth but about 4 km from the nearest shore, in the afternoon. The dolphins were joined by others at this time, and soon about seventy-five to one hundred animals were together in the general area. This feeding activity occurred about 7 km from our observation site.

Dusky dolphins generally rested within about 1 km of shore at night, and this nonforaging time was probably spent relatively close to shore in order for the animals to avoid predators such as killer whales (*Orcinus orca*) and possibly large sharks. We suspect that the surf zone makes it difficult for killer whales to locate and track their prey because of turbulence and air bubbles that reduce the efficiency of both visual and acoustic senses. On three occasions when killer whales and dusky dolphins were seen within 1 km of shore, the dolphins moved rapidly into the surf zone in successful attempts to evade the large predators (Würsig and Würsig 1980).

The diurnal pattern of movement described above occurred most often during spring, summer, and fall, when anchovy were present in varying degrees in the nearshore area. In winter, when anchovy were scarce, the dolphins hardly ever aggregated into groups larger than about 20 individuals and stayed within 3 km of shore at almost all times. Their behavior was dominated by slow movement near shore as they alternated rest with apparent feeding on nonschooling prey. Examples of representative radio tracks of two Argentine dusky dolphins in summer and one in winter are shown in figure 2.6. Apparently, there is enough food available for the dolphins to survive in winter in a rather restricted area. The generally low level of activity in winter and the lack of boisterous "large-group" socializing associated with feeding in other seasons, however, suggest that the animals may be conserving energy in a generally more food-limited environment. This hypothesis does not rule out other poten-

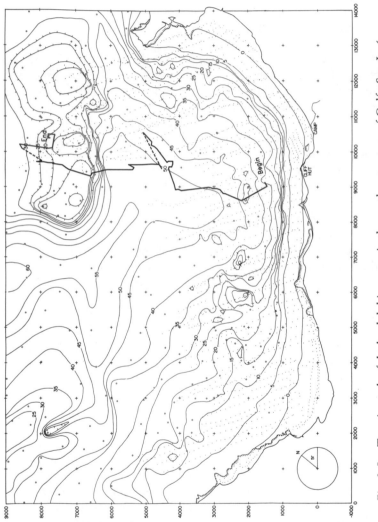

Fig. 2.5. Transit track of dusky dolphin group in the southeast corner of Golfo San José, Argentina. Small dots and circles represent positions of depth readings, used to draw depth contour lines, in meters at mean low water (MLW). Transiting was from "Cliff Hut" at 46.18 m altitude above MLW. x/y coordinate numbers are in meters. Track of dolphin group begins at 13:18:30 and ends at 15:15:10. Dashed lines indicate uncertain tracks. See figure 2.3 for sample plot data for first 11 positions and text for explanation of dolphin movements.

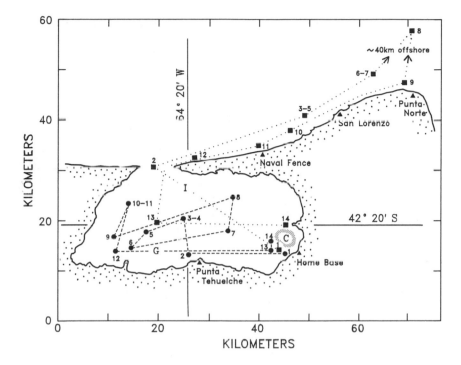

Fig. 2.6. Golfo San José, Argentina, and open ocean to the north. Three dolphin radio tracks, with numbers representing consecutive positions of dolphins at midday. Dolphin "I," adult male tracked November 30–December 14, 1974. Dolphin "G," adult female tracked January 3–January 16, 1975. Dolphin "C," adult male tracked July 31–August 6, 1974. "C" stayed in the confined area marked with small dots during the week of radio contact. Triangles on shore represent tracking stations.

tial factors, of course. Dolphin behavior is certainly also influenced by seasonal changes in courtship and breeding activity and perhaps by changes in behavior of prey as well.

Dusky Dolphins, New Zealand

In the Kaikoura area of the South Island of New Zealand, where we are currently studying the behavioral ecology of dusky dolphins (Cipriano 1985, Würsig et al. 1989), the physical environment is radically different from that of Argentina. The deep Kaikoura Canyon brings pelagic fishes and squid to within a few kilometers of shore. This canyon is a favorite winter feeding ground for sperm whales (*Physeter catodon*), which are known to feed largely on deep-water squid and bottom fishes in the area (Gaskin and Cawthorn 1967). Few schooling fishes except kahawai (*Arripis truta*), which occur in groups at a size too large to be taken by dusky dolphins, are found in these waters. Stomach analysis of fifteen dolphins incidentally killed during fishing operations shows that dusky dolphins in this area feed largely on mesopelagic fishes and squid (Cipriano 1989), like those commonly associated with the deep scattering layer (DSL) in many areas (Gibbs and Roper 1970, Hopkins and Baird 1977).

Dusky dolphins of the Kaikoura area do not show the same pattern of small group dispersal during foraging and of aggregation during fish herding and feeding as was observed in Argentina. Instead, it appears that subgroups are almost always closely associated within the larger school "envelope"; the entire school of several hundred dolphins, at times covering several square kilometers, usually travels as a unit. Subgroups within these larger schools of New Zealand dusky dolphins are sometimes observed to dive synchronously, and at times, almost the entire group may "disappear" for two or three minutes while being observed from high vantage points on shore. After such dives, subgroups returning to the surface are usually still widely separated, and on only a few occasions has there been any evidence that fish aggregations were associated with the surfacing dolphins. We suspect—from the generally loose schooling behavior of the mesopelagic prey taken by these dolphins—that cooperative herding of schooling fishes is much less important for New Zealand than for Argentine dusky dolphins. More field observations and specimens collected during spring and summer seasons are needed to substantiate the lack of coordinated feeding on schooling prey. Perhaps because the dolphins do not regularly aggregate after having been dispersed in many small groups (as occurs in Argentina), we have not observed conspicuously elevated levels of socializing at particular times.

Because the DSL tends to migrate vertically on a diel schedule, we might expect to see diel shifts in behavior of dolphins feeding on mesopelagic organisms that are found closer to the surface at night than during the day. During summer and fall months, the dolphins tend to be very close to land, often within several hundred meters of shore, in the early morning. In late afternoon and evening, they are usually found moving gradually into deeper waters, possibly to meet the rising DSL in pelagic waters during the night. The track of a group of about 160 dolphins shown in figure 2.7 illustrates this midday meandering in mid-depth waters, as far as 2 to 3 km from shore. Later in the evening, the dolphins are usually found at the fringes of the bay, in water 500 to 1,000 m deep. Once again, movement near shore while not feeding may be related to resting and predator avoidance, but we have less direct evidence for this than in the Argentine situation. The basic movement pattern from near-

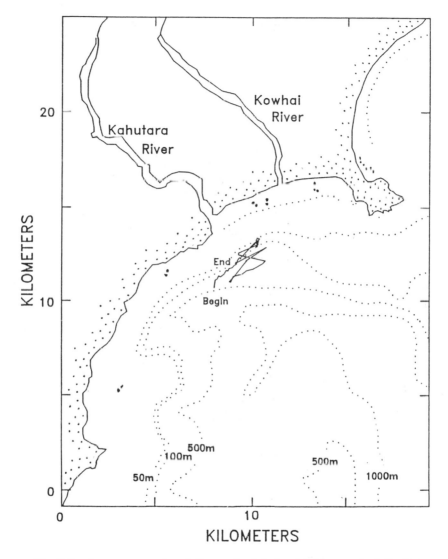

KILOMETERS

Fig. 2.7. Late summer theodolite track of dusky dolphins south of the Kai-koura Peninsula, South Island, New Zealand. Dolphin group positions were obtained from a station just south of the Kahutara River mouth, at an elevation of 100.84 m above MLW. Track begins at 1131, ends at 1417 on May 4, 1984, and shows nearshore movements along depth contours.

shore to offshore waters during the course of the day is similar in the two communities; but group sizes, timing and intensity of social activities, diurnality of feeding, and apparently the amount of social coordination during feeding are very different.

During winter months, groups of New Zealand dusky dolphins are

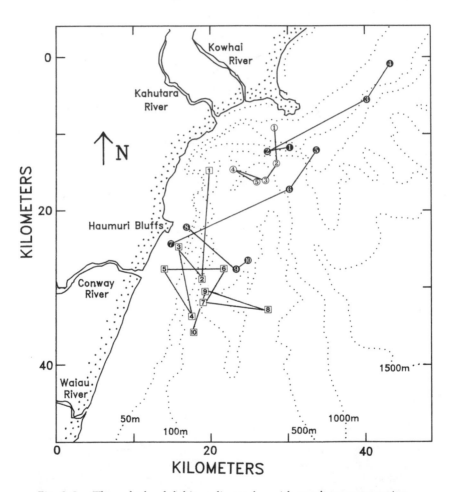

Fig. 2.8.   Three dusky dolphin radio tracks, with numbers representing positions at different times during the tracking period. Dolphin "A" (open circles), adult female tracked July 14–16, 1984. Dolphin "B" (filled circles), adult male tracked intermittently July 18–22, 1984. Dolphin "C" (open squares), juvenile female tracked July 20–24, 1984. Tracks show the large amount of movement and generally large distance from shore of New Zealand dusky dolphins in winter.

usually found much farther from shore than during summer, often 5 km or more off even the tip of the Kaikoura Peninsula where it projects to the east (fig. 2.8). Groups are observed to move quite quickly through the area, sometimes traversing the entire 30 km of coastline visible from high vantage points atop the peninsula in just a few hours. Some groups do enter the bay, which allowed capture and radio tagging of three individuals in different groups during July 1984 (fig. 2.8). These animals remained relatively far from shore during all times of day (8± s.d. 4.4 km, n = 25) and remained within reception range for only a few days each. One individual ("B": filled circles in fig. 2.8) moved out of range to the north in 1.5 days, reappeared a few days later, then moved out of range to the south in another two days. This fairly large range is like the pattern exhibited by Argentine dusky dolphins in summer, when the latter are searching widely for schooling fishes. This is quite unlike the situation for New Zealand dolphins in summer, when what is essentially the same large group can remain within the study area and be tracked from shore and photoidentified from boats repeatedly for up to one month. We do not know whether these patterns are related to daily and seasonal shifts in mesopelagic fish abundance or whether the total ranges of New Zealand dolphin groups are different between winter and summer. Further radio-tracking work is necessary, and use of instrumentation that provides depth-of-dive information, indicative of where and how dolphins are feeding, is planned for the near future.

Hawaiian Spinner Dolphins

Norris and Dohl (1980b) and Norris et al. (1985) studied the behavior and ecology of Hawaiian spinner dolphins on the west coast of Hawaii, concentrating most of their work around Kealake'akua Bay (fig. 2.9). This environment shares attributes of both the Argentine and New Zealand dusky dolphin habitats. Shallow areas near shore, although not nearly so extensive as those in Argentina, are clearly used by dolphins for daytime rest. Pelagic waters as close as one kilometer from the shore of this steep volcanic island are used for feeding on DSL-associated organisms (Norris and Dohl 1980b), similar to those consumed by dusky dolphins in New Zealand.

In the above discussion of diurnal shifts in dusky dolphin behavior in both habitats, we used somewhat undefined probability levels; for example, dolphins "usually" are in deeper water in the afternoon than in the morning. The diurnal pattern of Hawaiian spinner dolphins, however, follows a more strict regime, unaltered save for those occasional and relatively rare exceptions that teach observers of behavior never to see their data in absolute terms. Hawaiian spinner dolphins "always" enter near-

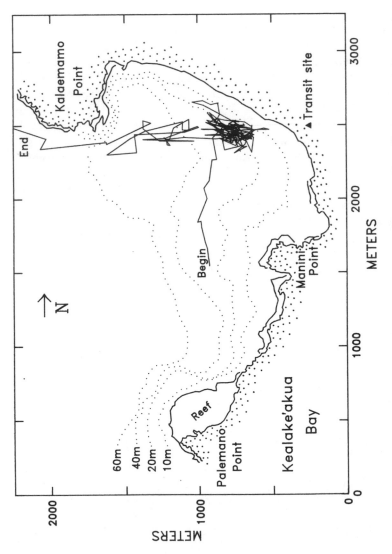

Fig. 2.9. Transit track of a Hawaiian spinner dolphin group in Kealake'akua Bay, Hawaii, July 30, 1979. Track begins at 06:35:38 and ends at 14:12:00. See text for explanation of dolphin movements.

shore shallow waters (usually protected bays) in the daytime, and they "always" move to deep, pelagic waters offshore at night. They need to feed in these waters since little food is available for them near shore, and they appear to need to go to shallow water to rest because of the dangers from pelagic sharks in deep tropical waters. In addition, their main food items, which are associated with the DSL in these clear tropical waters, sink to approximately 500 m during the day (Norris and Dohl 1980*b*) and are possibly not easily accessible for daytime foraging by small delphinids. Some of the DSL organisms of mid-latitudinal waters such as off the New Zealand coast may migrate down to only about 100 to 250 m during daylight hours (Dunlap 1970 and Zaneveld 1977 provide basic descriptions of DSL layers relative to this point), and dolphins there may occasionally be able to feed even during the day.

The diurnal pattern of movement and behavior of Hawaiian spinner dolphins near shore has been described in detail by Norris and Dohl (1980*b*) and Norris et al. (1985). The dolphins quickly and actively enter bays in the morning, alternate periods of rest with social and aerial activity, become restless in the afternoon, and finally move into deep water in the afternoon to aggregate with other groups and eventually to begin feeding in late evening.

The nearshore component of this diurnal behavior pattern has been documented in approximately fifty theodolite tracks, and an example is shown in figure 2.9. The group illustrated consisted of about thirty dolphins, which rapidly moved into Kealake'akua Bay in the early morning. They then descended into rest and stayed in a remarkably confined area less than 0.25 km^2 for about six hours, in waters only 10 to 20 m deep. In the afternoon, they began alternating movement toward the open ocean and back toward their resting spot, apparently as school members gradually reached a consensus to leave the bay (Norris and Dohl 1980*b*). Finally, the animals traveled out of sight in the evening, headed for the feeding grounds not far offshore.

An example of a radio track of a spinner dolphin traveling with a group illustrates the overall pattern of movement alongshore and in deep water (fig. 2.10). The tracked dolphin was tagged in the morning, in water less than 50 m deep. It did not cross the 100 m depth contour line until evening, and it moved into waters about 800 m deep at night. We do not know whether the group containing the dolphin tracked here joined other groups, as is common during nighttime feeding (Norris and Dohl 1980*b*). During early morning, it again approached 100 m depth, and during daylight, the dolphin moved into the shallow waters of Kealake'akua Bay. At no time was this animal farther than 2 km from shore, and yet it was presumably feeding (long "feeding dives" were registered during radio tracking at night) in truly pelagic waters. North of the area

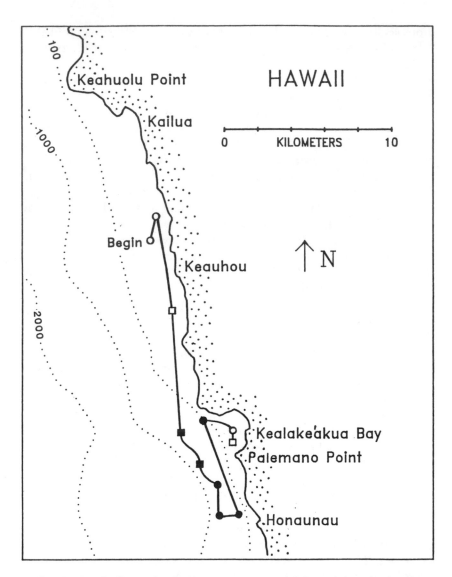

Fig. 2.10. Radio track of a Hawaiian spinner dolphin adult male, April 26–27, 1980. This dolphin was caught north of Keauhou and tracked to Kea-lakeʻakua Bay. Open circles and squares represent daylight hours, 6 to 12 and 12 to 18, respectively. Filled circles and squares represent nighttime track positions between 18–24 and 24–6 hours, respectively. See text for further explanations of dolphin movements.

shown in figure 2.10 there is less of an abrupt drop-off near shore, and there dolphins must travel 5 km or more to reach oceanic waters. In general, however, the proximity of feeding and resting areas for these island dolphins seems an ideal habitat.

There is little change in this diurnal pattern with the seasons, although in winter, the dolphins spend fewer hours per day in the protective near-shore areas. This is possibly because less food is available in deep water due to decreased larval recruitment of reef fishes during winter (Würsig and Würsig 1983, Norris et al. 1985). As a result, dolphins may need to spend more time feeding or traversing to feeding areas in winter and consequently have less time available for resting near shore. If this is true, winter could be an especially difficult time for these animals. It has also been suggested that Hawaiian spinner dolphins spend more time in pelagic waters in winter since nights are longer, and darkness also serves a protective function against sharks (Kenneth Norris pers. comm., University of California, Santa Cruz). This may well allow dolphins more time for feeding in winter than in summer, but we suspect that they would spend long nights outside of protective bays only if, indeed, they needed to forage for longer periods of time in winter than at other seasons.

Since both New Zealand dusky dolphins and Hawaiian spinner dolphins appear to feed primarily on mesopelagic DSL-associated organisms, the general pattern of food availability in the two regions may be quite similar. Their movements as large groups while feeding seem similar as well, although we know rather little about group sizes and behaviors of spinner dolphins while they are feeding at night. Both Argentine dusky dolphins and Hawaiian spinner dolphins split into smaller groups while they are not feeding. In the Argentine situation, many small groups can cover a large area while searching for food, and there appears to be rapid communication between subgroups when one locates schooling fish. In Hawaii, however, groups probably split up because the large feeding school would not efficiently "fit" as a social unit into the small bays that serve as protective and resting areas. Indeed, we find a rough correlation between dolphin group size and bay size and indications that groups are excluded from bays already occupied by a group (Norris et al. 1985). Instead of cooperation, groups of Hawaiian spinner dolphins may actually be practicing a form of competitive exclusion over access to near-shore resting areas.

DISCUSSION

Much has been learned about the behavioral ecology of nearshore dolphins by the simple and inexpensive techniques of theodolite and radio tracking. We show here that habitat type and general prey availability patterns are at least as important as phylogenetic relationships in determining basic patterns of daily and seasonal behavior. Thus, dusky dolphins and spinner dolphins inhabiting nearshore waters with close access to mesopelagic prey live in similar ways, while dusky dolphins without access to deep water but where appropriately sized schooling fishes are available live very differently.

Our broad comparisons must not lead us to a simplified view of these creatures, however. For example, dusky dolphins of Argentina appear to meander in search of food in winter in ways somewhat similar to those of their conspecifics in New Zealand year-round, albeit in smaller groups than in New Zealand. Dolphins feeding on DSL-associated organisms in deep water may be cooperating in similar but as yet unknown ways to the near-surface herding activities so prevalent in Argentina during spring and summer. Furthermore, New Zealand dusky dolphins do not exhibit the strict diurnal regime of Hawaiian spinner dolphins, possibly as a result of less predation pressure on the New Zealand than on the Hawaiian animals.

Photographic identification studies carried out during the work described here have shown that there is a measure of group fluidity, or changing interindividual affiliations, in all three populations. Such fluidity may be intrinsic to delphinid social systems in general (other than killer whales) and may be important in getting all (or at least, most) individuals of a population together periodically, so that animals that work together get to know each other well (e.g., De Waal 1982, on chimpanzees [*Pan troglodytes*]). We believe that this is probably an important reason for the high degree of social activity in after-feeding Argentine dusky dolphin aggregations. It may also be important for Hawaiian spinner dolphins to affiliate with different members during daytime rest and social activity periods in nearshore bays to enhance efficient and safer nighttime feeding.

Group fluidity is manifested in different ways. In Argentine dusky dolphins, small subgroups of approximately six to eight individuals appear consistently to stay together, and larger subgroups are composed of the almost kaleidoscopic arrangements of these small unchanging units. In groups of Hawaiian spinner dolphins, however, almost all dolphins can interchange (perhaps all but mother-calf pairs), and the "kaleidoscope" results from changes down to the individual level rather than only between stable subgroups. This may be because the Hawaiian spinner dolphins do not form the large after-feeding social groups seen in Argentine dusky dolphins, where all individuals can socialize together for a time, even though they are part of a more rigid subgroup. Instead, Hawaiian spinner dolphins come together as a large, diffuse group primarily while feeding at night. They segregate into smaller units after feeding, and their ever-changing affiliations while resting and socializing near shore must eventually allow all or most members of the community to get together on a social basis. This conjectured reason for our observations of differences in levels of affiliations rests on the assumption that dolphins who live together, and perhaps "work" together (by some form of cooperation), need to know each other well. Indeed, it appears that the dolphins know each other across the entire community of greater than 500 dolphins off the Kona coast of Hawaii, for fluid associations occur throughout the community (Norris et al. 1985). The details of New Zealand dusky dolphin group stability are intriguing, but at present, we know only that there are some longer-term and some more rapidly changing affiliations; as yet, we do not know at what social level changes in affiliations are most prevalent.

Bottlenose dolphins also appear to have different movement patterns in different habitats. Where they feed on locally abundant prey such as mullet (*Mugil* sp.) off the coast of west Florida, they have a well-defined and not very large range (Wells et al. 1980, Wells 1986). Where they feed primarily on reef fishes, which are probably not concentrated in any one area, they range over large stretches of coastline (Würsig and Würsig 1979). Bottlenose dolphins that feed on prey locally abundant near estuaries, but with intervening stretches with few prey, also range far as they travel from estuary to estuary in search of food (Ballance 1990).

Bottlenose dolphins also exhibit considerable group fluidity in all communities studied to date (e.g., Saayman and Tayler 1973; Shane 1977, 1980, 1990a, 1990b; Würsig and Würsig 1977; Wells et al. 1980; Gruber 1981; Ballance 1990), but potential differences in group structure between communities living in different habitats have not been adequately addressed. Bottlenose dolphins of Sarasota Bay, Florida, show less fluidity than Hawaiian spinner dolphins (Wells 1986, Wells et al. 1987) and Argentine bottlenose dolphins (Würsig and Würsig 1977). The differences in group fluidity between Florida and Argentine bottlenose dolphins may well be a result of one major feeding regime in the former and two different types of feeding—on nonschooling and on schooling prey—in the latter (described in Würsig 1979).

Wells (1986) recently compared Florida bottlenose dolphin groups to groups of African lions (*Panthera leo*), which also cooperate in obtaining food. The comparison of lions with relatively small schools of nearshore bottlenose dolphins is probably appropriate. We suspect that larger schools of the more pelagic dusky and spinner dolphins discussed above may have a general group structure and foraging patterns more akin to those of cooperating groups of chimpanzees. Chimpanzees in several study populations fluidly associate with many members of different troops, changing affiliations and troop sizes at least in part in response to environmental variables (Goodall 1965; Reynolds 1965; Hall 1968; Nishida 1968; De Waal 1982; Goodall 1983; Ghiglieri 1984, 1985). We still know rather little about social structure of delphinids, but we can surmise from frequent changes in affiliations and from the large testes size of males (Kenagy and Trombulak 1986) that the systems tend generally toward a promiscuous form of polygamy (see Wells 1986, for a detailed discussion of this assertion).

We suspect, from the different degrees and levels of dolphin group fluidity in different habitats, that habitat-related differences in social structure and mating strategies also exist, but this possibility has not been investigated.

REFERENCES

Ballance, L. T. 1990. Residence patterns, group organization, and surfacing associations of bottlenose dolphins in Kino Bay, Gulf of California, Mexico. In: The Bottlenose Dolphin, ed. S. Leatherwood and R. R. Reeves. San Diego: Academic Press.

Beaty, D. 1978. Antenna considerations for biomedical telemetry. Booklet for Montana Wildlife Society, Lewistown, Mont. 15 pp.

Bekoff, M., and M. C. Wells. 1980. The social ecology of coyotes. Sci. Amer. 242:130–148.

Cipriano, F. 1985. Dusky dolphins research at Kaikoura, New Zealand: A progress report. Mauri Ora 12:151–158.

Cipriano, F. 1989. Prey types, prey distribution and foraging ecology of dusky dolphins (*Lagenorhynchus obscurus*) off Kaikowa, New Zealand. Abstract. 8th Biennial Conference on the Biology of Marine Mammals, Pacific Grove, Calif.

Clutton-Brock, T. H., and P. H. Harvey. 1977. Primate ecology and social organization. J. Zool. Lond. 183 pp.

Davis, R. E., F. S. Foote, J. Anderson, and E. Mikhail. 1981. Surveying theory and practice. 6th ed. New York: McGraw Hill Books. 1,120 pp.

De Waal, Frans. 1982. Chimpanzee politics: Power and sex among apes. New York: Harper and Row.

Dietz, R. 1986. A feasibility study on satellite and VHF tracking of marine mammals. Report to Commission for Scientific Research in Greenland and Greenland Fisheries and Environmental Research Institute. Available from Danbiu Ap S. Henningsen Alle 58, DK-2900 Hellerup, Greenland. 93 pp.

Duffield, D. A. 1982. *Tursiops truncatus* genetics studies: Indian River 1980–1981. In: Live capture, marking, and resighting of bottlenose dolphins, *Tursiops truncatus*, ed. D. K. Odell and E. D. Asper. Final Report to NMFS, available from Rosenstiel School of Marine and Atmospheric Science, Univ. of Miami, 4600 Rickenbacker Causeway, Miami, FL 33149.

Duffield, D. A., and R. S. Wells. 1986. Population structure of bottlenose dolphins: Genetic studies of bottlenose dolphins along the central west coast of Florida. Final Report to NMFS, available from NMFS, SE Fisheries Center, 75 Virginia Beach Drive, Miami, FL 33149.

Dunlap, C. R. 1970. A reconnaissance of the deep scattering layers in the eastern tropical Pacific and the Gulf of California. In: Proceedings of an international symposium of biological sound scattering in the ocean, ed. G. B. Farquhan. Superintendent of Documents, Washington, D.C., stock no. 0851-0053, 395–408.

Erickson, A. W. 1978. Population studies of killer whales (*Orcinus orca*) in the Pacific Northwest: A radio-marking and tracking study of killer whales. Nat'l Tech. Inform. Serv. Rep. No. P.B. 285615. 34 pp.

Evans, P. G. H. 1987. The natural history of whales and dolphins. New York: Facts on File Publ.

Evans, W. E. 1971. Orientation behavior of delphinids: Radiotelemetric studies. Ann. of the N.Y. Academy of Sci. 188:142–160.

———. 1974. Radiotelemetric studies of two species of small odontocete cetaceans. In: The whale problem, ed. W. Schevill. Cambridge: Harvard Univ. Press.

Evans, W. E., J. D. Hall, A. B. Irvine, and J. S. Leatherwood. 1972. Methods for tagging small cetaceans. U.S. Fish. Bull. 70:61–65.

Gaskin, D. E. 1982. The ecology of whales and dolphins. London: Heinemann Press. 459 pp.

Gaskin, D. E., and M. W. Cawthorn. 1967. Diet and feeding habits of the sperm whale (*Physeter catodon* L.) in the Cook Strait region of New Zealand. N.Z. J. of Mar. and Freshwater Sci. 2:156–179.

Gaskin, D. E., G. J. D. Smith, and A. P. Watson. 1973. Preliminary study of movements of harbour porpoises (*Phocoena phocoena*) in the Bay of Fundy using radiotelemetry. Can. J. Zool. 53:1466–1471.

Ghiglieri, M. P. 1984. The chimpanzees of Kilale Forest. New York: Columbia Univ. Press.

———. 1985. The social ecology of chimpanzees. Sci. Amer. 252:102–113.

Gibbs, R. H., and C. F. E. Roper. 1970. Ocean acre preliminary report on vertical distribution of fishes and cephalopods. In: Proceedings of an international symposium on biological sound scattering in the ocean, ed. G. B. Farquhar. Superintendent of Documents, Washington, D.C., stock no. 0851-0053, 119–408.

Goodall, J. 1965. Chimpanzees of the Gombe Stream Reserve. In: Primate behavior, ed. I. De Vore. New York: Holt, Rinehart, and Winston.

————. 1983. Population dynamics during a 15-year period in one community of free-living chimpanzees in the Gombe National Park, Tanzania. Z. Tierpsychol. 61:1–60.

Gruber, J. A. 1981. Ecology of the Atlantic bottlenosed dolphin (*Tursiops truncatus*) in the Pass Cavallo area of Matagorda Bay, Texas. M.Sc. thesis, Texas A&M University.

Hall, K. 1968. Social organization of the Old World monkeys and apes. In: Primates: Studies in adaptation and variability, ed. P. Jay. New York: Holt, Rinehart, and Winston.

Hopkins, T. L., and R. C. Baird. 1977. Aspects of the feeding ecology of oceanic midwater fishes. In: Oceanic sound scattering prediction, ed. N. R. Andersen and B. J. Zahuranec. New York: Plenum Press.

Irvine, A. B., M. D. Scott, R. S. Wells, and J. H. Kaufman. 1981. Movements and activities of the Atlantic bottlenose dolphin, *Tursiops truncatus*, near Sarasota, Florida. Fish. Bull. U.S. 79:671–688.

Irvine, A. B., and R. S. Wells. 1972. Results of attempts to tag Atlantic bottlenose dolphins, *Tursiops truncatus*. Cetology 13:1–5.

Irvine, A. B., R. S. Wells, and M. D. Scott. 1982. An evaluation of techniques for tagging small odontocete cetaceans. U.S. Fish. Bull. 80:135–143.

Jarman, P. J. 1974. The social organization of antelope in relation to their ecology. Behaviour 48:215–267.

Jennings, J. G. 1982. Tracking marine mammals by satellite: Status and technical needs. Oceans (Sept.):751–754.

Jennings, J. G., and W. F. Gandy. 1980. Tracking pelagic dolphins by satellite. In: A handbook on biotelemetry and radio tracking, ed. C. J. Amlaner, Jr., and D. W. MacDonald. Oxford: Pergamon Press.

Kenagy, G. J., and S. C. Trombulak. 1986. Size and function of mammalian testes in relation to body size. J. Mammal. 67:1–22.

Kirby, V. L., and S. H. Ridgway. 1984. Hormonal evidence of spontaneous ovulation in captive dolphins, *Tursiops truncatus* and *Delphinus delphis*. In: Reproduction in whales, dolphins, and porpoises, ed. W. F. Perrin, R. L. Brownell, Jr., and D. P. De Master. Report of the International Whaling Commission, Special Issue no. 6, Cambridge, England.

Kooyman, G. L. 1965. Techniques used in measuring diving capacities of Weddell seals. Polar Rec. 12:391–394.

Kooyman, G. L., J. O. Billups, and W. D. Farwell. 1983. Two recently developed recorders for monitoring diving activity of marine birds and mammals. In: Experimental biology at sea, ed. A. G. MacDonald and I. G. Priede. New York: Academic Press.

Kooyman, G. L., R. L. Gentry, and D. L. Urquhart. 1976. Northern fur seal diving behavior: A new approach to its study. Science 193:411–412.

Leatherwood, S., and D. K. Ljungblad. 1979. Nighttime swimming and diving behavior of a radio-tagged spotted dolphin, *Stenella attenuata*. Cetology 34:1–6.

Leatherwood, S., and R. R. Reeves. 1982. Bottlenose dolphin (*Tursiops truncatus*) and other toothed cetaceans. In: Wild mammals of North America, ed. J. A. Chapman and G. A. Feldhamer. Baltimore: Johns Hopkins University Press.

Leatherwood, S., R. R. Reeves, and L. Foster. 1983. The Sierra Club handbook of whales and dolphins. San Francisco: Sierra Club Books. 302 pp.

Marshall, N. B. 1966. The life of fishes. New York: Universe Books. 402 pp.

Mate, B. R. 1989. Watching habits and habitats from earth satellites. Oceanus 32:14–18.

Mech, L. D. 1970. The wolf: The ecology and behavior of an endangered species. Garden City, N.Y.: Doubleday Press.

———. 1983. Handbook of animal radio-tracking. Minneapolis: University of Minnesota Press. 107 pp.

Nishida, T. 1968. The social group of wild chimpanzees in the Mahali Mountains. Primates 9:175–198.

Norris, K. S., and T. P. Dohl. 1980*a*. The structure and functions of cetacean schools. In: Cetacean behavior: Mechanisms and functions, ed. L. M. Herman. New York: John Wiley and Sons.

———. 1980*b*. The behavior of the Hawaiian spinner dolphin, *Stenella longirostris*. Fish. Bull. U.S. 77:821–849.

Norris, K. S., and K. W. Pryor. 1970. A tagging method for small cetaceans. J. Mamm. 51:609–610.

Norris, K. S., B. Würsig, R. S. Wells, M. Würsig, S. M. Brownlee, C. Johnson, and J. Solow. 1985. The behavior of the Hawaiian spinner dolphin, *Stenella longirostris*. Final Report for NMFS, Southwest Fisheries Center. 213 pp.

Perrin, W. F. 1975. Distribution and differentiation of populations of dolphins of the genus *Stenella* in the eastern tropical Pacific. J. Fish. Res. Bd. Can. 32:1059–1067.

Read, A. J., and D. E. Gaskin. 1985. Radio tracking the movements and activities of harbor porpoises, *Phocoena phocoena* (L.), in the Bay of Fundy, Canada. Fish. Bull. U.S. 83:543–552.

Reynolds, V. 1965. Some behavioral comparisons between the chimpanzee and the mountain gorilla in the wild. Am. Anthropol. 67:691–706.

Saayman, G. S., and C. K. Tayler. 1973. Social organization of inshore dolphins (*Tursiops truncatus* and *Sousa*) in the Indian Ocean. J. Mamm. 54:993–996.

Scott, M. D., R. S. Wells, A. B. Irvine, and B. R. Mate. 1990. Tagging and marking studies on small cetaceans. In: The bottlenose dolphin, ed. S. Leatherwood and R. R. Reeves. San Diego: Academic Press.

Shane, S. H. 1977. The population biology of the Atlantic bottlenose dolphin, *Tursiops truncatus*, in the Aransas Pass area of Texas. M.Sc. thesis, Texas A&M University.

———. 1980. Occurrence, movements, and distribution of bottlenose dolphins, *Tursiops truncatus*, in southern Texas. Fish. Bull. U.S. 78:593–601.

———. 1990*a*. Behavior and ecology of the bottlenose dolphin at Sanibel Island, Florida. In: The bottlenose dolphin, ed. S. Leatherwood and R. R. Reeves. San Diego: Academic Press.

———. 1990*b*. Comparison of bottlenose dolphin behavior in Texas and Florida, with a critique of methods for studying dolphin behavior. In: The bottlenose dolphin, ed. S. Leatherwood and R. R. Reeves. San Diego: Academic Press.

Shane, S. H., R. S. Wells, and B. Würsig. 1986. Ecology, behavior, and social organization of the bottlenose dolphin: A review. Mar. Mamm. Sci. 2:34–63.

Tanaka, S., K. Takao, and N. Kato 1987. Tagging techniques for bottlenose dolphins (*Tursiops truncatus*). Nippon Suisan Gakkaishi 53:1347–1325.

Tyack, P. L. 1982. Humpback whales respond to sounds of their neighbors. Ph.D. dissertation, Rockefeller University.

Tyack, P. L. 1986. Population biology, social behavior, and communication in whales and dolphins. Trends Ecol. and Evol. 1:144–150.

Uda, S., and Y. Mushiake. 1954. Yagi-Uda antenna. Sendai, Japan: Sasaki Press.

Watkins, W. A. 1985. The use of radio tags in tracking large whales: A review of present knowledge. The Third Conference on the Biology of the Bowhead Whale, *Balaena mysticetus*, Abstract. Available from Alaska Eskimo Whaling Commission, Barrow, Alaska.

Wells, R. S. 1984. Reproductive behavior and hormonal correlates in Hawaiian spinner dolphins, *Stenella longirostris*. In: Reproduction in whales, dolphins, and porpoises, ed. W. F. Perrin, R. L. Brownell, Jr., and D. P. De Master. Report of the International Whaling Commission, Special Issue no. 6, Cambridge, England.

———. 1986. Structural aspects of dolphin societies. Ph.D. dissertation, Univ. of California, Santa Cruz. 234 pp.

Wells, R. S., A. B. Irvine, and M. D. Scott. 1980. Social ecology of inshore odontocetes. In: Cetacean behavior: Mechanisms and functions, ed. L. M. Herman. New York: John Wiley and Sons.

Wells, R. S., M. D. Scott, and A. B. Irvine. 1987. The social structure of free-ranging bottlenose dolphins. In: Current mammalogy, ed. H. Genoways. Vol. 1:247–305, 519 pp.

Wells, R. S., and B. Würsig. 1983. Patterns of daily movements and distribution of Hawaiian spinner dolphins, *Stenella longirostris* Fifth Biennial Conference on the Biology of Marine Mammals, Abstract. Available from Wells, Long Marine Laboratory, Univ. of California, Santa Cruz, CA 95060.

Wilson, E. O. 1975. Sociobiology: The new synthesis. Cambridge: Harvard University Press.

Würsig, B. 1978. On the behavior and ecology of bottlenose and dusky porpoises in the south Atlantic. Ph.D. dissertation, State University of New York at Stony Brook. 335 pp.

———. 1979. Dolphins. Sci. Amer. 240:136–148.

———. 1982. Radio tracking dusky porpoises in the south Atlantic. Mammals in the Seas, FAO Fisheries Series no. 5, 4:145–160. FAO, United Nations, Rome, Italy.

Würsig, B., F. Cipriano, and M. Webber. 1985. Movement and dive patterns of radio-tagged dusky dolphins (*Lagenorhynchus obscurus*) in New Zealand. Sixth Biennial Conference on the Biology of Marine Mammals, Abstract.

Würsig, B., and T. A. Jefferson. In press. Methods of photoidentification for small cetaceans. Report of the International Whaling Commission (Special Issue 12). Cambridge, England.

Würsig, B., and M. Würsig. 1977. The photographic determination of group size, composition, and stability of coastal porpoises (*Tursiops truncatus*). Science 198:755–756.

———. 1979. Behavior and ecology of the bottlenose dolphin, *Tursiops truncatus*, in the south Atlantic. Fish. Bull. U.S. 77:399–412.

———. 1980. Behavior and ecology of the dusky dolphin, *Lagenorhynchus obscurus*, in the south Atlantic. Fish. Bull. U.S. 77:871–890.

———. 1983. Association, movement, and daily activity patterns of the spinner dolphin (*Stenella longirostris*) in Hawaii. Fifth Biennial Conference on the Biology of Marine Mammals, Abstract. Available from Würsig, Moss Landing Marine Labs, Moss Landing, CA 95039.

Würsig, B., M. Würsig, and F. Cipriano. 1989. Dolphins in different worlds. Oceanus 32:71–75.

Zaneveld, J. R. V. 1977. Optical parameters that may affect vertical movement of animals in the ocean. In: Oceanic sound scattering prediction, ed. N. R. Andersen and B. J. Zahuranec. New York: Plenum Press.

Killer whales, resident in Puget Sound and the waters of British Columbia, are observed by researchers from The Whale Museum, on San Juan Island in Washington state. (Photo by Fred Felleman.)

CHAPTER THREE

THE FEEDING ECOLOGY OF KILLER WHALES (*ORCINUS ORCA*) IN THE PACIFIC NORTHWEST

Frederic L. Felleman, James R. Heimlich-Boran, and Richard W. Osborne

INTRODUCTION

Numerous isolated accounts in the scientific and popular literature describe predation by the killer whale, *Orcinus orca*. Foremost is D. F. Eschricht's (1862) often misquoted account of the remnants of thirteen

We would like to thank M. A. Bigg, his co-workers, and Ken Balcomb for initiating this study and for sharing the data and insights they gathered from their research. In addition, we would like to thank our co-workers Nancy Haenel, Sara Heimlich-Boran, Camille Goebel-Diaz, and Rick Chandler, who also contributed to the collection of data presented here. We received constructive criticism from our friends and graduate advisors: Gary Thomas, Bernd Würsig, Ken Norris, and Sam Wasser. Jim Heimlich-Boran produced most of the figures presented. This work was funded by grants to the senior author from the Environmental Internship Program (EIP/NW), the Bailey Sussman Foundation, the Select Program at the University of Washington, and the Houston Underwater Club. The Washington Cooperative Fishery Research Unit at the University of Washington provided a research vessel and computer time. Additional research vessels were provided by Michael Tillman of the National Marine Mammal laboratory in Seattle, the Whale Museum, and Moss Landing Marine Laboratory. Biosonics, Inc., generously permitted the hydroacoustic data to be processed on their equipment. Jim Heimlich-Boran's research was supported, in part, by a grant from the Hewlett-Packard Foundation. The Whale Museum has provided Rich Osborne with an operating budget and much of the logistical support for the observations reported in this chapter. This work was initiated by grants to Ken Balcomb from the National Marine Fisheries Service and the Marine Mammal Commission.

porpoises and five seals in the stomach of an orca, as if it had eaten them all on one occasion. In addition, the nature of the wear observed on the orca's uniquely interlocking oval teeth has suggested "ferocious and highly predatory feeding behavior, or the functional modifications of their cranial anatomy which support this behavior" (Caldwell and Brown 1964 : 139). Such anecdotes and occasional observations of predation on large whales, which typically include graphic descriptions of the selective removal of the prey's lips and tongue (Tomilin 1957), have traditionally provided the basis for considering the killer whale primarily a predator of marine mammals. In this chapter, the common name "orca" will be used in lieu of killer whale, as it does not imply a simplistic characterization of its feeding behavior.

E. Hoyt (1984) has compiled the most current list of known orca prey, which included 24 species of cetaceans, 14 species of pinnipeds, 31 species of fish, 9 species of birds, 2 species of cephalopods, 1 species of turtle, and 1 species of otter. Orca foraging strategies have been described as "opportunistic" (Martinez and Klinghammer 1970; Defran and Pryor 1980; Dahlheim 1981), but summaries of orca prey from around the world can be misleading when addressing the feeding strategies of any one population. The summation of stomach content data only provides an indication of what orcas can eat, not what a discrete population does eat. Distinguishing between the ability and behavior of a predator, M. L. Rosenzweig (1985) uses the term *generalist* to describe a predator that feeds on many kinds of prey and the term *specialist* for a predator that uses some subset of the available prey. Rosenzweig suggests using the term *selector* to describe the behavior of a specialist and *opportunist* to describe a generalist predator. We suggest that orcas have opportunistic abilities but behave as selectors with the flexibility to follow fluctuations in their preferred prey.

Seasonal movements of orcas are clearly responsive to prey distributions. Orca occurrence has been found to be positively correlated with the presence of herring in the North Atlantic (Jonsgaard and Lyshoel 1970), elephant seal pups in the southwest Indian Ocean (Condy et al. 1978), sea lion and elephant seal pups along the Patagonia coast (Lopez and Lopez 1985), adult sea lions, pups, and elephant seal pups along the west coast of North America (Norris and Prescott 1961, Norris and Dohl 1980), gray whales with calves along the west coast of North America (Morejohn 1968, Baldridge 1972), minke whales in the Southern Ocean (Mikhalev et al. 1981, Budylenko 1981), and returning adult salmon in Pacific Northwest inland water (Spong and White 1970; Spong et al. 1971, 1972; Balcomb et al. 1980; Heimlich-Boran 1986).

Orcas exhibit varying degrees of cooperative foraging behavior depending on the type of prey selected (for reviews, see Martinez and

Klinghammer 1970; Hoyt 1984; Felleman 1986). Cooperative foraging in orcas has been identified by observations of group movements, from synchronous respirations while chasing and encircling prey (Ljungblad and Moore 1983, Steltner et al. 1984) to divisions of labor in the attack (Tarpy 1979) and the sharing of prey (Lopez and Lopez 1985). The ability to capture prey larger than the predator is the most commonly cited selective advantage of cooperative foraging in terrestrial social carnivores (Kleiman and Eisenberg 1973, Beckoff et al. 1984).

It is not intended to imply, however, that natural selection need be functioning above the individual level. For example, Schaller (1972) has shown that cooperatively hunting lions have a success rate of 30 percent as compared to 15 percent exhibited by solitary lions. In addition, social carnivores are typically territorial and often highly related so that individuals in a population are very familiar with each other (Beckoff et al. 1984). Consequently, cooperative foraging can be maintained in a population through the benefits accrued to individuals (Williams 1966) and kin selection (Hamilton 1964).

Specialization in prey choice among parapatric orca populations was found in the Soviet catch of 906 orcas (785 stomach samples taken) during the 1979–80 whaling season in the Antarctic Ocean (Berzin and Vladimirov 1983). Two types of orca populations were found: offshore and inshore. The offshore orca population consisted of groups of ten to fifteen individuals whose stomach contents contained 89.7 percent marine mammals, 7.1 percent squid, and 3.2 percent fishes. The inshore population consisted of groups of 150 to 200 individuals whose stomach contents contained 98.5 percent fishes, 1.1 percent squid, and 0.4 percent marine mammals. The individuals in the inshore population were approximately one meter smaller on average than individuals in the offshore population. These distinctions were striking enough to the Soviet authors to tentatively define the inshore population as a new species, *Orcinus glacialis*.

Prey choice also appears to vary between individuals within an orca population. Orca stomach content data from populations in the eastern North Pacific (Rice 1968), western North Pacific (Nishiwaki and Handa 1958), and northeast Atlantic (Jonsgaard and Lyshoel 1970) indicate that prey choice varies between age and sex classes, with larger individuals eating a greater percentage of marine mammals. The peripheral involvement of males observed during attacks on cetaceans (Tarpy 1979, Arnbom et al. 1987) apparently contradicts the few stomach samples or may suggest that males feed after females have immobilized the prey, as observed in lions (Schaller 1972).

In the Pacific Northwest, Bigg has defined three socially isolated populations of orcas, which he called northern resident, southern resident, and transient, in the inland waters of Washington and British Columbia, Canada (Bigg 1982, Bigg et al. 1987). It has been suggested that resident and transient orcas have different diets, judging by direct observations of predation and two analyses of stomach contents (Balcomb et al. 1980).

This chapter addresses the question of distinctions in the feeding ecology of resident and transient pods. We describe a series of field techniques that are being used to distinguish the prey choice and foraging strategies of the three orca communities.

METHODS

Oceanographers have shown that fronts and eddies in the open ocean can cause patterned patches of organisms that predators are able to exploit predictively (Owen 1981). Japanese whaling fleets have known how this information can be used successfully to locate and exploit cetaceans commercially (Nasu 1966). Recently, oceanographic information has been incorporated into field studies of cetacean ecology (Gaskin 1976; Saayman and Tayler 1979; Norris and Dohl 1980; Würsig and Würsig 1980; Wells et al. 1980; Felleman 1986; Shane 1987; Heimlich-Boran 1987, 1988). These studies, among others, have begun to show that an understanding of the spatial and temporal characteristics of prey availability provides indications of how feeding behavior may be responsive to prey distributions. Ecological research on cetaceans is severely limited by the fact that the majority of a whale's life is spent under the surface of the water. The standard methods of feeding habit analysis, including direct observation of the search, pursuit, and capture of prey, fecal analysis, and the collection of stomach contents, are often impossible or impractical to use with cetaceans.

A fisheries-oceanographic approach to the study of feeding ecology was employed in this study to determine how the temporal distribution of orcas correlates with abundance of food. Orca prey choice in the Pacific Northwest was deduced from surface observations of behavior, echoranging on prey fish, group size, and inference from the oceanographic and geographic features of the habitat as well as temporal and spatial distribution of prey gleaned from fishery data.

The Study Area

Orcas were observed in the inland waters of the Pacific Northwest coast of North America between 47°N and 51°N (fig. 3.1). The primary study area occupied by the southern resident community, referred to as "Greater Puget Sound," lies between 47°N and 49°N and 122°W and

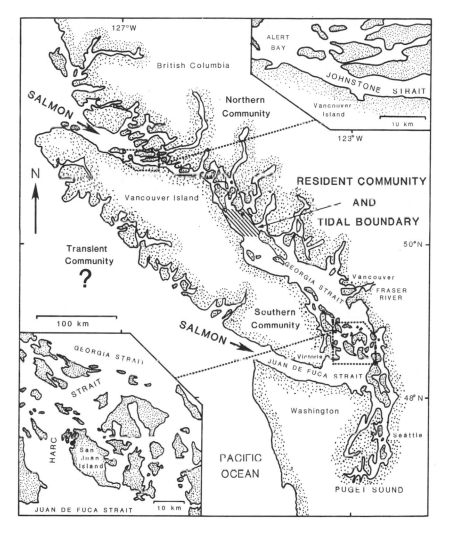

Fig. 3.1.   Home range of Pacific Northwest orca communities. Cross-hatching indicates the boundary area between the northern and southern resident communities (after Bigg and Wolman 1976). This is also the area of tidal convergence (Tully and Dodimead 1957). Core areas of the two resident communities are shown in detail.

124°W. This region includes the glacially carved waters of Puget Sound, the Strait of Juan de Fuca, the San Juan archipelago, and the southern Strait of Georgia. A separate northern resident community is centered in Johnstone Strait, which is approximately 50.5°N and between 126.3°W and 126.9°W.

The region is characterized by a varied bathymetry. Average depth is about 100 to 150 m, with maximum depths of over 400 m in central Georgia Strait and over 500 m in Johnstone Strait. The tides are of mixed semidiurnal type with two highs and lows of unequal magnitude. Average maximum tidal current speeds are approximately 50 cm/sec., with maximum current speeds of up to 250 cm/sec. in Haro Strait (Thomson 1981).

The study area is a productive region with abundant prey resources for orca. The Fraser River (fig. 3.1) is the single largest source of fresh water and salmon in the region. A list of potential fish prey is dominated by several species of seasonally abundant salmon (*Oncorhynchus*) that migrate through the area from June through September, steelhead trout (*Salmo gairdneri*) and cutthroat trout (*Salmo clarkii*) during the winter, and several other genera of schooling and bottom fish. Potential pinniped prey include a few thousand resident harbor seals (*Phoca vitulina*) and seasonally occurring species such as elephant seals (*Mirounga angustirostris*), California sea lions (*Zalophus californianus*), and Stellar's sea lions (*Eumetopias jubatus*) (Spalding 1964; Everitt et al. 1979; Angell and Balcomb 1982). Potential cetacean prey include the two resident porpoise species, Dall's porpoise (*Phocoenoides dalli*) and harbor porpoise (*Phocoena phocoena*), and two baleen whales, seasonally resident minke whales (*Balaenoptera acutorostrata*) and migratory gray whales (*Eschrichtius robustus*) (Scheffer and Slipp 1948; Everitt et al. 1979; Dorsey 1983).

Pacific Northwest Orca Population

The orcas of Washington and British Columbia offer exceptional opportunities for the study of a cetacean species in the wild. Most can be easily identified by naturally occurring markings. Sexual dimorphism is marked. They live in cohesive social groups in nearshore waters. However, little was known about them until after extensive aquarium captures in the 1960s and 1970s when concern for the population grew and field studies were initiated (Bigg and Wolman 1976). These orcas live in stable family units called pods (Bigg et al. 1976; Bigg 1982; Balcomb et al. 1982; Bigg et al. 1987). From extensive photographic documentation of pod associations from spring through fall over the last decade, it appears that a pod is an extended family unit composed of mothers and their offspring and associated males. The adult males are thought to be sons or brothers of the matriarchs (S. L. Heimlich-Boran 1986, J. R. Heimlich-Boran 1988). It is generally believed that in most cases, both sexes spend their entire life spans in maternal subgroups, remaining within their natal pods.

The southern resident community is composed of 81 individuals in three pods, the northern resident community is composed of 172 individuals in sixteen pods, and the transient community is composed of 79 individuals in thirty pods (Bigg et al. 1987). The two resident communities have nonoverlapping ranges that correspond to two distinct tidal regions (Felleman 1986): a northern region with southward-flowing flood currents and northward-flowing ebb currents, and a southern region with opposite ebb and flood current directions (Tully and Dodimead 1957). The transient community travels throughout the range of both northern and southern resident communities but does not socially interact with them (Bigg 1982). At least one of these transient pods has been observed as far away as southeast Alaska (Leatherwood et al. 1984). Resident and transient orcas appear to be reproductively isolated based on the morphologic distinctions in their dorsal fins (Bigg et al. 1987), distinct call dialects (Ford and Fisher 1982, 1983), characteristic utilizations of their habitat (Felleman 1986; Heimlich-Boran 1988), and differing degrees of social interactions with other pods (Osborne 1986, Jacobsen 1986).

Identification

Pods were recognized by visual identification of known members (fig. 3.2). Photographs of the dorsal fin and back of all individuals, using 35 mm cameras with 200 to 300 mm lenses, allowed confirmation of identification of all pod members.

Observations

Two primary types of observations were used: sightings and encounters. Sightings were reports of whales received from the public through a toll-free telephone reporting system that extended our seasonal and geographic coverage and helped researchers locate whales. Encounters were observations by researchers in which pod identity was determined. Behavioral and ecological data were collected from shore observation points and from research vessels. Vessel encounters were conducted in accordance with National Marine Fisheries Service permit regulations.

Seasonal Occurrence and Correlation with Fisheries

All observations were logged chronologically and plotted on charts. For analyses, sightings were grouped by day and by fishery reporting areas;

Fig. 3.2. Members of a subgroup from L pod displaying unique variations in white "saddle" and shape of dorsal fin. From left to right: L-35 (the mother of the others in the photograph), L-50, L-65 (L-35's current calf), L-54, and L-1 (adult male). The dorsal fin of an adult male such as L-1 may reach 2 m in height; those of females are usually less than 1 m. (Photo by Fred Felleman.)

one or more sightings on any given day was an "area sighting day." Data were then tabulated to provide the frequency of sighting days per week. These data were compared with salmon sport fishery data using a correlation test for paired variables (Zar 1984).[1] We also calculated a seasonally corrected correlation coefficient using the difference between data for corresponding weeks in consecutive years to compensate for any common seasonal variation.

We attempted to maximize the time spent with whales. Almost one quarter of the observations were made in Haro Strait; over half of the observations were made from July through September. Sighting effort was not quantified during the study. The large number of observations probably indicates fairly the kinds of activities of these whales in their core area of distribution from spring to fall.

Behavioral Sampling

Behavior data from 1976 to 1982 were collected by a continuous sampling method (Altmann 1974). In subsequent years, data were collected by a scan sampling method at fifteen-minute intervals (ibid.). A fifteen-minute sampling period was overlaid on the continuous data from 1976 to 1982 for the analysis of habitat use by both resident and transient whales.

The methodology used to define the behavior categories was described by Osborne (1986). Behaviors such as a breach, spy hop, tail lob, or observations of prey were noted directly onto data sheets. Coded sentences were used to describe the most common surface formations. Osborne defined eight functional categories of behavior as forage, percussive forage, mill, travel, percussive travel, rest, play, and intermingle. This chapter will address only the feeding behaviors that most directly indicate prey selection: foraging, percussive foraging, milling, and marine mammal predation.

Habitat Use

Whale locations were determined from compass triangulations on nearby shorelines and marker buoys, using a Morin Opti2 hand-bearing compass. The study area was divided into a grid of 441 quadrats. Each quad-

---

[1] Fisheries catch data are published in annual statistical reports by the Washington State Department of Fisheries. We used data from the salmon sport fishery because it was more complete than data for other fishes or fishing methods, but we do not thereby imply orcas only prey on salmonids.

rat was approximately 4.6 km square. Whale routes were plotted on this grid, and occurrences in particular quadrats were recorded at fifteen-minute intervals along with behavioral and tidal data.

Whale routes and behavior patterns for each encounter were sorted by quadrat and behavior, thus pooling all observations of a given behavior in a given quadrat throughout the months for which we have data. In some cases, observations were then pooled into eight major regions. For statistical purposes, the intrinsic null hypothesis was that the behavior distribution for each quadrat should be the same as that for the study area as a whole (i.e., that behaviors were distributed uniformly throughout the study area). The frequent occurrence of specific behavior patterns in certain quadrats allows the hypothesis that the whales used specific areas for specific behavioral purposes.

Tidal Data

The whales' behavior and direction of travel in relation to the tidal current direction provide further indication of prey choice. We recorded, at fifteen-minute intervals, whether whales were moving with, against, or nonoriented to the current.[2] A half-hour prior to and following slack current was used as the time of minimal current, during which time all behavior was called nonoriented.

These data were also used to test if the whales' direction changes were independent of tidal changes using a Pearson's chi-square test with Cochran's correction for continuity (Zar 1984). Autocorrelation analysis (Box-Jenkins) revealed that the tendencies of the whales to make direction changes were independent events. (A direction change was defined as a complete reverse of direction resulting in a new orientation to the tidal current; consequently, a direction change would have to occur during or just after the hour slack interval to be considered associated with the change in the tidal current.)

Hydroacoustic Data

During 1984 and 1985, the 5 m boat was equipped with echo-ranging acoustic devices to detect and quantify the presence of prey fishes.[3] The

---

[2] The direction of the tidal current was extracted from U.S. and Canadian government publications, while the extrapolation of phase differences between tide stations and the locations of whales were taken from Mojfeld and Larsen (1984), Crean (1983), and Thomson (1981).

[3] Hydrophones were made by Lab-Core or Aquadyne. Tape recorders used aboard vessels were Sony TC-152 SD or Sony TC-D5. Fixed array recorders were Sony SRA-3, Akai

majority of recording was done when whales were actively milling because of noise-induced limitations at the whales' higher speeds of movement during other behaviors.

Both single and multiple targets, indicating individual fish and schools of fish, were present in the study area. Species identification was inferred by observations of nearby fishing effort, depth distribution of targets, associated substrates, and the size and schooling tendency of the targets.

RESULTS

Feeding Ecology of Southern Resident Pods

Seasonal Occurrence and Correlation with Fisheries The seasonal sightings of southern resident whales from 1977 to 1983 are shown in figure 3.3. The distribution in time and space of public sightings closely parallels the distribution of research encounters. Whales were seen during all months of the year but most frequently during August. Over half of the observations occurred between June and September.

The occurrence of the whales during 143 weeks from April 1976 to December 1978 correlated significantly with the occurrence of salmon caught by sportfishermen (R = 0.74, P < 0.001; J. R. Heimlich-Boran 1986). When there was an increase in the number of salmon caught, there was a similar increase in the frequency of orca observations. The seasonally corrected correlation coefficient was also significant (5 = 0.45, P < 0.001), indicating that any variation in the number of salmon caught from year to year matched a similar variation in the number of orca sighting days.

The weekly distribution of sightings shows small-scale winter peaks in whale occurrence (ibid.). The peaks correspond to catches of juvenile chinook salmon that move downstream from their natal rivers into the estuarine waters during this time as well as to catches of winter runs of adult cutthroat and steelhead trout that are moving into fresh water at this time (Hart 1973). In addition to the correlation between whale occurrence and salmon catch in the overall study area, significant correla-

---

1730D-SS, or Akai GX-365-D. The frequency response of the systems ranged from 30 Hz to 16 kHz. The system consisted of a 70 kHz Simrad EYM scientific echo-sounder with a pulse length of 0.6 msec. The half-angle of the transducer (at −3 dB points) was 14 degrees. The transducer was mounted on the transom of the boat. The returning echoes were digitized through a Sony PCM-1 digitizer, and echoes were stored for later analysis on a Sony TCDM-5 cassette or Panasonic Rv-100 videotape recorder. The signal strength of fish targets and calibration signals were monitored on a Tektronics DDM 305 oscilloscope. All components operated from a 12-volt DC power supplied by deep-cycle marine batteries.

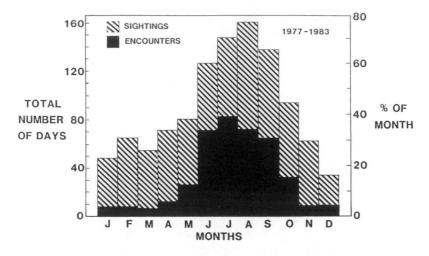

Fig. 3.3. Seasonal occurrence of southern resident orcas. "Sightings" are reports from the public to the Whale Museum's toll-free telephone system. "Encounters" are documented observations of identified orcas.

tions were found for most of the subdivisions of the study area as well. For example, the presence of whales in the narrow and confined area of southern Puget Sound was strongly correlated with the arrival of salmon that spawn in that region.

Geographic Distribution   Travel routes of the southern resident whales were mapped using 239 vessel encounters for 994.5 hours (mean 4.1 hrs./encounter) from 1976 through 1983 (Heimlich-Boran 1987, 1988). They were tracked for 6,660 km through 177 (40%) of the 441 study quadrats. The number of encounters per quadrat varied from one to 141. The Haro Strait region at the center of the study area accounted for almost two-thirds of the observations, while the remainder were spread nearly equally throughout the other regions (table 3.1).

Feeding Behavior   Feeding behaviors—forage, percussive forage, milling, and marine mammal predation—occurred in 47 percent of 985 hours of observations from 1976 to 1983 (fig. 3.4). Foraging was characterized by loose forward-oriented travel interspersed with brief instances of nondirectional active milling by subgroups or peripheral individuals (Osborne 1986). This sporadic milling was presumed to indicate the detection and pursuit of prey. Observations of salmon herding have been made (Felleman 1986, Osborne 1986), but direct observations of consumption are rare (see Balcomb et al. 1980, Bigg et al. 1987). The only

*Table 3.1.   Regional Distribution of Southern Resident
Whale Observations*

Region	Hours	Percent Occurrence
Haro Strait	617.5	62.1
Georgia Strait	113.2	11.4
Juan de Fuca Strait	108.0	10.9
San Juan Islands	79.0	7.9
Puget Sound	76.8	7.7
Total	994.5	100.0

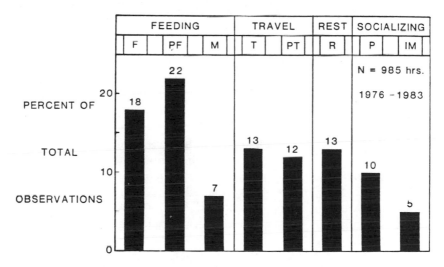

Fig. 3.4.   Distribution of behavior for southern resident whales: foraging
(F), percussive foraging (PF), milling (M), traveling (T), percussive traveling
(PT), rest (R), play (P), and intermingling (IM).

stomach contents collected from a resident pod came from a dead,
beached, adult male (L-8). His stomach contained only fish remains, including salmon and bottom fishes (Balcomb et al. 1980).

Typically, a basic foraging pattern extended over a 3 to 10 km range
with pod members traveling abreast in flank formations, oriented at right
angles to the shoreline or to underwater ridges (fig. 3.5). In the southern
resident community, females and calves often traveled together in loose

Fig. 3.5.   Members of J pod, an adult orca male, three females, and a calf, traveling in flank formation.

or tight subgroups, while adult males and females without calves traveled individually or in pairs. During foraging behavior, adult female and calf groups commonly engaged in percussive splashing just prior to milling. Percussive behaviors included slapping tails, dorsal fins, or pectoral fins on the water and breaching (fig. 3.6). All whales present sometimes engaged in continuous milling behavior, for a minimum duration of 0.2 hour, a maximum of 4.0 hours, and a mean of 1.5 hours (Osborne 1986).

Individuals or subgroups often exhibited foraging behavior peripheral to the pod. In this peripheral strategy, adult males typically foraged individually as much as 3 km from the nearest individual, though still traveling in the same direction as the pod. Adult females foraging at a distance from the pod were normally in pairs. These peripheral individuals and subgroups often exhibited rapid sharp turns at the surface probably indicative of the rapid turning avoidance tactics of salmon which we have observed in the presence of foraging whales. Pairs of peripheral females frequently oriented toward each other as they dove, which possibly enabled them to herd fish. In contrast, females with calves traveling close inshore remained parallel to each other and rarely changed direction sharply.

Predation on marine mammals by southern resident pods was ob-

Fig. 3.6. Members of L pod exhibit percussive behavior while foraging; left: breach; right: tail lob. (Photo by Fred Felleman.)

served on only three occasions, which amounts to less than 1 percent of recorded sightings. All instances involved a female and calf subgroup in L pod. The first instance took place during August 1976 in Haro Strait and involved 20 minutes of coordinated chase and/or play with a neonate harbor porpoise (*Phocoena phocoena*) followed presumably by consumption (Chandler et al. 1977, Balcomb et al. 1980). The other instance occurred in July 1982 and involved a Dall's porpoise. Again an L pod subgroup of females and calves was involved, and again they engaged in about 20 minutes of coordinated chase/play before the porpoise disappeared. Kenneth Balcomb has told us of a third instance that took place in summer 1987 and also involved a subgroup in L pod feeding for approximately 20 minutes on a harbor seal.

On twenty-eight other occasions, we observed marine mammals, including porpoise, minke whales, and harbor seals, in the vicinity of resident whales without any noticeable predatory reaction by the orcas or their potential prey. On two occasions, the whales appeared to follow a minke whale that passed by, but no attack was observed. Twelve of the cases involved observations of pinnipeds that were usually within 50 m of the whales.

Habitat Use   The majority of observations on feeding were in the thirteen quadrats of Haro Strait (Heimlich-Boran 1987, 1988), especially in three areas that are characterized by high-relief subsurface topography and the most varied bathymetry in the area. (Other areas were used for other behavior such as rest and socializing.)

The first area is off the southwest shore of San Juan Island, a canyon 260 m deep with numerous seamounts rising up over 100 m from the canyon floor. Whales were observed in two quadrats there for 106.5 hours during 279 encounters; foraging was the predominant feeding behavior. A second area of frequent feeding was found just north of San Juan Island where 170 m depths were ringed by a number of kelp-covered seamounts. Milling and percussive foraging predominated here. The third area of Haro Strait with significant amounts of feeding was the mouth of the Strait of Georgia. In this area, a prominent sill is oriented at right angles to the strait, with steep slopes at either side. Whales were observed here for 18.5 hours. Pecussive foraging was the predominant feeding behavior.

Feeding behavior also occurred in two larger regions, Puget Sound and Georgia Strait, especially along the Fraser River Delta. This delta is well known as a feeding and lingering area for salmon migrating upstream (Groot et al. 1986); steep slopes drop down from the 5 m deep banks to 100 m of depth within 1 km of the bank edge. The predominant feeding behavior in Georgia Strait was percussive foraging.

Milling was the predominant feeding behavior in Puget Sound. Although the subareas of Puget Sound did not have large enough samples to test independently, milling was observed consistently throughout. The presence of whales inside Puget Sound correlated with increased salmon catches by sportfishermen (P < 0.001; J. R. Heimlich-Boran 1986).

Tidal Orientation   Southern resident orcas had a tendency to move with the flood and against the ebb current. Milling occurred in 68.7 percent of all observations at slack current, almost ten times as often as during flood and ebb currents from 1983 through 1985 (Felleman 1986). Sonically tagged salmon have also been shown to mill at slack currents (Stasko et al. 1973, 1976).

The whales often confined the area of their daily range by changing their direction of travel after milling at slack current in areas of salmon abundance. Orcas changed their direction of travel within an hour of slack current seven times more frequently than would be expected by chance (P < 0.001). They were also likely to change their behavior within an hour of a slack current, especially if they did not change direction (P < 0.001).

Acoustic Behavior   Acoustic recording of southern resident pods revealed the occurrence of general vocal activity during 83 percent of the observations of foraging behavior and 91 percent of the observations of milling behavior (Osborne 1986). Milling behavior had the highest vocalization rate for any behavior (Hoelzel and Osborne 1986).

Hydroacoustics   Sonar recordings made in the presence of milling orcas from the southern resident community revealed a predominance of large single targets in the upper water column (above 20 m) which were probably salmon, based on comparisons with previous studies of salmonid stock assessment, knowledge of fish distributions in the region, and concurrent observations of the commercial salmon fishery in the region (Thomas and Felleman 1988).

Northern Resident Pods

Seasonal Occurrence and Correlation with Fisheries   The northern resident community consists of 16 pods numbering approximately 172 individuals (Bigg et al. 1987). Northern resident pods were observed for 38.5 hours during a pilot study conducted by the senior author in July 1985 to compare northern resident pods' movements and behavior relative to the tidal current with that of the southern resident and transient pods. Three subgroups of A pod, one subgroup of I pod, and all of B and C pods were observed during the pilot study.

Sightings of northern resident orcas appear to coincide with peaks of salmon abundance, similar to southern resident whales. The northern resident community usually travels along a 25 km stretch of Johnstone Strait during the summer months (Ford and Fisher 1982; Jacobsen 1986) but ranges from northern Vancouver Island to northern Georgia Strait. As in the southern resident community, a core area exists which overlaps with the commercial salmon fishery (Hilborn and Ledbetter 1979), located at a point of aggregation for returning adult Fraser River salmon. Paul Spong and his co-workers were the first to recognize the association between tide and season on orca and salmon occurrence in the northern resident community (Spong and White 1970; Spong et al. 1971, 1972). In Johnstone Strait, there is a predictable five-day ebb current corresponding to monthly neap tides. During these times, the flood current is not strong enough to counteract the estuarine flow of river water at the surface moving toward the ocean. Consequently, the amount of fresh water at the surface is at a maximum during these times. All Fraser River salmon runs were late in 1985, and so was the arrival of most of the northern resident pods. Only three pods were observed in Johnstone Strait just prior to and during the five-day ebb period, July 23 through July 28. Sockeye salmon catches for that week were estimated to reach their maximum of 8,100 fishes on July 28. The following day, when the tides began their normal reversing process, the estimated sockeye catch increased to 262,000. This coincided with the arrival of three additional northern resident pods into the region for the first time that season.[4] Apparently, some of the pods waited for the majority of salmon to return before they entered the inshore waters.

Feeding Behavior

Feeding behavior was observed in the northern resident community 31.4 percent of the total 38.5 hours of observations. Milling behavior occurred in 9.8 percent of the observations. Foraging behavior was similar to that of southern resident pods in most respects. However, less high-speed

---

[4]It is suggested that the adult salmon were waiting offshore for olfactory cues to lead them back to their natal streams (Felleman 1986). These cues are presumably associated with the fresh water, which was abundant during the prolonged ebb current. It also seems likely that the salmon could not progress far inshore until the flood currents resumed. A similar association between the timing of returning adult Atlantic salmon and freshwater influences was observed by Huntsman (1936).

traveling or "porpoising" was observed in the northern resident community than in the southern resident community, which agrees with observations by Ford and Ford (1981).

Predation on marine mammals by northern resident pods has never been documented. However, some chasing interactions have been noted between northern resident pods and Dall's porpoise (Jacobsen 1986). In general, as with the southern resident community, the northern resident orcas do not appear to select available marine mammal prey (Jeffersen 1987).

Tidal Orientation

The northern resident pods were observed to travel with the current 31.7 percent of the time and against it 37.9 percent. Milling behavior occurred during 13.9 percent of the slack current observations (table 3.2). Direction changes in the northern resident community, like those of the southern resident community, were seven times more likely to occur within an hour of slack current than would be expected by chance (P < 0.001). This corresponds with Spong's observations of northern resident orcas. In addition, behavior changes also tended to occur within an hour of slack current.

Feeding Ecology of Transient Pods

Seasonal Occurrence and Correlation with Prey Availability   Observations of transient pods presented here are from eighteen encounters totaling 42.8 hours of data collected between 1976 and 1985 on six pods

Table 3.2.   Occurrence of Milling Behavior as Percent of
Tidal Observation

Pod Community	Flood	Ebb	Slack
S. Resident, 1976–1979	4.8	2.1	11.5
S. Resident, 1983–1985	7.8	7.2	68.7
N. Resident, 1985	7.5	8.8	13.9
Transient, 1976–1985	13.9	33.3	6.9

Note: Increased occurrence of milling during slack current in the 1983–1985 resident data base as compared to the 1976–1979 data base may be due in part to the improved accuracy of making in situ tidal calculations.

totaling 13 individuals. These data are supplemented by 181 hours of observation of two radio-tagged whales from O pod (Erickson et al. 1987). These same pods have been documented at the northern end of Vancouver Island by Bigg and his colleagues, along with twenty-four other transient pods (Bigg et al. 1987). Transient pods are small, with fewer than 6 individuals (mean = 3); some sightings consist of lone adults (ibid.).

The occurrence of transient pods is sporadic throughout all regions of Washington and British Columbia. Baird and Stacey (1987) have found that transient whales can be encountered more frequently than elsewhere along the southwest shore of Vancouver Island. Their ranges may extend throughout the North Pacific, including southeast Alaska (Leatherwood et al. 1984) and possibly Oregon and northern California.

Although there is a seasonal bias in the sighting effort for transient pods, they never occur as frequently in Puget Sound as the southern resident pods. Since transient pods often occur in different parts of the habitat than resident pods, their occurrence may go unnoticed. They occur less frequently than resident pods in Haro Strait throughout the year.

The sample size of transient sightings was too small to conduct a statistical assessment of the correlation between their occurrence and that of potential prey. They have never been seen during June and July (fig. 3.7). Sightings in spring and fall may be related to the peaks in abundance of adult Pacific herring (Mar.–Apr.) and returning Fraser River adult salmon (Aug.–Oct.). However, it appears that their primary prey species include harbor seals and various bottom fishes that are available year-round in the inshore water (Everitt et al. 1979, Hart 1973). Consequently, transient pods are not limited to a specific season for finding these prey.

Geographic Distribution and Habitat Use   Transient whales were observed in 26 of the 441 quadrats of the southern community study area (Heimlich-Boran 1988). Ten of those 26 quadrats (38%) are areas in which resident whales were never recorded in almost 1,000 hours of observation. Transient pods were always sighted in relatively shallow and nearshore areas away from the major tidal channels; in many cases, these correspond to major harbor seal haul-out areas (Everitt et al. 1979). This analysis was recently applied to a study that radio tracked two transient whales from O pod for ten days in 1976 (Erickson et al. 1987). The whales traveled through 88 quadrats, 34 (38.6%) of which have been exclusively used by transient pods and are regions of high pinniped abundance. During the late summer and fall, sighting reports indicate that transient pods enter the inland waters from Juan de Fuca Strait at the

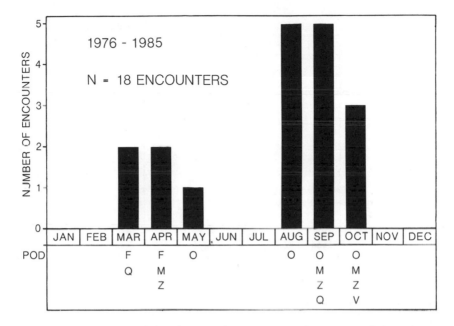

Fig. 3.7. Seasonal distribution of encounters with transient whale pods.

southern tip of Vancouver Island. This area is known to be a sea lion haul-out (Angell and Balcomb 1982).

Feeding Behavior Feeding behavior occurred during 81 percent of the transient pod observations (fig. 3.8). Transient pod fish-foraging behavior was very similar to peripheral adult foraging among resident pods. There was a general directional trend in the movement of the pod, interspersed with periods of individual nondirectional milling. However, transient pods exhibited a greater tendency to follow coastlines and to spend long periods milling in one area, often in bays, before moving on. Percussive behavior was observed only rarely during these foraging patterns.

Stomach contents have been examined in only one adult male known to be transient, O-1. The stomach contained only marine mammal remains: elephant seal claws and harbor seal bones (Balcomb et al. 1980). M. A. Bigg has told us that the stomach contents from two other solitary males that were probably transients contained primarily harbor seal remains but also included remains of sea lions, an elephant seal, and unidentified cetacean skin.

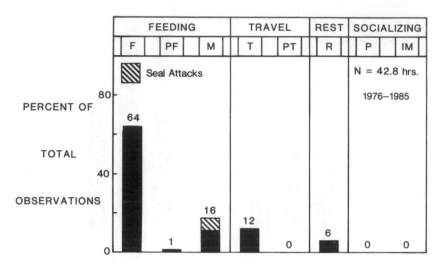

Fig. 3.8. Distribution of behavior for transient whales: foraging (F), percussive foraging (PF), milling (M), traveling (T), percussive traveling (PT), rest (R), play (P), and intermingling (IM).

Marine mammal predation by transient pods was observed on four occasions, involving individuals from four of the six transient pods documented in the study area. In every instance, the whales were preying on free-swimming harbor seals. This represented 22 percent of all transient encounters, a much higher percentage than was seen in southern resident pods. Harbor seal predation included both cooperative and individual hunting strategies. In the cooperative strategy, the seal was surrounded by circling whales, and the whales took turns swimming rapidly at the seal, striking it with their bodies, flukes, dorsal fins, or pectoral fins. In the solitary strategy, a single individual within the pod repeatedly slapped its tail on the seal (fig. 3.9). Birds were often attracted to the site of the kill, suggesting that the seal was not consumed whole. In all instances, after 18 to 24 minutes of attack, the seal disappeared and the pod milled in a small area, possibly sharing the kill.[5] In one instance, 10 minutes after the seal disappeared, a whale surfaced with what appeared to be entrails draped on its dorsal fin. In addition, four harbor seals were apparently

[5] Lopez and Lopez (1985) have observed orcas sharing pinniped prey in Patagonia, Argentina. However, Rice (1986) has also reported that small pinniped prey is often swallowed whole.

Fig. 3.9. A transient male orca repeatedly strikes a harbor seal with his flukes. This is the only context in which percussive behavior is regularly exhibited by transient whales. (Photo by Fred Felleman.)

consumed during the ten-day radio tracking of O pod. Predation of sea lions by transient orcas has been observed by other investigators near haul-outs located along the southwest end of Vancouver Island (Baird and Stacey 1987).

Tidal Orientation   In contrast to the usually directional movements of resident pods, nondirectional milling took place in 60 percent of all transient whale observations. Although transient pods have been known to travel approximately 550 km in six days (Ford and Ford 1981), once inshore, they tend to move quickly across major straits and then to remain in restricted areas for extended periods. They were observed to mill during only 6.9 percent of slack observations (table 3.2), which suggests that their prey do not exhibit the same tidal dependence as resident whale prey or that they use the tide differently to catch their prey. Protected bays and channels will often remain essentially slack throughout the tidal cycle, but migratory species rarely enter such areas.

Changes in the tidal current did not significantly affect transient whale behavior or direction of travel or behavior. However, these data must be considered preliminary due to the small sample of tide changes and limited amount of in situ tidal calculations. It would be expected that transient whales would utilize slack current as resident whales do if they also feed on salmon, a hypothesis currently being tested in the retrospective tidal analysis of the radio tracking of O pod.

Although transient whales did not seem to respond to changes in tidal currents, all observations of seal predation or presumed predation during radio tracking occurred during positive tide heights. Seals are most likely to be found off their haul-outs during high tides because many of these rocks are only exposed at low tides (Everitt et al. 1979).

Acoustics and Hydroacoustics   Transient pods were silent in the majority of instances when they were acoustically monitored (Ford and Fisher 1983, Osborne 1986). On one occasion, a few vocalizations were recorded during daylight foraging; this represents less than 2 percent of the daylight acoustic sample of transient whales (N = 4.6 hrs.).

Recordings were made in September 1984 of Q pod while milling in San Juan Channel. Analysis of these data revealed the presence of large single targets and schooled targets at the surface, indicative of salmon and schooling bait fish. However, there were almost three times the percentage (27%) of schooled targets than with the resident pod recordings. Since salmon and bait fish are predator and prey, respectively, it is not surprising to find them co-occurring in the recordings.

DISCUSSION

Orcas are long-term inhabitants of the Pacific Northwest (Scheffer and Slipp 1948), undoubtedly highly familiar with the environment. Habitat selection theory suggests that an animal must make decisions about how much time to spend in different resource patches of the habitat (MacArthur and Pianka 1966, Rosenzweig 1985). Since prey distributions vary spatially and temporally, distinctions between when and where resident and transient pods find and obtain food within the same region may be good indications of differences in their prey choice.

Similarly, the way in which a whale forages can be a predictor of prey choice. Each community had its own methods of foraging. The transient community orcas foraged very differently from the resident community, even while in the same habitat, which indicates they were seeking different prey.

Resident orcas commonly swim in flank formation when searching for prey but typically break into smaller subgroups for its capture. Norris and Dohl (1980) suggest that a group of odontocetes swimming in flank formation can effectively scan more water than lone individuals in search of prey. Percussive activities were also regularly observed during feeding behavior. This may be a form of communication, may facilitate fish herding (Würsig and Würsig 1980), and may also serve to coordinate the maneuverings of the whales (Würsig 1986).

After a period of coordinated percussive foraging, the resident orcas usually break into apparently random movements of milling, lacking any semblance of a coordinated group effort. However, the whales may still be cooperating to some degree. Although peripheral males were often observed to be milling alone, it was common for subgroups of two to three females to dive oriented toward each other, which may be a form of herding.

Orca vocalization rates are greatest during milling. Playback experiments of orca vocalizations on potential orca prey species have demonstrated that only mammalian prey exhibit avoidance responses (beluga whales: Fish and Vania 1971; gray whales: Cummings and Thompson 1971; southern right whales: Herman and Tavolga 1980), whereas fishes showed no responses (Pacific herring: Schwarz and Greer 1984; salmonid spp.: Abbot 1973). Consequently, the degree to which orcas vocalize while foraging may provide a further indication of their prey choice.

Silence provides the predator on marine mammals with the benefits of a sneak attack. Ljungblad and Moore (1983) recorded no vocalizations during their observations of orcas chasing gray whales in the northern Bering Sea. It appears that group coordination in the detection and capture of marine mammal prey may be accomplished without vocalizations. Norris (1967) recognized the importance of passive listening in the feeding behavior of cetaceans. Yablokov (1966) has suggested that the importance of vision in orca group coordination is indicated by the high degree of contrast in their pigmentation patterns.

The erratic surfacing observed in transient whales may be partly an artifact of efforts to avoid boats. Transient whales seemed distinctly less habituated to close approaches by small vessels than resident whales, who appear surprisingly tolerant of multiple observation vessels tracking them simultaneously. However, the tendency for transient whales to remain in confined regions for extended periods suggests a sit and wait search strategy (Webb 1984) adapted for finding seals off their haul-out. The foraging strategy of transient whales feeding on harbor seals was characterized by extended handling of the prey. Additional observations are needed to discern if these behaviors represent cautious predation, cat and mouse behavior described as "malicious joy" by Norris and Prescott (1961), or serve some other function such as teaching the young to kill prey, as observed in orcas of Patagonia, Argentina (Lopez and Lopez 1985). Further observations are also needed to discern the relative importance of fishes in the transient diet. Their winter occurrence corresponds to peak aggregations of Pacific herring in Georgia Strait (Hart 1973), and preliminary hydroacoustic recordings made amid a foraging transient pod showed many schooled bait fishes as well as large single targets, presumably salmon.

A unique feature of the habitat use by transient whales was that they were observed in shallow, nearshore quadrats never frequented by resident pods. Their use of these quadrats suggests a feeding strategy consistent with the distribution of bottom fishes and seals along rocky, kelp-covered shores, rather than predation on salmon that occur more commonly midstrait (Stasko et al. 1973, 1976; Simenstad et al. 1979; Long 1983). The most frequently sighted transient pod (O pod) has been repeatedly seen near Eliza Island in Bellingham Bay. During the ten-day radio tracking of two members from this pod, they returned to this area on three separate occasions and joined transient male F-1 there. Eliza Island is the center of the Lummi Indian traditional sealing grounds. The other region most frequented by the radio-tagged whales was Protection Island, which is home to the second-largest aggregation of harbor seals in Greater Puget Sound. Seal predation was observed in both of these locations.

The feeding areas of the resident pods in Haro Strait were characterized by the most highly varied bathymetry of the entire southern study area. Two possible benefits to feeding in these areas of high-relief topography are increased density of prey due to physical limitations on their movements and the availability of underwater barriers to facilitate the herding of prey (Heimlich-Boran 1988). The locations of feeding areas also closely correspond with regions utilized by the commercial salmon fisheries (Hilborn and Ledbetter 1979).

Resident communities are also keenly responsive to changes in the direction of the tidal current as shown by corresponding changes in their behavior and direction of travel. The coordination of these group changes in direction was often associated with percussive behaviors. Occasionally, when two or more pods were traveling together, only one of them would change their direction within an hour of slack tide.

It is conceivable that the whales could swim with the current continuously by maintaining the appropriate orientation to the current when they change directions at slack. However, resident communities spend approximately equal amounts of time swimming with and against the current. Since swimming against the current may slow down the rate the whales travel or increase the energetic cost of travel, enhanced feeding success may account for the persistence of this behavior.

Results from studies that tracked sonically tagged salmon through the study area suggest that Fraser River sockeye (*O. nerka*) and pink (*O. gorbuscha*) progress north through the estuarine waters by moving north with the flood current and holding ground during the ebb current (Stasko et al. 1973, 1976). Resident whales tended to move with flood currents and against ebb currents. While whales could effectively scan more water for salmon when oriented against the flood current, because they would be approaching the prey in the opposite direction, there may be a trade-off for the whales between the encounter rate and ease of prey capture. When the whales are moving against the flood current, they would be approaching the salmon head-on, which may enable the salmon to see and avoid the orcas. In addition, the whales would be approaching the salmon at a higher velocity than if they were traveling with the salmon. Consequently, they would be reducing the time they have to respond to the presence of prey, which might reduce their hunting efficiency.

By maintaining the same relative orientation to the current as returning adult salmon, however, the whales would be able to approach their prey from behind. This might enable the orcas to approach their prey closely without being seen. Also, the salmon might be easier to catch because the orcas would have more time to respond to the fishes' movements than when they are traveling in opposite directions.

Another example of tidally influenced behavior was the tendency for the resident whales to mill during slack currents. This nondirectional behavior may reflect the lack of oriented movements by salmon, which have been shown to mill at slack currents presumably because they rely on currents to impart directional cues for homing (Hasler and Wisby 1951, Dodson and Dohse 1984).

Transient pods tend to attack harbor seals at positive tide heights. A similar tidal utilization has been observed in the predation of sea lions and elephant seals by orcas in Patagonia, Argentina (Lopez and Lopez 1985). However, rather than having to wait for the pinnipeds to leave their rocky haul-outs, the orcas of Patagonia wait for the tide to rise high enough for them to swim up a tidal channel and take the pinnipeds off the smooth sloping beach.

In summary, from spring to fall, large resident pods behave as cruising predators, scanning vast amounts of water for the detection and capture of their seasonally and tidally predictable salmonid prey. During these seasons, they confine their core area to where these and other prey fishes can be found consistently. They appear to feed along specific routes that correspond to the predictable paths of migratory salmon, utilizing areas of high-relief subsurface topography that may aid them in collecting prey into higher densities. Resident pods utilize specific tidal conditions, especially slack tides, to capture their prey. Their selective use of tidal currents suggests they favor maintaining the same orientation as the salmon, enabling them to approach the fishes from behind as stable targets. The resident pods appear to forage as slack maximizers by timing their arrival at areas of fish abundance to coincide with slack current, when salmon are known to aggregate.

In contrast, transient whales occur sporadically in the study area, foraging in small groups on easily localized pinnipeds and bottom fishes. They appear to behave as sit-and-wait predators, relying on the site tenacity and year-round residence of their prey for its detection and using either individual or cooperative techniques for its capture. They go into shallow, protected areas where seal prey are abundant and where resident whales have never been observed. They appear to forage at high tide, utilizing the tides to predict the availability of their pinniped prey.

Differences between resident and transient pods' behavior and ecology are evident, yet they are members of the same species, living in overlapping habitats. Resident and transient orcas look different, vocalize differently, travel in groups of different sizes, and socially interact within separate, closed communities. They follow traditional community-specific geographic routes and employ different foraging strategies. Norris and Dohl (1980) state in reference to dolphins that "learning capabilities provide for a high level of behavioral flexibility in nature and this is translated into local variations in group behavior that we might call culture (p. 253)." The discrete nature of a pod and/or a community of pods provides ample opportunity for preferences or traditions to become established.

REFERENCES

Abbot, R. R. 1973. Acoustic sensitivity of salmonids. M.S. thesis, School of Fisheries, University of Washington, Seattle. 113 pp.

Altmann, J. 1974. Observational study of behavior: Sampling methods. Behaviour 49:227–265.

Angell, T., and K. C. Balcomb. 1982. Marine birds and mammals of Puget Sound. Seattle: University of Washington Press. 123 pp.

Arnbom, T., V. Papastavrou, L. S. Weilgart, and H. Whitehead. 1987. Sperm whales react to an attack by killer whales. J. Mamm. 68(2): 450–453.

Baird, R. W., and P. J. Stacey. 1987. Foraging behavior of transient killer whales. Cetus 7(1):33.

Balcomb, K. C., J. R. Boran, and S.L. Heimlich. 1982. Killer whales in Greater Puget Sound. Rep. Intl. Whal. Comm. 32:681–686.

Balcomb, K. C., J. R. Boran, R. W. Osborne, and N. J. Haenel. 1980. Observations of killer whales (*Orcinus orca*) in Greater Puget Sound, state of Washington. Final Report to Marine Mammal Commission, NTIS PB80-224728, U.S. Dept. of Commerce, Springfield, Va.

Baldridge, A. 1972. Killer whales attack and eat a gray whale. J. Mamm. 53(4):898–900.

Beckoff, M., T. J. Daniels, and J. L. Gittleman. 1984. Life history patterns and comparative social ecology of carnivores. Ann. Rev. Ecol. Syst. 15:191–232.

Berzin, A. A., and V. L. Vladimirov. 1983. A new species of killer whale (Cetacea, Delphinidae) from Antarctic waters. Zool. Zh. 62(2): 287–295.

Bigg, M. A. 1982. Assessment of killer whale (*Orcinus orca*) stocks off Vancouver Island, British Columbia. Rep. Intl. Whal. Comm. 32:655–666.

Bigg, M. A., G. M. Ellis, J. B. Ford, and K. C. Balcomb. 1987. Killer whales: A study of their identification, genealogy and natural history in British Columbia and Washington State. Nanaimo, B.C.: Phantom Press and Publishers, Inc. 79 pp.

Bigg, M. A., and A. Wolman. 1976. Live capture killer whale (*Orcinus orca*) fishery in B.C. and Washington. J. Fish. Res. Board Can. 32(7):1213–1222.

Budylenko, G. A. 1981. Distribution and some aspects of the biology of killer whales in the south Atlantic. Rep. Int. Whal. Comm. 31 paper SC/32/SM3.

Caldwell, D. K., and D. H. Brown. 1964. Tooth wear as a correlate of described feeding behavior by the killer whale with notes on a captive specimen. Bull. So. Calif. Academy Sciences, vol. 63, pt. 3, 128–140.

Chandler, R. D., C. A. Goebel, and K. C. Balcomb. 1977. Who is that killer whale? A new key to whale watching. Pacific Search 11(7): 25–35.

Condy, P. R., R. J. Van Ararde, and M. N. Bester. 1978. The seasonal occurrence and behavior of killer whales, *Orcinus orca*, at Marion Island. J. Zool. 184:449–464.

Crean, P. B. 1983. The development of rotating, non-linear numerical models (GF2, GF3) simulating barotropic mixed tides in a complex coastal system located between Vancouver Island and the mainland. Canadian Technical Report of Hydrography and Ocean Sciences, no. 31. 65 pp.

Cummings, W. C., and P. O. Thompson. 1971. Gray whales, *Eschrichtius robustus,* avoid the underwater sounds of killer whales, *Orcinus orca.* Fish. Bull. 69(3):525–530.

Dahlheim, M. E. 1981. A review of the biology and exploitation of the killer whale, *Orcinus orca,* with comments on recent sightings from Antarctica. Rep. Intl. Whal. Comm. 31:541–550.

Defran, R. H., and K. Pryor. 1980. Social behavior and training of eleven species of cetaceans in captivity. In: Cetacean behavior: Mechanisms and functions, ed. L. Herman. New York: Wiley-Interscience. 319–362.

Dodson, J. J., and L. A. Dohse. 1984. A model of olfactory-mediated conditioning of directional bias in fish migrating in reversing tidal currents based on the homing migration of American shad (*Alosa sapidissima*). In: Mechanisms of migration in fishes, ed. J. D. McCleave, G. P. Arnold, J. J. Dodson, and W. H. Neill. New York: Plenum Press. 263–281.

Dorsey, E. 1983. Exclusive adjoining ranges in individually identified minke whales (*Balaenoptera acutorostrata*) in Washington State. Can. J. Zool. 61:174–181.

Erickson, A. W., M. B. Hanson, and F. L. Felleman. 1987. Habitat utilization of two radio tagged transient killer whales (*Orcinus orca*). (Abstract) Seventh Biennial Conference on the Biology of Marine Mammals, Miami, Fla.

Eschricht, D. F. 1962. On the species of the Genus Orca, inhabiting the northern seas. In: Recent memoirs on the cetacea, ed. W. H. Flower. London: Royal Society. 151–188.

Everitt, R. D., C. H. Fiscus, and R. L. DeLong. 1979. Northern Puget Sound marine mammals. Report in partial fulfillment of EPA Interagency Agreement #D6-E693-EN. NMML, NMFS. NOAA, Seattle, Wash. 139 pp.

Felleman, F. L. 1986. Feeding ecology of the killer whale (*Orcinus orca*). M. Sc. thesis, University of Washington, Seattle. 163 pp.

Fish, J. F., and J. S. Vania. 1971. Killer whale, *Orcinus orca,* sounds repel white whales, *Delphinopterus leucus.* Fish. Bull. 69(3):531–535.

Ford, J. K. B., and H. D. Fisher. 1982. Killer whale (*Orcinus orca*) dialects as an indicator of stocks in British Columbia. Rep. Intl. Whal. Comm. 32:671–680.

———. 1983. Group-specific dialects of killer whales (*Orcinus orca*) in British Columbia. In: Communication and behavior of whales, ed. R. S. Payne. Boulder: Westview Press. 129–161.

Ford, J. K. B., and D. Ford. 1981. The killer whales of B.C. Waters 5(1):1–32.

Gaskin, D. E. 1976. The evolution, zoogeography and ecology of cetacea. Oceanogr. Mar. Biol. Ann. Rev. 14:247–346.

Groot, C., T. P. Quinn, and T. J. Hara. 1986. Responses of migrating adult sockeye salmon (*Onchorynchus nerka*) to population specific odors. Can. J. Zool. 64:926–932.

Hamilton, W. D. 1964. The genetical theory of social behavior, I, II. J. Theoretical Biol. 12(1):12–45.

Hart, J. L. 1973. Pacific fishes of Canada. Ottawa: J.F.R.B.C., Bull. 180. 740 pp.

Hasler, A. D., and W. J. Wisby. 1951. Discrimination of stream odors by fishes and its relation to parent stream behavior. Am. Nat. 85:223–238.

Heimlich-Boran, J. R. 1986. Fishery correlations with the occurrence of *Orcinus orca* in Greater Puget Sound. In: Behavioral biology of killer whales, ed. B. C. Kirkevold and J. S. Lockard. New York: A. R. Liss. 113–131.

———. 1987. Habitat use patterns and behavioral ecology of killer whales (*Orcinus orca*) in the Pacific Northwest. M.Sc. thesis, Moss Landing Marine Laboratory. 61 pp.

———. 1988. Behavioral ecology of killer whales (*Orcinus orca*) in the Pacific Northwest. Can. J. Zool. 66:565–579.

Heimlich-Boran, S. L. 1986. Cohesive relationships among Puget Sound killer whales. In: Behavioral biology of killer whales, ed. B. C. Kirkevold and J. S. Lockard. New York: A. R. Liss. 251–284.

Herman, L. M., and W. N. Tavolga. 1980. The communication systems of cetaceans. In: Cetacean behavior: Mechanisms and functions, ed. L. M. Herman. New York: John Wiley Interscience. 149–209.

Hilborn, R., and M. Ledbetter. 1979. Analysis of the British Columbia salmon purse seine fleet: Dynamics and movement. J. Fish. Res. Board Can. 36:384–391.

Hoelzel, A. R., and R. W. Osborne. 1986. Killer whale call characteristics with implications for their role in cooperative foraging strategies.

In: Behavioral biology of killer whales, ed. B. C. Kirkevold and J. S. Lockard. New York: A. R. Liss.

Hoyt, E. 1984 [1st ed. 1981]. The whale called killer. New York: E. P. Dutton. 287 pp.

Jacobsen, J. 1986. Behavior of the killer whale (*Orcinus orca*) in the Johnstone Strait, B.C. In: Behavioral biology of killer whales, ed. B. C. Kirkevold and J. S. Lockard. New York: A. R. Liss. 135–185.

Jefferson, T. J. 1987. A study of the behavior of Dall's porpoise (*Phocoenoides dalli*) in the Johnstone Strait, British Columbia. Can. J. Zool. 65:736–744.

Jonsgaard, A., and P. B. Lyshoel. 1970. A contribution to the biology of the killer whale, *Orcinus orca* (L.). Norwegian J. Zool. 18(4): 1–48.

Kleiman, D. G., and J. F. Eisenberg. 1973. Comparisons of canid and felid social systems from an evolutionary perspective. Animal Behavior 21:637–659.

Leatherwood, S., K. C. Balcomb, C. O. Matkin, and G. Ellis. 1984. Killer whales (*Orcinus orca*) of southern Alaska: Results of field research 1984. Preliminary Report, Hubbs-Sea World Research Institute, Technical Report No. 84–175.

Ljungblad, D. K., and S. E. Moore. 1983. Killer whales (*Orcinus orca*) chasing gray whales (*Eschrichtius robustus*) in the northern Bering Sea. Arctic 36(4):361–364.

Long, E. R. (ed.). 1983. A synthesis of biological data from the Strait of Juan de Fuca and northern Puget Sound. NOAA, Office of Marine Pollution Assessment, Seattle, Wash. 295 pp.

Lopez, J. C., and D. Lopez. 1985. Killer whales of Patagonia and their behavior of intentional stranding while hunting near shore. J. Mamm. 66(1):181–183.

MacArthur, R. H., and E. R. Pianka. 1966. On optimal use of a patchy environment. Am. Nat. 100(916):603–609.

Martinez, D. R., and E. Klinghammer. 1970. The behavior of the whale, *Orcinus orca:* A review of the literature. Zeitschrift für Tierpsychologie 27:828–839.

Mikhalev, Y. A., M. V. Ivashin, V. P. Savusin, and F. E. Zelenaya. 1981. The distribution and biology of killer whales in the southern hemisphere. Rep. Int. Whal. Comm. 31:551–565.

Morejohn, G. V. 1968. A killer whale-gray whale encounter. J. Mammal. 96(2):341–342.

Nasu, K. 1966. Fishery oceanographic study on the baleen whaling grounds. Sci. Rep. Whales Res. Inst. 20:157–210.

Nishiwaki, M., and C. Handa. 1958. Killer whales caught in the coastal waters of Japan for recent 10 years. Sci. Rep. Whales Res. Inst. 13:85–96.

Norris, K. S. 1967. Some observations on the migration and orientation of marine mammals. In: Animal orientation and navigation, ed. R. M. Storm. Corvallis: Oregon State University Press. 101–125.

Norris, K. S., and T. P. Dohl. 1980. The structure and function of cetacean schools. In: Cetacean behavior, ed. L. Herman. New York: John Wiley Interscience. 211–262.

Norris, K. S., and J. H. Prescott. 1961. Observations on Pacific cetaceans of Californian and Mexican waters. Univ. Calif. Publ. Zool. 63(4): 229–402.

Osborne, R. W. 1986. A behavioral budget of Puget Sound killer whales. In: Behavioral biology of killer whales, ed. B. C. Kirkevold and J. S. Lockard. New York: A. R. Liss. 211–250.

Owen, R. W. 1981. Fronts and eddies in the sea: Mechanisms, interactions and biological effects. In: Analysis of marine ecosystems, ed. A. R. Longhurst. New York: Academic Press. 197–233.

Rice, D. W. 1968. Stomach contents and feeding behavior of killer whales in the eastern North Pacific. Norsk Hvalfangst-Tidende 2:35–38.

Rosenzweig, M. L. 1985. Some theoretical aspects of habitat selection. In: Habitat selection in birds, ed. E. V. Hoyt. New York: Academic Press. 517–541.

Saayman, G. S., and C. K. Tayler. 1979. The socio-ecology of humpback dolphins (*Sousa* sp.). In: Behavior of marine animals, vol. 3: Cetaceans, ed. H. E. Winn and B. L. Olla. New York: Plenum Press. 165–226.

Schaller, G. B. 1972. The Serengeti lion. Chicago: Univ. of Chicago Press. 480 pp.

Scheffer, V., and J. W. Slipp. 1948. The whales and dolphins of Washington State with a key to the cetaceans of the west coast of North America. American Midland Naturalist 39(2):257–337.

Schwarz, A. L., and G. L. Greer. 1984. Responses of Pacific herring, *Clupea harengus pallasi,* to some underwater sounds. Can. J. Fish. Aquat. Sci. 41:1183–1192.

Shane, S. 1987. The behavioral ecology of the bottlenose dolphin. Ph.D. dissertation, Univ. of Calif., Santa Cruz. 97 pp.

Simenstad, C. A., B. S. Miller, C. F. Nyblade, K. Thornburgh, and A. J. Bledsoe. 1979. Food web relationships of northern Puget Sound and the Strait of Juan de Fuca. Springfield, Va.: NTIS, EPA-600/7-79-256.

Spalding, D. J. 1964. Comparative feeding habits of the fur seal, sea lion, and harbor seal on the British Columbia coast. Fish. Res. Board Can. Bull. 146. 52 pp.

Spong, P., J. Bradford, and D. White. 1970. Field studies of the behavior of the killer whale (*Orcinus orca*) I. Proc. 7th Ann. Conf. on Biol. Sonar and Diving Mammals. 169–174.

Spong, P., H. Michaels, and L. Spong. 1971. Field studies of the behavior of the killer whale (*Orcinus orca*) II. Proc. 8th Ann. Conf. Biol. Sonar and Diving Mammals. 181–185.

Spong, P., L. Spong, and Y. Spong. 1972. Field studies of the behavior of the killer whale (*Orcinus orca*) III. Proc. 9th Ann. Cong. Biol. Sonar and Diving Mammals. 187–192.

Stasko, A. B., R. M. Horrel, and A. D. Hasler. 1976. Coastal movements of adult Fraser River salmon (*Oncorhynchus nerka*) observed by ultrasonic tracking. Trans. Am. Fish. Soc. 105(1).

Stasko, A. B., R. M. Horrel, A. D. Hasler, and D. Stasko. 1973. Coastal movements of mature Fraser River pink salmon (*Oncorhynchus gorbuscha*) as revealed by ultrasonic tracking. J. Fish. Res. Board Can. 30(1).

Steltner, H., S. Steltner, and D. E. Sergeant. 1984. Killer whales, *Orcinus orca*, prey on narwhals, *Monodon monoceros:* An eyewitness account. Canadian Field-Naturalist 98(4):458–462.

Tarpy, C. 1979. Killer whale attack. National Geographic 155(4): 542–545.

Thomas, G. L., and F. L. Felleman. 1988. Acoustic measurement of the fish assemblage beneath killer whale pods in the Pacific Northwest. Rit Fiskidvelder 11:276–284.

Thomson, R. E. 1981. Oceanography of the British Columbia coast. Ottawa: Dept. Fish. Oceans, CSPFAS #56. 281 pp.

Tomilin, A. G. 1957. Mammals of the USSR and adjacent countries, Cetacea. Vol. 9: Jerusalem. 605–626.

Tully, J. P., and A. J. Dodimead. 1957. Properties of the water in the Strait of Georgia, British Columbia, and influencing factors. J. Fish. Res. Board Can. 14:241–319.

Webb, P. W. 1984. Body form, locomotion and foraging in aquatic vertebrates. Amer. Zool. 24:107–120.

Wells, R. S., A. B. Irvine, and M. D. Scott. 1980. The social ecology of inshore Odontocetes. In: Cetacean behavior, ed. L. Herman. New York: John Wiley Interscience. 263–317.

Williams, G. C. 1966. Adaptation and natural selection: A critique of some current evolutionary thought. Princeton: Princeton University Press. 307 pp.

Würsig, B. 1986. Delphinid foraging strategies. In: Dolphin cognition and behavior: A comparative approach, ed. R. Schusterman, P. Thomas, and F. G. Wood. Hillsdale, N.J.: Lawrence Erlbaum Press. 347–360.

Würsig, B., and M. Würsig. 1980. Behavior and ecology of the dusky dolphin, *Lagenorhynchus obscurus,* in the south Atlantic. Fish. Bull. 77:871–890.

Yablokov, A. V. 1966. Variability of mammals. Springfield, Va.: NSF/ Smithsonian Inst., NTIS. 350 pp.

Zar, J. H. 1984. Biostatistical analysis. 2d ed. Englewood Cliffs, N.J.: Prentice-Hall. 718 pp.

Killer whales being closely approached by several boats in Johnstone Strait. (Photo by Jeff Jacobsen.)

CHAPTER FOUR

THE INTERACTIONS BETWEEN KILLER WHALES AND BOATS IN JOHNSTONE STRAIT, B.C.

Susan Kruse

Introductory Comments: When I first met Susan Kruse, she was an undergraduate at the University of California, Santa Cruz. She wanted to observe killer whales in Canada. Together, we evolved the notion that she should attempt to use the rather new technique of theodolite tracking to see if the numerous local fishing and recreational boats were disturbing the whales.

I showed her the theodolite, an expensive and delicate instrument, and warned her to take good care of it. We found someone to train her in its use. Then she set off for lonely Cracroft Island. Her next months on the muddy, steep slopes of the island amid clouds of mosquitoes would have tested anyone. She brought the theodolite back intact. She also brought back reams of data and then accomplished the not inconsiderable task of reducing the data to tracks on a chart.

Susan Kruse's apparently simple killer whale data are far from inconsequential. In salmon fishing season, killer whales in Canadian waters have to wend their way through a maze of nets and hordes of commercial fishing boats as they go about their daily rounds. Other kinds of vessel traffic—sportsmen, tourists, sailboats, even whale watchers and whale researchers—are abundant in and out of salmon season and are increasing yearly. Do the whales mind?

Kruse's simple tracks are solid indications that matters are not simple for the whales. They do indeed react strongly when boats come near, boats of any size, large or small. There is now no question about that, thanks to her summer on the muddy cliff with the bugs. If the conflict worsens still further, hers may be the only informed voice speaking for the whales.

—*K.S.N.*

INTRODUCTION

Over the last fifteen years, British Columbia has become well known as a place to see free-ranging killer whales. During the summer salmon run,

I heartily thank Kenneth Norris and Peter Tyack for unending patience and help from the study's inception through data analysis and early drafts of this manuscript. Randy Wells was an invaluable source of help and encouragement throughout the data analysis. Don Snyder of Bowman Williams, Inc., Santa Cruz, kindly provided his surveying expertise. I thank David Bain, Shannon Brownlee, Jim Darling, Jeff Jacobsen. Mark and Dorothy

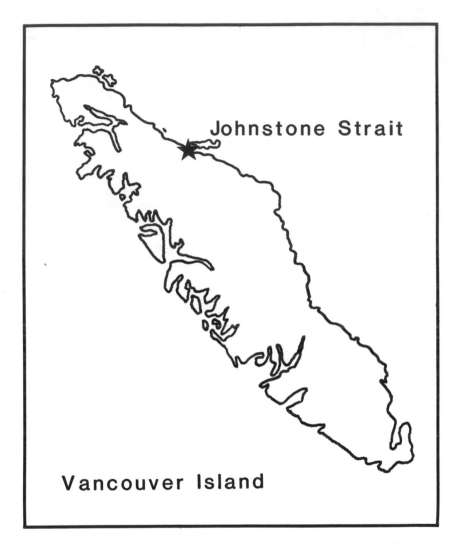

Fig. 4.1. The coastline of Vancouver Island, B.C. My study area, Johnstone Strait, is indicated by the star.

killer whales frequent Johnstone Strait, which borders the northeast coast of Vancouver Island (fig. 4.1). The strait is easily accessible to the public and has grown in popularity as a killer whale viewing area. The number

Kruse, Steve Swartz, and Bernd and Melany Würsig for their many and varied contributions to this study. Stubbs Island Charters, Telegraph Cove, B.C., provided support for this project.

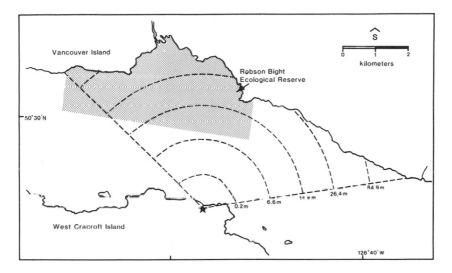

Fig. 4.2. My study area in detail. The star represents the theodolite station. Concentric lines radiating at 1-km intervals from the station indicate ranges of error with increasing distance from the observation point.

of commercial whale watching businesses increases each year. In addition to burgeoning tourist traffic, Johnstone Strait supports large logging and salmon fishing industries and is commonly used by barges and ferries passing between Alaska and the south.

Local residents and researchers have become concerned that the potential for human impact on the killer whales is mounting. In 1981, a small section of Johnstone Strait, including Robson Bight, was designated an ecological reserve by the Canadian Department of Lands, Parks, and Housing (fig. 4.2). Although it was initially established to prevent the construction of a logging dump at the Tsitika River estuary, Robson Bight Ecological Reserve soon became recognized as a killer whale sanctuary. Boaters are advised to stay at least 300 m from the whales while in the reserve. Researchers and commercial photographers may apply to the Department of Lands, Parks, and Housing for permits to approach the whales more closely.

Despite these restrictions, people and killer whales come into contact frequently (frontispiece). In 1982, it was common to see a group of whales with an entourage of up to four boats following them through the strait. Animals rarely passed through the Robson Bight Ecological Reserve without drawing a crowd of onlookers. Although there was concern about how boats might have been affecting the whales, no one had actually tried to assess these effects.

In summer 1983, I returned to British Columbia to conduct a study of the interactions between boats and killer whales in the Robson Bight area. Using a theodolite from onshore, I could track the movements of the boats and whales quite precisely (see chap. 2, this vol.). Because I observed from shore, my presence was unlikely to affect the whales' behavior. In addition, I could easily and instantly survey the entire study area from the clifftop observation station and could get a sense of "the big picture" of the interactions between the whales and boats.

METHODS

From July 20 through September 1, 1983, I tracked killer whales and boats from a land-based station overlooking Johnstone Strait. I used a Nikon model NT2A theodolite with a precision of ± 10 seconds of arc. The station was located on the north shore of Cracroft Island, 58.8 m above mean low tide. My view included 10.5 km of the Vancouver Island shoreline 3.3 to 6.4 km away (fig. 4.2). The Cracroft Island theodolite station provided an expansive view of the Robson Bight Ecological Reserve and its surrounding waters, an area where both whales and boats could commonly be observed.

Of the 4,147 theodolite readings made during the study, 74.5 percent of them were within 3 km of the observation station. Sightings made within one km of the station had a calculated precision of 0.2 m, given the 10-second precision of the vertical angle measurements. At 3 km, the calculated precision was 14.8 m, or about two whale lengths.

Observations were made between 0700 and 2000. During an observation period, whales were serially numbered as I began tracking them. Groups of whales were identified as a single unit if they surfaced relatively synchronously and were within about a whale's length of each other. Single whales were tracked when they were traveling alone or were associated with a dispersed group.

Theodolite readings were taken at rates varying from five times a minute to as infrequently as every 10 to 15 minutes. Every time an animal was located through the theodolite, I dictated into a tape recorder the time of observation, the animal's identifier, the vertical and horizontal bearings, and behavioral notes. If boats were operating in the same area, I identified and tracked them using the same procedure.

Peter Tyack of Woods Hole Oceanographic Institution provided the data reduction and analysis programs that converted the raw data to rectangular coordinates and provided the measures that I used in this study: cumulative speed, milling index, and course bearing. I calculated cumulative speed by measuring the total length of the track from beginning to end and dividing it by the elapsed time. The milling index (MI) is a measure of the linearity of the route taken by the whale between any two points. Thus, MI=1 when a whale swims in a straight line; values less than one indicate how far an animal has deviated from a straight-line path. If a whale has a milling index score of zero, it ended up exactly where it had started (Tyack 1982). Course bearing is the compass bearing of the track's course relative to the x/y coordinate system, where 0 degrees corresponds to south (the positive x axis) and 270 degrees to west (the positive y axis) (Malme et al. 1983).

I summarized cumulative speeds, milling indexes, and course bearings of a track for each 15 minutes of observation. These 15-minute summaries were treated as separate data points in the analyses. If there were gaps in theodolite readings of whales' positions > 15 minutes, the observation session was dropped from the data base. I classified observations during which boats came within 400 m of whales as potentially disturbed conditions (Baker et al. 1982). When boats did not come within 400 m of the whales, I classified observations as presumably undisturbed, control conditions. I divided tracked boats into two size classes, < 7 m and > 7 m, to separate small runabouts from the cabin cruisers and commercial fishing boats. Later, the whales' responses to different boat size classes and engine types were compared.

Michael Bigg of the Pacific Biological Station, Nanaimo, British Columbia, and his colleagues have identified 30 pods, or social groups, of killer whales in British Columbia (Bigg 1982; see also chap. 3, this vol.). Pods typically contain 5 to 20 animals and may include all age and sex classes. Bigg has identified two types of pods in Johnstone Strait: residents and transients. Resident pods are seen frequently throughout the summer and in some cases, all year long. Their movements are predictable, and they generally travel continuously up and down the strait. The regularity of the behavior of resident pods made it easier to judge whether their movements and activities were altered by boats.

RESULTS

I collected data on 31 of 43 field days. A minimum of fifteen whales moved through the study area each day. A total of 155 15-minute whale observation periods were analyzed in this study. In 84 observation periods, boats were operating within 400 m of whales. These periods were classified as potentially disturbed, while 71 were identified as undisturbed conditions. Natural diurnal periodicity of "undisturbed" killer whale

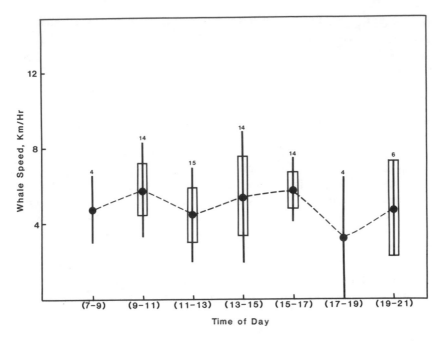

Fig. 4.3. "Undisturbed" killer whales showed no diurnal variation in swimming speed and maintained a moderate pace (X = 5.19 km/hr., χ^2 = 5.38, p<0.5) throughout the day. (N is indicated at the top of each bar; the mean, standard deviation, and 95% confidence intervals are included on the graph).

movement patterns was examined, as well as the effects of boats on their movements. The principal findings are described below.

Daily Movement Patterns

Undisturbed killer whales swam at an average speed of 5.19 kilometers per hour (sd = 2.52 km/hr.). They maintained this pace throughout the day, suggesting that activity levels did not fluctuate in a diurnal pattern (Kruskal-Wallis, χ^2 = 5.38, p. <0.5; Zar 1984, fig. 4.3).

Whale Responses to Boats

Potentially disturbed conditions (boats within 400 m) existed during 54.2 percent of the tracks. Killer whales showed a clear response to the presence of boats by swimming, on average, 1.4 times as fast as whales in the undisturbed category (X̄ = 6.37 km/hr; sd = 3.48). This reaction did not

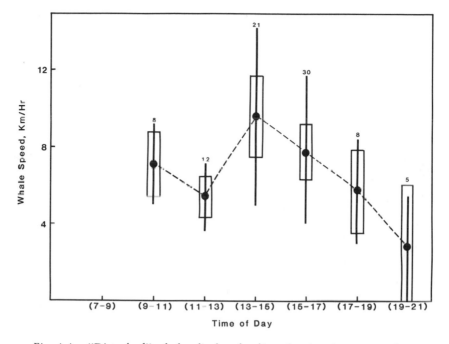

Fig. 4.4. "Disturbed" whales displayed a diurnal swimming pattern characterized by a peak in speed between 1300 and 1500 when numbers of recreational and "whale-watching" boats were highest. (X = 6.37 km/hr.; χ^2 = 20.34, p<0.001). (N is indicated at the top of each; the mean, standard deviation, and 95% confidence intervals are included on the graph.)

diminish over the course of the summer, indicating that the killer whales did not habituate to the presence of boats.

Unlike the undisturbed whales, disturbed animals showed a distinct diurnal pattern in swimming speed which was characterized by a peak between 1300 and 1500 (fig. 4.4). Boat numbers also peaked during this time (31.4% of all boats counted). Twenty-five percent of the disturbed whale observations were made within this three-hour period. Perhaps the hours between 1300 and 1500 represent times when most of the "whale watchers" arrive at the Robson Bight area, which is about a one-hour boat ride from the nearest public boat launching ramp.

Once in the study area, undisturbed whales did not appear to have preferred swimming directions. Disturbed whales, however, favored a westerly course when moving within the study area (Kruskal-Wallis, χ^2 = 6.45; p. < 0.025; Zar 1984). Westward movement may lead whales out of narrow Johnstone Strait into the open waters of Queen Charlotte Strait.

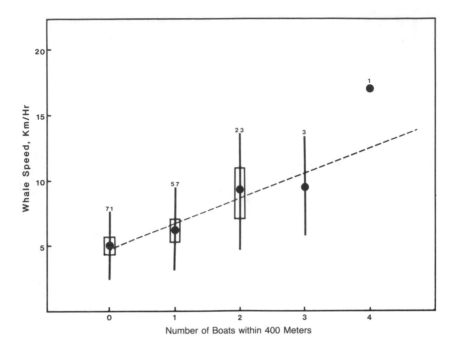

Fig. 4.5. Swimming speeds tended to increase as the number of boats present within 400 m increased (n = 155, r = 0.442, H = 21.17, p>0.001). (n is indicated at the top of each bar; the mean, standard deviation, and 95% confidence intervals are indicated on the graph.)

Swimming speed was positively correlated with the number of boats operating within 400 m of killer whales; as more boats approached, the whales swam faster (fig. 4.5). Disturbed whales typically had from one to four boats operating within 400 m of them. However, single boats were most common (67.9%).

Boats were present during all but 12 of the 155 killer whale tracking sequences. Within the study area, hourly boat counts were made on 26 days between August 3 and September 1. During a total of 271 census hours, the average boat density was 17.86 boats per hour and ranged between 0 and 107 boats per hour. Traffic was greatest during the commercial salmon fishing openings.

I attempted to determine which features of approaching boats evoked responses from the whales. In particular, were whales simply reacting to the presence and approach of boats, or were some recognizable features such as boat size or engine type responsible for the whales' responses? Boat size had been observed to be related to varied responses by humpback whales in southeast Alaska (Baker et al. 1982). Moreover, Stewart et al. (1982) noted that beluga whales responded differently to outboard-powered vessels than they did to boats with diesel engines. Were killer whales sensitive to these boat characteristics as well?

Boats that approached whales closer than 400 m were divided into two size classes, smaller than 7 m and larger than 7 m. Swimming speeds of whales were compared with respect to these size classes. Killer whales did not respond differently to varied boat sizes (Mann-Whitney U Test, $Z = -0.68$; p. > 0.05; Zar 1984). Nor did the whales respond differently to outboard motors and inboard engines ($Z = -0.28$; p. > 0.05; Zar 1984).

Killer whales commonly responded to boats by increasing their swimming speeds but not by changing their courses. Milling indexes were about the same for both disturbed and undisturbed whales, indicating that whales usually did not radically change their courses when approached by boats. Indeed, I did not observe many obvious avoidance responses by whales. However, one group of four animals radically changed its course when simultaneously approached by several boats from different directions.

DISCUSSION

The approach of boats clearly affected the movement patterns of the killer whales in Johnstone Strait. Short-term effect by approaching boats was fairly high and consistent during this study. Killer whales responded to approaching boats by increasing their swimming velocities and by tending to swim toward open water. Increased swimming speed is a common cetacean response to boat disturbance. Richardson et al. (1985) note strong and consistent increases in swimming speeds of bowhead whales in response to approaching boats. Humpback whales in Alaskan waters swim faster when boats draw near (Baker et al. 1983).

Other odontocete species respond to boat disturbance by fleeing from restrictive waters. Stewart et al. (1982) reported that beluga whales responded to outboard motor sounds by heading out of rivers into open water. Irvine et al. (1981) reported that bottlenose dolphins fled from shallow grass flats into a deep channel apparently in response to the engine sounds of a boat that had been used to capture the dolphins the year before. Westward movement may be a killer whale escape response, as it allows whales to move out of narrow Johnstone Strait into the open waters of Queen Charlotte Strait. Open waters provide whales with greater freedom of movement that may allow them to avoid boat disturbances.

The lack of swimming speed periodicity supports the notion that the Robson Bight area might primarily be a resting area for killer whales (G. Ellis pers. comm. 1986). Norris and Reeves (1978) described harassment as having two types of effects, short-term and long-term. Short-term harassment is stress related and may be linked with human activities that result in avoidance, flight, or aggression. Long-term effects as a result of human activities may reduce the biological fitness of a population. Fitness-related harassment might include habitat destruction or repeated disruption of energy budgets and "critical behaviors" such as reproduction, feeding, and rest. Such changes may be detrimental to animal populations, especially if their undisturbed habits are optimal for survival. Although the results of chronic harassment have not been described for any whale population, if the killer whales I studied are suffering from chronic stress caused by human harassment throughout the summer, there may be long-term effects on the population's biological fitness. This Canadian population of killer whales has low mortality rates (2.8%/yr. for males and 0.7%/yr. for females). Calving intervals range from 8.59 to 12.47 years (Bigg 1982). Because of these low birth and mortality rates, the animals may be subjected to continued stress related to human activities for years before noticeable changes in the population's fitness could become evident.

REFERENCES

Baker, C. S., L. M. Herman, B. G. Bays, and G. B. Bauer. 1983. The impact of vessel traffic on the behavior of humpback whales in southeast Alaska—1981 season. Contract No. 81-ABC-00114. Report to the National Marine Fisheries Service, Seattle, Washington. 39 pp., figs., tables.

Baker, C. S., L. M. Herman, B. G. Bays, and W. F. Stifel. 1982. The impact of vessel traffic on the behavior of humpback whales in southeast Alaska. Contract No. 81-ABC-00114. June 1, 1982, to the National Marine Fisheries Service, Seattle, Washington. 39 pp., figs., tables.

Bigg, M. A. 1982. An assessment of killer whale (*Orcinus orca*) stocks off Vancouver Island, British Columbia. Paper SC/Jn81/KW4. In: Rept. Int. Whal. Comm. 32(1982):655–666.

Irvine, A. B., M. D. Scott, R. S. Wells, and J. H. Kaufmann. 1981. Movements and activities of the Atlantic bottlenosed dolphins (*Tursiops truncatus*). Cetology 13:1–5.

Malme, C. I., P. R. Miles, C. W. Clark, P. Tyack, and J. E. Bird. 1983. Investigations of the potential effects of underwater noise from petroleum industry activities on migrating gray whale behavior. Final rept. for the period of 7 June 1982–31 July 1983. Contract No. AA851-CT2-39. Prepared by Bolt Beranek and Newman, Inc., Cambridge, Mass.

Norris, K. S., and R. R. Reeves (eds.). 1978. Report on a workshop on problems related to humpback whales (*Megaptera novaeangliae*) in Hawaii. Final report to the U.S. Marine Mammal Commission in fulfillment of Contract No. MM7AC018.

Richardson, W. J., M. A. Fraker, B. Würsig, and R. S. Wells. 1985. Behaviour of bowhead whales, *Balaena mysticetus,* summering in the Beaufort Sea: Reactions to industrial activities. Biological Conservation 32: 195–230.

Stewart, B. S., W. E. Evans, and F. T. Awbrey. 1982. Effects of man-made waterborne noise on behavior of belukha whales (*Delphinapterus leucas*) in Bristol Bay, Alaska. HSWRI Technical Report No. 82-145.

Tyack, P. L. 1982. Humpback whales respond to the sounds of their neighbors. Ph.D. dissertation, The Rockefeller University. 204 pp.

Zar, J. H. 1984. Biostatistical analysis. 2d ed. Englewood Cliffs, N.J.: Prentice-Hall. 718 pp.

Karen Pryor and Ingrid Kang Shallenberger on the bridge of the tuna purse seiner Queen Mary, *just prior to a set. A dolphin-tuna aggregation has been sighted; speedboats have been lowered and are traveling beside the ship until they are close enough to herd the dolphins into position so the net can be set around them. (Photo by Ralph Silva, Jr.)*

CHAPTER FIVE

SOCIAL STRUCTURE IN SPOTTED DOLPHINS (*STENELLA ATTENUATA*) IN THE TUNA PURSE SEINE FISHERY IN THE EASTERN TROPICAL PACIFIC

Karen Pryor and Ingrid Kang Shallenberger

INTRODUCTION

Along the west coast of South America, reaching north from the Galapagos and 1,500 miles out toward Tahiti, lies a million-square-mile area of ocean known as the eastern tropical Pacific, or ETP. A peculiarity of the ETP is that while in this region the sea floor is thousands of meters below, the warm, productive surface water is shallow, sometimes only 10 m deep. This surface layer, in which temperatures are typically greater than 24.5°C, is cut off from the water beneath by a thermocline, a "floor"

This study was undertaken as a portion of the 1979 Dedicated Vessel Program, under the joint direction of the Marine Mammal Commission, the National Marine Fisheries Service, and the U.S. Tuna Foundation. The research was supported by National Marine Fisheries Service Contract No. 01-78-027-1043.

We would like to thank William Perrin and Warren Stuntz, of the Southwest Fisheries Center, Ethyl Tobach, American Museum of Natural History, and Rosamund Gianutsios, Adelphi University, for assistance in designing the data acquisition system and the initial analysis and interpretation of the data (SFC Admin. Rept. LJ-80-11C). We are very grateful for the kindness, cooperation, and support of Captain and Mrs. Ralph Silva, Jr., and the crew of the *Queen Mary,* without which this work would have been impossible. Our special thanks to Philippe Vergne, Living Marine Resources, who managed our small boat and swam as shark guard. This chapter was reviewed, and greatly improved thereby, by Kenneth Norris, William Perrin, Kevin Lohman, and Erich Klinghammer.

or discontinuity, below which the water is not only colder but higher in salinity and unusually low in oxygen. Therefore, the sea life that in other tropical oceans is often spread through a much deeper water column, is here concentrated in the shallow surface layer (Au et al. 1979). Vertical discontinuities and confluences of water bodies along and within this surface layer attract enormous concentrations of prey and predator, most conspicuously, dolphins, tuna, and tuna fishing vessels.

Schools of dolphins in the ETP, principally of the genus *Stenella,* are often accompanied by schools of yellowfin (*Thunnus albicares*) and skipjack tuna (*Katsuwonus pelamis*). The association, which appears to be involuntary on the part of the dolphins, is probably due to the behavior of the tuna, which tend to join up with dolphin schools in the daylight hours. Presumably, this increases their foraging efficiency; while dolphins do not eat everything tuna eat (e.g., pelagic portunid crabs), tuna eat dolphin prey (Perrin et al. 1973).[1]

Tuna fishermen, using hooks and poles, had long known that the "porpoise," as they call all species of small cetaceans, could be an indicator of the presence of tuna. After World War II, the American tuna fleet, based in San Diego, developed techniques for using large purse seine nets to surround tuna and dolphins both, gathering in the tuna and releasing the dolphins. Initially, dolphin mortality was extremely high: pelagic animals are not behaviorally adapted to avoiding or surmounting obstacles, and the nets were fatal to hundreds of thousands (Pryor and Norris 1978).

Although the ETP dolphin populations were postulated to consist of many millions, they could not sustain this kind of loss for long (Perrin et al. 1982). Techniques were developed and adopted by the industry for safe release of encircled dolphins; the federal government established detailed regulations governing the fishing procedure; and by 1982, mortality caused by the U.S. fleet dropped to biologically acceptable levels (unfortunately, annual mortality caused by domestic and foreign fleets had climbed again to an estimated 120,000 by 1986). Meanwhile, the National Marine Fisheries Service undertook extensive research into the biology of the dolphins. In 1978, a consortium was formed between the U.S. tuna industry, the National Marine Fisheries Service, and other interested groups to use a tuna purse-seining vessel exclusively for scientific research on the tuna-porpoise problem. The *Elizabeth C.J.,* a large seiner, was chartered experimentally. Then a middle-sized seiner, the *Queen Mary,* (150 ft., 500-ton capacity) was refitted to accommodate up

[1]The association may also include several species of seabirds, sharks, and other fast-swimming predators such as marlin. Such mixed-species foraging groups are a phenomenon related to areas of rich but patchy resources (McArthur and Pianka 1966).

to six scientists for a full year of research during 1979, the "Year of the Dedicated Vessel." Six research voyages of four to six weeks each were accomplished in the eastern tropical Pacific. During these voyages, the vessel located and encircled schools of tuna and dolphins, in the usual manner called fishing "on porpoise," but scientific tasks and requirements took precedence over fishing success.

In January 1979, the authors contracted with the Southwest Fisheries Center to join one voyage of the Dedicated Vessel to study the behavior of encircled dolphins by making underwater observations within the net. The principal species of dolphins involved in the tuna fishery are the Pacific spotted dolphin (*Stenella attenuata*) and the spinner dolphin (*Stenella longirostris*, so-called for its behavior of leaping into the air and spinning on its long axis). Spotted and spinner dolphins have been maintained successfully in captivity only at Sea Life Park in Hawaii. Karen Pryor was head trainer and curator at this oceanarium from before its opening, in 1963, until 1971. Ingrid Kang Shallenberger was second-in-command from 1965 until 1971 and was head trainer and curator from 1971 until 1990. Both authors were also trained as ethologists, Shallenberger at the University of Stockholm and the University of Washington, Pryor at Cornell, New York University, and Rutgers University.

Between 1963 and 1979, the authors personally adapted to captivity, maintained, and trained many individuals of several pelagic species of cetaceans, including over forty spinner and spotted dolphins (Pryor 1973, 1975). Spinner and spotted dolphins were maintained as performing and research animals, in groups of two to eight individuals, throughout this period.

Training as a Tool for the Ethologist

George Schaller has been widely quoted as saying that to describe the behavior of any species of animals in the field, one needs "5,000-hour eyeballs," meaning that it takes considerable looking before one can understand or indeed even notice the crucial but often small events that constitute social communication. If this is true of terrestrial animals, whose signals—a growl, a threatening posture—are often at least partly familiar to us, it is much truer of the cetaceans. For example, the lay literature abounds with egregious accounts of dolphins "laughing," "smiling," and "acting playful," when what the author witnessed were the gaped jaw and head movements that in dolphins signify aggressive intent (Nollman 1987).

We had, individually, considerably more than Schaller's requisite

5,000 hours of looking at the behavior of spinner and spotted dolphins as well as several other species (Pryor 1973, 1975; Defran and Pryor 1980). Furthermore, we had looked at these animals from the special viewpoint of the trainer.[2] Training animals in a group, as we did with spinner and spotted dolphins, provides excellent opportunities for learning about social relationships and the nature of social signaling (Lorenz 1975; Pryor 1981, 1987). For example, we dolphin trainers have a straightforward technique for identifying an animal's place in the dominance hierarchy: if a fish falls between two dolphins, who gets it? The dominant animal (barring the rare gesture: a dominant male spotted dolphin at Sea Life Park, named Hoku, sometimes deliberately gave a fish to his female conspecific, Kiko; see Pryor 1975).

Knowledge of the dominance hierarchy is of practical importance to the trainer.[3] It also opens a window to the ethologist. Knowing which animal in a pair or group is dominant, one can then learn to identify aggressive and submissive displays and other indications of relationships, often in rather fine detail. As an example, a common threat gesture in spotted dolphins (not seen in bottlenose dolphins) is a rapid nodding of the head, sometimes with jaws open, and sometimes accompanied by burst-pulse sounds (dubbed "snitting" by Sea Life Park trainers). The male spotted dolphin mentioned above, Hoku, routinely used this threat to force an adult female false killer whale (*Pseudorca crassidens*) ten times his weight to give him some of her fish. When we looked at wild spotted dolphins underwater, at sea, and saw groups of males "snitting" at each other, we knew this to be an exchange of threats, not, say, affiliative greetings, however much it might have looked like "nodding and smiling."

Social Organization in Captivity

Studying our captive animals, we learned to identify many individual behavioral events, from postures and gestures to stylized bubble releases, and to recognize their significance at least in part (Defran and Pryor 1980, Pryor and Kang 1980). Of major importance in our captive school was the dominance hierarchy, to which the animals devoted much time. Behavioral indications of this hierarchy are many: we have observed that

[2] For detailed discussions of dolphin training methods, see Karen Pryor's *Lads Before the Wind* (Harper & Row, 1975) and *Don't Shoot the Dog!* (Bantam Books, 1985).

[3] All too often, novice trainers dismiss a nonperforming animal as "untrainable" when it is merely at the bottom of the hierarchy, being prevented from earning reinforcement by more dominant animals. These animals may actually go hungry if not given special attention.

the dominant animal often swims slightly in advance of the others, or above them; subordinate animals in a group tend to swim closer to equals and a bit farther away (wider interanimal distance) from animals dominant to them; subordinate animals move aside from dominant animals and can be displaced from feeding or work stations. Gregory Bateson found that our *Stenella* group sometimes rested in "rank order," with the superior animals cruising at the surface and the most subordinate at the bottom of the school (Bateson 1974). He also observed that changes in the hierarchy (introduction of new animals) or transgressions by subordinates (stealing fish, for example) sometimes produced open conflict, blows, and ramming attacks that could be quite severe.

In any established group of captive dolphins, a network of affiliative connections is, as it were, thrown across the dominance hierarchy. Affiliative relationships can be expressed by unison swimming and respiration; by close interanimal distances; by frequency and duration of swimming together; and most conspicuously, by contact, especially by animals petting, touching, or rubbing each other with fins, flukes, rostrum, or body (Defran and Pryor 1980, Norris et al. 1985). We observed repeatedly that in the Sea Life Park *Stenella* group, individuals newly introduced into the school were not included in affiliative exchanges. In fact, a stranger, unless highly dominant, was likely to be shunned and avoided for days or even weeks before becoming part of the affiliative network (Defran and Pryor 1980; Pryor 1973, 1975).

Research Goals

Of primary interest to the National Marine Fisheries Service in our research at sea were evidences of stress that might or might not be evinced by encircled animals. Their hypothesis was that animals encircled by tuna nets might be undergoing hazardous levels of stress, perhaps leading to mortalities additional to those observed during the fishing process.

Of primary interest to us, however, was the question of social organization. At sea, these species can be found in aggregations of hundreds or even thousands of animals. It seemed likely that some sort of finer-grained social organization continues to be maintained in these large schools. Subgroups of two to perhaps six or eight animals or so could sometimes be seen even in aerial photographs of these aggregations. But what might the nature of those subgroups be? And how might they relate to each other? The brief confinement of whole schools of these ever-traveling pelagic animals in a purse seine was an unprecedented opportunity for behavioral observation. We hoped our familiarity with the interactions between individual captive animals would help us investigate the next step up the ladder of social behavior: the nature of the subgroups and the interactions and relationships between them.

METHODS

The Fishing Method

A tuna purse seiner cruises until a school of dolphins is sighted (some larger vessels carry helicopters to spot schools). Several speedboats are then lowered overboard which chase, turn, and herd the dolphins, like cattle, back toward the ship. Tuna accompanying the school usually remain with it during this process. The captain directs the herding by radio from the top of the mast.

The seine net, which is 200 m deep and sometimes nearly a mile long, is stacked on the back deck of the ship, fan-folded so that it will run into the water smoothly. The location of winches and other gear requires the net to be set on the left or port side of the vessel. When the dolphins reach the port side of the ship, a heavy open boat, the net skiff, is dropped off the stern of the vessel, pulling one end of the seine net after it. The fishing vessel then travels in a circle until the net is paid out, or set, around the animals.

Once the circle of net is complete, a cable running through steel rings around the bottom of the net is drawn tight, thus pursing or closing the net underwater. The top of the net, or cork line, is kept open in a circle, by the way in which the net is set relative to the wind and current and if necessary, by speedboats pulling the cork line outward; this allows the dolphins room to swim and breathe. The vessel is now temporarily immobilized by the inertia of the net in the water.

The dolphins usually cluster at or near the water surface and as far from the vessel as possible. The tuna, meanwhile, move constantly about in the net throughout the time the net is in the water. Unlike the object-shy dolphins, if the tuna find a hole in the net, they will rapidly escape.

The net is then "rolled," or pulled back aboard the vessel, through a power winch at the top of the boom. Crew members restack the net below the winch on the deck. When approximately three-quarters of the net is back aboard, the vessel starts its engines and begins backing up, hauling the remaining bowl-shaped net through the water. The net elongates, and in due course, the cork line in the portion of the net farthest from the boat is pulled underwater. This area of the net is lined with special fine-meshed net, the "Medina panel," which prevents accidental entanglement of dolphin fins or beaks. When the cork line sinks, the backward movement of the vessel slips the net out from under the dolphins, releasing them unharmed (fig. 5.1). Many dolphins appear to have learned to

Fig. 5.1. Taken from the mast of the fishing vessel, this photograph shows the completion of the backdown procedure. The vessel has pulled the net until the cork lines at the far end sank, passing under the dolphins. White splashes in the distance are made by the released dolphins. A school of tuna "boil" inside the cork line, which is now at the surface again. The speedboat driver is examining the area to make sure all dolphins have been released. (Photo courtesy of United States Tuna Foundation.)

expect this procedure, called "backing down"; they wait quietly in the backdown area and swim out or allow themselves to be sluiced out.[4]

When all the dolphins have been released, the remainder of the net is rolled aboard. The tuna are "sacked up" in the last of the net and scooped aboard the vessel to be frozen in tanks of brine in the hold. Sharks and other unwanted animals in the catch are disposed of overboard. Cleaning and other processing of the fish is done later by canneries onshore. A single set of the net may take two or more hours to complete, requires a crew of about twenty, and may catch no tuna, a few fish, or 30 tons or more (Orbach 1977).

Investigators' Procedures during a Set

The fishing procedure allowed a period of one to two hours, during pursing and rolling of the net, when the seiner was stationary and underwater observations were possible. Once the seiner had halted and the two ends of the net had been drawn together, we lowered an inflatable, outboard-powered rubber raft over the side, climbed into it, and crossed the cork line into the net. It was usually possible to approach the school very closely by moving slowly and by flanking the school rather than heading right at it (fig. 5.2). A third member of the scientific party, Phillippe Vergne, launched and handled the raft for us and dove with us as shark guard. First mate Ralph Silva, Jr., also dove with us when duties permitted, and other members of the scientific party sometimes assisted us from the surface.

We then entered the water using snorkeling gear and collected data and photographs and made observations until backdown was about to begin. The clarity of the water and the approachability of the dolphins made SCUBA gear unnecessary. The water temperature, approximately 80°F, was not a limitation on observation time (fig. 5.3).

During backdown, as the dolphins were released from the net, we tied the raft to the cork line in the backdown area, next to the release point. We continued observations underwater until ordered into the raft (to avoid being sluiced out into open sea; Pryor did get backed out of the net once). We then returned to the vessel, transferred and recorded our data and observations, and prepared for the next set.

---

[4]Care must be taken if the tuna come into the backdown area while the dolphins are being released. Speedboat drivers are assigned to signal if the tuna head toward the opening, whereupon the vessel halts temporarily and the cork lines pop to the surface until the tuna swim away. Crew members also assist any dolphins that may be in difficulties and inspect the net underwater for stragglers (fig. 5.1).

Species, Age, and Sex Recognition

Spotted and spinner dolphins in the ETP are easily separated by coloration, spotters having a dark cape and in adult animals, light and dark speckles or spots, while spinners are unspotted and either all gray or gray with white bellies. Also, spotted dolphins have falcate or sickle-shaped

Fig. 5.2. Investigators make observations in a school of spotted dolphins, near the cork line. One shark guard swims behind them while another watches from the surface. (Photo courtesy of National Marine Fisheries Service.)

Fig. 5.3. Ingrid Kang Shallenberger taking data underwater, using a plastic slate and ordinary lead pencils. The pair of spotted dolphins directly behind her are "columning," or diving vertically after surfacing to breathe. Bubbles, at left, were released by a whistling animal. (Photo by Karen Pryor.)

dorsal fins, and spinner dolphins have triangular fins, making them easily separable even in silhouette or at a distance.

Conveniently for the investigator, spotted dolphins change their color pattern as they age. William Perrin, studying specimens taken during tuna fishing, divided spotted dolphins into five maturation stages, based on size and on the degree of spotting: neonatal (gray with ivory belly), two-toned (dark gray above, light gray below), speckled (with a few dark ventral spots), mottled (spotted all over, with spots overlapping below), and fused (heavily spotted, with a black mask, and spots below fused and faded; Perrin 1969). We divided the spotted dolphins in the nets, by size, into five age groups—neonate, calf, juvenile, young adult, and large or fused-pattern adult—that corresponded to the differences in coloration as defined by Perrin (table 5.1). While transitional animals undoubtedly

Table 5.1.   Color Pattern Changes with Age in Pacific Spotted Dolphins

Age group[a]	Size[b]	Coloration	Pattern type[c]
Neonate 1–2 weeks	80–110 cm	gray and ivory	neonatal
Calf 2 weeks–1 year or ¾ adult	100–160 cm	dark gray above, light gray below	"two-tone"
Juvenile 1–3 years	150–180 cm	dark above, light below with a few ventral spots	"speckled"
Young adult 3–6 years	160–190 cm	spotted all over	"mottled"
Large adult 6 or 7+ years	170–200 cm	black mask, heavily spotted, spots fused and faded below	"fused"

[a] Age estimates are based on sizes and color patterns of captive spotted dolphins of known ages.

[b] Size estimates are adapted from Perrin et al. 1976, based on measurements of animals killed during fishing.

[c] From Perrin 1969.

existed, in practice, we found we could readily assign animals to their age groups based on Perrin's five color pattern groups and on size relative to nearby animals.

Sex identification was usually possible in large adult spotted dolphins. In fused-pattern adults, the sexes are dimorphic: the most conspicuous difference is that males have a postanal bulge, or keel, and a thickened caudal peduncle (Perrin 1975). Since this keel is more pronounced in ETP spotted dolphins than in Hawaiian spotted dolphins, the keel on large male dolphins in the nets seemed quite conspicuous to our eyes (fig. 5.4a); it is not, however, as conspicuous as the keel in large male ETP spinners, which is almost grotesque (fig. 5.4b). Nevertheless, to verify that the individuals we identified as males were indeed males, we tested our assumptions on a group of fused-pattern animals that was resting at the surface (rafting, see fig. 5.4a) by verbally agreeing on the presumed sex of an animal and then taking turns visually checking its genital area. Those with postanal keels were indeed males, and those without were females. In addition, adult animals closely associated with a calf were usually assumed to be female, based on behavior.

Fig. 5.4. (*above, a*) Surfacing to breathe, two fused-pattern Pacific spotted dolphins exhibit the postanal keels of mature males. The third animal accompanying them is an adult female. Rafting animals are visible in the background. (Photo by Karen Pryor.)

(*right, b*) In a group of spinner dolphins of mixed ages and sexes, the nearest animal shows the "backward" dorsal fin and exaggerated postanal keel of mature male spinners in the eastern tropical Pacific. (Photo by Karen Pryor.)

Solitary animals and younger animals could not be sexed unless circumstances provided a close look at the genital area (males have two visible swellings, at the anus and penis, while females have a single genital slit, flanked, in adults, by two small mammary slits). We never succeeded in sexing a calf or a juvenile.

Recording Data

We recorded data underwater on lightly sanded plastic slates made of 3/16-inch Lucite, cut into 9-inch-by-12-inch sheets and painted white on the reverse side to improve legibility (the white surface was then painted over again with black to reduce visibility to sharks). The writing tools were ordinary no. 2 soft lead pencils. Each slate was drilled with two 1/4-inch holes, one for a string to tether the slate to the investigator, and one to tether the pencil to the slate. About twenty extra pencils and slates were carried in the raft for each dive.

We used Focal Animal Sampling as one data collection method (Altmann 1974). Based on our experience with captive animals, we selected five minutes as a representative sample of an animal's activity when viewed underwater. Insofar as possible, we spread our focal animal selection over both sexes, all age groups, and over active, inactive, solitary, and grouped animals. To record focal animal samples without taking our eyes off the animals, we listed events vertically, moving the hand down the edge of the slate after each entry. If two animals were interacting or swimming in unison, we could sometimes watch two focal animals at a time.

Each set also provided a wealth of observational data other than the behavior of focal animals. We tape recorded our other observations and discussion immediately on returning to the ship after each set. Written field notes and taped observations were also made whenever possible during the chase and the early part of the set, from shipboard on sets in which we did not dive, and during and after backdown. Weather, location, time at start of chase, and shipboard school size estimates were taken from the ship's log.

Recognition and Labeling of Behavior

Before going to sea, we constructed a "dictionary" of all the behavioral events known by us to occur in spotted dolphins in captivity and the social or communicative significance of these events when known (based on a previous survey of senior trainers: Defran and Pryor 1980). We used standard terms where they existed and coined terms if they did not, especially for behavior specific to these species, such as certain leaps and aggressive displays. Each behavior was assigned a two-letter, mnemonic code (BR for "breathe," DV for "dive," TS for "tailslap," etc.) to facilitate data taking underwater and subsequent computer analysis of focal animal samples. We augmented the dictionary and codes in the field as needed (Pryor and Kang 1980.)

Sets Made

Between July 23 and August 13, 1978, the M.V. *Queen Mary* made seventeen sets of the net in the eastern tropical Pacific (fig. 5.5). The investigators dove in the net on fourteen sets (omitting two night sets and one set in which sharks were visible in the net). In two more sets, data could not be collected under water because of rough seas, though the attempt was made. Details of each set are given in table 5.2.

A total of over 4,000 dolphins were encircled and released during the voyage; three mortalities occurred. Spotted dolphins were present in all

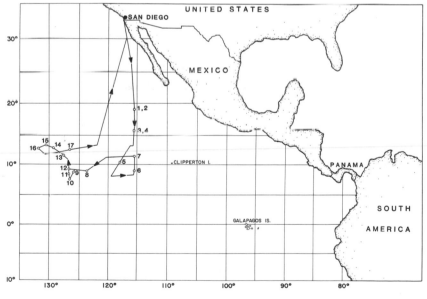

FIGURE 1: CRUISE TRACK AND SET LOCATIONS

Fig. 5.5. Cruise track and set locations.

Table 5.2. Set-by-Set Overview

Set	Spinners present	School size	Time (at chase start)	Agitation levels[a]	Social activity[b]
1		23	12:36 P.M.	1,1,2	III
2		60	3:30 P.M.	2,2,4	II
3		60+	9:45 A.M.	1,1,1	III
4		45	1:00 P.M.	1,1,1	II
5	√	180	7:12 A.M.	1,1,3	III
6		300+	4:58 P.M.	4,4,4–5	II
7	√	300–500	2:21 P.M.	2,3,3	—
8	√	—	11:19 A.M.	2,1,3	—
9	√	300+	11:22 A.M.	1,1,2–3	III
10	√	500+	11:18 A.M.	1,1,2	IV
11	√	100	6:23 P.M.	2,1,2	—
12	√	220	7:59 A.M.	2,1,2	IV
13		120	12:51 P.M.	1,1,2	—
14	√	350	9:15 A.M.	1,1,2	III
15		50–100	12:37 P.M.	2,3,3	—
16	√	1000	4:40 P.M.	3,3,3	—
17		450+	7:44 A.M.	3,3,2	III

[a] Types of surface activity during pursuing, rolling, and backdown. See text for details.
[b] Degree of social interaction underwater. See text.

Table 5.3. Focal Animal Samples

	Spotted Dolphins	Spinner Dolphins
Adult male	26	13
Adult female	16	2
Adult, sex unknown	5	1
Juvenile	23	—
Calf	5	—
Females and calves in pairs	22	6

seventeen sets; spinner dolphins were present only in nine sets. A total of 119 focal animal samples were completed (table 5.3).

FINDINGS AND DISCUSSION

Dolphin Aggregations Underwater

Excellent visibility prevailed on most dives during our cruise (30–75 m, gauged by the distance at which the net walls and floor were visible). We could often view the aggregation of animals as a whole, at least until the crowded conditions before backdown.

We follow Norris in defining a school as any aggregation of animals that swim together as a unit, separated from other aggregations (Norris and Dohl 1980). If spotted and spinner schools were both present, they occupied the same general area in the net but did not mingle; no social interactions were observed between species. The aggregations of each species in the net might be further divided into several separate groups containing from 20+ to 100+ individuals. For example, Set 3 initially contained three groups of spotted dolphins that stayed separate for half an hour. Set 9 contained two separate groups of spotted dolphins, a cluster of about 30 mostly stationary animals and another of about 60 actively swimming animals, as well as three separate groups of spinners. In these multiple-school sets, the groups kept to themselves as neatly as if they were contained in invisible plastic bags, what Norris aptly calls "the school envelope" (Norris et al. 1985).

Once within the net, spinners have been seen to group themselves in a circle, with animals in the middle engaged in "rafting" (floating in a somewhat vertical posture near or at the surface) and more active animals moving protectively in a "ring of aggression" around the central group—what one might call the musk-ox model. But spotted dolphins did not, in our underwater observations, appear to arrange themselves in this manner. In nine of the eleven sets in which we could see the whole school, spotted dolphins were arranged in a truncated cone (likened by Norris to an upside down tea cup). The top of the cone, the smallest portion, was the area at the surface, where rafting animals, if present, were indeed gathered; rafting animals might be of any age, including large adult males, except neonates. Here, also, other animals in the school surfaced to breathe. Meanwhile, most of the rest of the group, including large and small adults and juveniles, circled and cruised below this apex, in a pyramidal mass that might extend 20 or more m downward and widen to 30 m or more at the base. In the center of this mass, a column of animals were moving upward and downward vertically, rising to breathe and diving again in a column, and then joining the animals cruising horizontally at various depths.[5]

Rather than staying in the middle of the school à la musk-ox, females with calves under one-year size tended to remain, during rolling and pursing of the net, at the outside perimeter of the bottom of the mass of animals. Female-calf pairs on which focal animal samples were taken in this location surfaced to breathe, of course, but then returned to the bottom rim of the school, staying, as it were, out of the crowds.[6]

Fear and Stress

While social organization and the role of males in spotted dolphin schools is our principal topic, it seems appropriate to discuss, first, the question of fear and stress in the nets, so as to respond to the not uncommon supposition that one could observe little or no normal social behavior in such frightening circumstances.

When each dolphin school was finally driven into the net by the pursuing speedboats, the animals were certainly agitated and, if the chase had been a long one, probably severely fatigued. All schools, when set on, took up a similar position in the net, described as a "node" of minimum fear, as far from the ship as possible without coming into actual contact with the net (Norris et al. 1978). Whether aggregation in the net consisted of a single group or of two or more separate groups, all the animals responded to any alarming stimulus, such as a speedboat motor, by moving away.

At the beginning of some sets (9 out of 17), we saw signs of agitation at the surface, such as headslaps, tailslaps, thrashing, and "bunching" (animals traveling rapidly, grouped tightly together, and breathing in sharp puffs). In captive *Stenella* spp., these are all signs of fear, stress, excitement, or frustration (Pryor 1973, 1975; Norris and Dohl 1980). In most sets, these agonistic displays diminished partially or entirely during pursing and rolling but sometimes increased again just prior to backdown (table 5.2).

Presence or absence of behavioral events can be used as an index of the state of an organism, as is done in many physiological assessment scales (King-Thomas and Hacker 1987). Norris and his associates observed

---

[5] The picture was very different in spinners (present in 9 of 17 sets), which seldom rafted and which surfaced to breathe in long, horizontal arcs rather than vertical columns. Spinners also moved faster and over a wider area throughout the sets (Pryor and Kang 1980).

[6] We saw only two neonates, identifiable by visible fetal folds, on this cruise; in both cases, the mother and neonate, viewed repeatedly by both observers during the set, were moving rapidly, continuously, and well beyond the perimeter of the school.

Table 5.4. Agitation Levels

Level 1:	No aerial behavior. Animals moving quietly.
Level 2:	Rapid swimming; rolling or leaping across surface ("porpoising"); loud exhalations ("chuffing").
Level 3:	Level 2 plus some headslaps, tailslaps, rostrums in air.
Level 4:	Level 2 and 3 activity plus fluking, tailwaves, sideswipes, charging; whistling audible in air.
Level 5:	Level 2, 3, and 4 activity plus panicky dashing about, charging the net, struggling.

Note: Assignment of behaviors to a given level are based on observations of captive animals under known circumstances of stress.

that some types of surface behavior in spinner dolphins increased in frequency prior to the school's traveling from a rest area to offshore feeding sites. Ranking of these behaviors enabled them to predict when the departure was imminent (Norris et al. 1985). Similarly, we created an agitation scale, using surface-visible behavior related to fear and stress in *Stenella* in captivity, to rank the evidences of stress in the schools (table 5.4). We rated each school on the Agitation Scale three times, at the beginning, middle, and end of the set (Pryor and Kang 1980). Schools varied from being highly agitated throughout (Set 6) to being very calm, with no surface displays even at backdown (Sets 3 and 4; see table 5.2).

Except for the animals in Set 6, we saw no panicky dashing about or blundering into the nets, scenes that had been described by many observers in previous years. We also saw no "sleepers," animals lying on the bottom of the backdown channel (Norris et al. 1978), though we were able to scan the backdown area through face masks in most sets. It is our surmise that this difference was a result of experience on the part of the animals we happened to observe. There seems to be no doubt that some dolphins in the ETP have learned what to expect when set on and can develop accommodating behavior (Stuntz and Perrin 1979). We think it probable that those schools showing the most agitation were those with the least experience; highly agitated schools also showed the most incompetence at backdown, going out of the net sideways, swimming back into the net, and so on, whereas calmer schools sometimes left the net very efficiently, some individuals even slithering over the cork lines before they sank.

As the intensity of fear and agitation varied from school to school, so did evidences of fear and stress in individuals. Some cruised slowly as captive animals do when resting or engaged in social activity; others showed signs of stress, such as sinking briefly or whistling repeatedly. One individual, a large adult male, caught our eye by lashing out in all directions for about 10 seconds. The other animals gave him a wide berth (this was the only event of its kind that we observed in 11 underwater sets).

Rafting

The behavior of rafting, or hanging at the surface with just the blowhole exposed, was recorded in animals of both sexes and in all age groups except neonates and small calves (which, according to Sea Life Park staff, never lie still at the surface in captivity, either). Percentage of animals rafting varied from none to over half the animals in a given set; there was no correlation between agitation levels and the percentage of animals rafting.

It is our guess that rafting may combine elements of both fear and adaptation in individually varying amounts. Nearby rafting animals appeared alert and could be seen inspecting us and our equipment attentively; yet in several sets, we swam among rafting animals, touched them, and even moved them about. At Sea Life Park, newly captured *Stenella* specimens are often extremely passive, allowing themselves to be caught, force-fed, and even inoculated without resistance, until they learn to feed, whereupon they become hard to catch again. Some rafting animals may be exhibiting this "learned helplessness." However, in captivity, spotted dolphins that are not otherwise occupied sometimes loaf or rest in the rafting position. It is possible that some animals were doing so by choice. And some animals may have been rafting simply in mimicry of others. One focal animal case describes a juvenile repeatedly nudging its rafting mother (a play invitation). When the female gaped (a threat display), the calf finally gave up and rafted beside her.

Social Activity Levels

The *Stenella* school at Sea Life Park tended to engage in abundant social interaction in intermittent bouts, alternating with rest periods (in spinners, aerial displays often occurred at night and were reported by night watchmen). Bouts of leaping and social interaction in the dusky dolphin (*Lagenorhynchus obscurus*) commonly occur following feeding episodes (Würsig and Würsig 1981). The social activity observed by us in spotted dolphins in the net also appeared to be occurring in bouts. Either a lot was going on or very little. Whatever the state of socializing in a particular set happened to be, that level of activity remained constant throughout the set until just before backdown.

Table 5.5. Social Activity Levels

Level I:	No apparent social activity; animals *not* in visible subgroups.
Level II:	Animals in subgroups, rafting, columning, and/or cruising. Social interactions not conspicuous.
Level III:	Level II activity plus social interaction between individuals, such as affiliative exchanges, aggressive display, and adult-calf play.
Level IV:	Level II and III activity, plus extensive social interactions of long duration or involving several animals.

While actions of individual animals—leaps, tailslaps, and so on—can be seen from the surface, one needs to be underwater to see and record the small but significant interactions between two or more animals, such as a pectoral pat or a gape and the reaction to it. Therefore, we can report on spotted dolphin social activity only in the eleven sets we observed underwater. We created a scale for assessment of the intensity of social activity, similar to our scale for agitation levels, as shown in table 5.5.

Whatever the state of socializing in a particular set happened to be, that level of activity remained constant throughout the set until just before backdown. In three sets, social activity was at a minimum. In six sets, social activity was commonplace. The remaining two sets were very high in social activity. The social activity level for each set seen underwater is given in table 5.2.

Social Activity Levels, Agitation Levels, and Time of Day

If spotted dolphins feed at dawn and dusk, and if social activity normally follows feeding as in the dusky dolphin, we might expect time of day to affect social activity levels. All sets observed underwater before 1:00 P.M. were ranked for social activity at Levels III or IV; the two sets observed after 1:00 P.M. were low (Level II) in social activity. Given our limited data, there is nevertheless a significant visible trend toward social activity in the morning (fig. 5.6a). We also compared agitation levels for all seventeen sets to time of day and to size of school and found a random scatter and no correlation (Fig. 5.6b).

It has been suggested that some of what we considered to be normal social activity, particularly male-male aggression, was a by-product of fear and stress. We compared social activity levels to agitation levels for the eleven sets observed underwater (fig. 5.6c). We found a highly significant inverse correlation of social activity to agitation levels: sets with higher levels of agitation showed less incidence and less variety of social

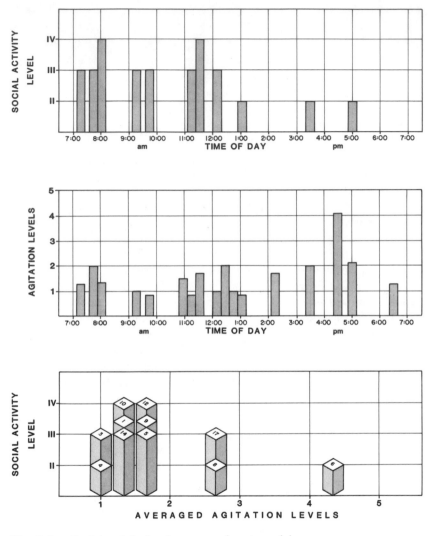

Fig. 5.6a. Social activity levels compared to time of day.

    5.6b. Agitation levels compared to time of day.

    5.6c. Social activity levels compared to agitation levels.

activity; sets with high levels of social interaction (including male-male conflict) were, conversely, low in surface signs of agitation (Spearman rank correlation = −0.84; n = 11; 0.002 < P < 0.005; Zar 1984).

Spotted Dolphin Subgroups: General Observations

We define subgroups as a relatively small (<20) group of animals oriented and traveling in the same direction, maintaining close interanimal distances (usually <2 m), and, typically, separated from their neighbors by a wider gap (>3 m). Subgroups usually surfaced to breathe all together. Some subgroups moved in physical contact and in complete unison, tailbeat matching tailbeat. In the schools we observed, subgroups constituted about half the animals; most of the remaining half moved in a single loose congregation.[7]

Female-Calf Pairs   Focal animal sampling was the technique that helped us tease apart the subgroup structure. We began by taking five-minute study cases on the most obvious and identifiable animals, the female-calf pairs that stayed in one location on the perimeter of the school. These pairs were easily recognizable not only by their size difference but also by behavior; rather than cruising or resting side by side as adult pairs might, they often oriented toward each other, rubbing heads or fins or bodies, even when cruising or surfacing to breathe. Studies of the reproductive tract indicate that female spotted dolphins do not normally have a new calf every year but may keep a single calf with them for two, three, or even four years. Therefore, we could expect that some females would be accompanied by quite large young, even juveniles in the speckled state (Kasuya et al. 1974; Hohn et al. 1985). In fact, we did occasionally see the somewhat irregular swimming pattern and the frequency and types of contact that characterized females and calves in pairs consisting of a mottled or fused-pattern adult and a speckled juvenile. Based on these behavioral patterns, we considered these to be mother-young pairs, even though the size difference might no longer be appreciable. It is possible, however, that such pairs may also be siblings (Perrin pers. comm.).

Female-Young Subgroups   Female-calf pairs did not always remain alone; sometimes they joined other animals or were joined by them. We

---

[7] A few animals moved about individually but rarely for long: of 14 focal animals selected for being solitary, only 4 failed to join another animal or group even in the brief five minutes of case study.

Fig. 5.7. A female-young subgroup. The calf in the center is traveling from the adult female at left to the two female-calf pairs at right. The net floor is visible below these animals. (Photo by Karen Pryor.)

noted the animals female-calf pairs associated with and thus identified another type of subgroup—female-young. These groups, when inspected animal by animal, consisted of adult females in the mottled or fused pattern, each accompanied by a calf or two-tone or speckled juvenile. Sometimes female-young groups did include a single, large, fused-pattern female without a youngster. Spotted dolphin schools sometimes contain a few (>1%) truly postreproductive females (Myrick et al. 1986). It seems at least possible that, as in many terrestrial animals, older females that are neither pregnant nor lactating may remain in social association with their daughters and their offspring.

Female-young groups were by no means static organizations: the numbers in such groups changed constantly as animals joined and left. However, the separation of the group from other nearby animals and the composition of the group remained constant (fig. 5.7). Female-young groups never seemed to contain any other classes of animals, such as independent juveniles, young adults, or large adult males, and the presence of even one calf seemed diagnostic. Whenever we inspected all the animals in a group with a calf in it, all appeared to be adult females with associated young.[8]

Juvenile Subgroups   Juvenile subgroups, which were not seen in every set, consisted of three to six two-tone individuals, smaller than adult size and quite without spots, characterized by minimum interanimal distances and near-perfect synchrony of breathing and swimming. In fact, the behavior of these juvenile subgroups was strikingly similar to that of adult male subgroups: both swam shoulder to shoulder in precise formation, both moved through the schools at a constant speed and on their own course, and both appeared to remain intact throughout sets, neither gaining nor losing individuals. Also, a peculiar behavior we called rafting head down was seen only in juvenile and adult male subgroups (fig. 5.8; Pryor and Kang 1980).

Juvenile subgroups were never seen to join other subgroups, although once a juvenile subgroup was joined by an adult female that exchanged body rubs and pectoral pats with one of the juveniles. Juvenile subgroups did not appear to gain or lose animals; focal animals in a juvenile subgroup studied in Set 10 were seen repeatedly, their behavior and numbers unchanged, until backdown. We had no explanation for why some juveniles formed these closely associated subgroups, while others, as large or larger, were still paired with adult females.

---

[8] In examining photographs of schools in the nets, Norris et al. (1978) found that any group with a calf in it had smaller interanimal distances, overall, than groups without calves.

Fig. 5.8.   A juvenile subgroup "rafting head-down" near the bottom of the net. These five juveniles persisted in this behavior, except when surfacing to breathe, throughout the set until backdown. Two adult male subgroups were seen rafting head-down in other sets. (Photo by Karen Pryor.)

Adult Male Subgroups   The most conspicuous subgroups, seen in every school, were made up of adult males. These were groups of three to eight large, fused-pattern males, all with heavy keels, dark coloration, black facial masks, and white rostrum tips, the fully developed mature pattern of the largest animals in Perrin's scale (table 5.1, Perrin 1969). These animals characteristically moved in unison, often shoulder to shoulder with pectoral fins overlapping and no individual in advance of another. They cruised slowly through the school without swerving or altering speed, while a path opened up before them; even rafting animals bestirred themselves to move aside when these dominant animals passed through (fig. 5.9).[9]

Every school observed contained at least one adult male subgroup. If the aggregation in the net was divided into separate schools, each such school contained one or more of these conspicuous groups. In Set 12, among 200+ spotted dolphins, we counted eight adult male subgroups.

These subgroups did not seem to fluctuate in numbers as, for example, female-young subgroups did. We sometimes recorded a fused-pattern female (slender and keelless) joining an adult male subgroup to exchange affiliative gestures (patting, body rubbing) with one (or in one case, two) of the males, but the visit was always brief; conversely, an individual male might leave its subgroup, perhaps to interact with another individual or to investigate one of us, but in all recorded observations these individuals returned to the subgroup they came from, usually in less than one minute. In three sets, we noted an adult male subgroup containing a marked or scarred individual that could be positively identified throughout the set; in each case, the numbers of males in that subgroup remained the same.

Young Adults　The rest of the school, the amorphous mass of animals that were not in clear-cut subgroups, proved to consist of speckled juveniles and mottled-pattern young adults. We never saw fused-pattern animals or calves in these larger aggregations.[10] Young adult gatherings differed from the smaller subgroups in size, up to fifty or more individuals, but they were similar to the smaller groups in that interanimal distances were small (about 1 body width), and exchanges of affiliative displays were abundant.

Other Associations

In addition to the subgroup types recognized above, we observed many kinds of transient associations between two or more animals that we did not consider as subgroups, such as male-female pairs, animals engaged in aggressive interchanges, and triads, two adults with a young calf sandwiched tightly between them.

---

[9] We identified similar subgroups of large, heavily keeled males in some (not all) spinner schools; they did not exhibit the tight unison of spotted male groups but swam in loose clusters and echelons, as has been reported for Hawaiian spinners (K. Norris pers. comm.).

[10] Again, the picture was very different in spinners. In some spinner schools we saw large, heavily keeled males, females, half-grown animals, and calves jumbled together in a big mass of animals throughout the set.

Fig. 5.9. An adult male subgroup cruising through the school. An individual in another male subgroup rising from below gapes his jaws in a threat gesture at the group passing over him. (Photo by Karen Pryor.)

The Behavior of Adult Male Subgroups

The behavior of adult male subgroups differed from the behavior of all other animals in the school perhaps most conspicuously in the way they carried out so much of their activity in synchrony. Synchronized behavior is a particular skill of dolphins, and episodes of unison breathing, leaping, and so on, are not uncommon. Usually, however, this unison behavior is a transient event. These adult male subgroups took it to extremes. In unison, they rose to breathe, and dived again. Together, they investigated us; an adult male subgroup was often our first sight, positioned between us and the school, when we entered the water (fig. 5.10). If one adult male

Fig. 5.10. An adult male subgroup positions itself between the main part of the school and the investigators. Another adult male subgroup of four animals cruises behind these three. Two of the first group cock their heads stiffly to study us with both eyes; these pelagic dolphins do not have the head-neck mobility of coastal bottlenose dolphins. Note the conspicuous white rostrum tips of the mature males, also visible on a large, fused-pattern female in a female-calf subgroup, lower right. (Photo by Karen Pryor.)

subgroup met another, they made threat displays in unison (see fig. 5.9).[11]
Often, we could identify adult male subgroups, even at considerable distance underwater, by the stereotyped precision of their swimming formation. This "military" unanimity, which has been remarked on by other
observers, seems to us to be peculiar to this species (James Coe pers.
comm., Brower and Curtsinger 1981).

Social Aggression

We feel that it is important to differentiate between noninteractive agonistic behavior (such as headslaps and breaching), which is often fear or
stress related, and agonistic behavior directed at another specific individual animal (such as a threat display or actual blow), which constitutes
social aggression.

Social aggression in captive animals is usually related to dominance
disputes. As such, we consider it normal social behavior. We were not
surprised, therefore, to see episodes of social aggression occurring, along
with many other sorts of social behavior, in schools engaged in social activity. The principal aggressive behavior we recorded in focal animal sampling was gaping. Females sometimes gaped at calves. Solitary animals,
whether adults or juveniles, sometimes gaped at other animals investigating or approaching them. Animals in young adult subgroups occasionally
gaped at newcomers (but in reviewing data from all sets, young adults
were about three times more likely to greet animals joining them with a
pectoral pat). We recorded only one instance of threat between males in
an adult male subgroup; two males in a subgroup of five exchanged two
rounds of gapes and head noddings. Even including this modest exchange
of threats, in data compiled from all sets, aggressive signaling was recorded less often within adult male subgroups than in female-young and
young adult subgroups.

Outside of focal animal sampling, however, we collected numerous
observations of more extensive episodes of social aggression, almost entirely between adult males. There were three exceptions: once a female
reprimanded a lively calf in the backdown area by knocking it into the air
with her head three times before it settled at her side; in Set 10, a female
drove another female out of an adult male subgroup with gapes; and in
Set 12, three pairs of young adults engaged in noisy and bubbly bouts of
toothraking, which may or may not have been aggressive behavior.

---

[11] Senior male subgroups even scouted out the unknown together. In Set 13, just before
backdown, we observed a group of four make a simultaneous high-speed excursion 200 m
back toward the ship to investigate some research gear being hauled up from the net floor;
together, they circled the gear tightly, twice, before returning to the school.

We saw three interchanges between spotted dolphins which we would describe as fights, each time between a fused-pattern male and a younger individual (a juvenile in the speckled phase in Set 1 and mottled-phase young adults in Sets 3 and 8). The animals exchanged threat signals and burst-pulse sounds and whirled around each other head to tail, until the younger animal swam away (with the older male, in one case, in aggressive pursuit). These episodes took place in clear water and about 4 to 5 m from the observer, so the color pattern differences were distinct, but the sex of the younger animal could not be identified positively.

In Set 12, Pryor recorded a confrontation between two adult male subgroups. An adult male subgroup of four animals and another of three met face to face, came to a halt, and exchanged open-jawed burst-pulse threat displays for about 10 seconds before the smaller subgroup ducked under and both continued on their previous courses. In Set 14, two fused-pattern males, with a fused-pattern female between them, fenced with open jaws across her bows as they cruised; the clash of teeth was audible.

Little aggression was manifested toward divers. On seven occasions, a male spinner or spotted dolphin directed a loud click train and sharp eye contact at an investigator, a signal pairing that in captivity is sometimes followed by a feint or a blow. Once, in Set 17, the nearest animal in a subgroup of four adult males threatened an observer with jaw shaking and burst-pulse sounds but deflected his course when threatened back.

Affiliative Behavior in Spotted Dolphin Schools

Much of the typical behavior of individuals within subgroups, such as unison breathing, unison swimming, close interanimal distances, and body contact, could be characterized as affiliative. Therefore, the very existence of subgroups might be considered to be affiliative behavior. In captive spotted dolphins, such affiliative activities are engaged in largely or exclusively by individuals that recognize each other and have established relationships. One may deduce that the animals seen in spotted dolphin subgroups in the nets are well acquainted with each other.

In addition, we observed many occurrences of affiliative behavior between animals that were not in the same subgroups. The most frequently recorded events might be termed greeting behavior and consisted of an individual briefly joining another individual or group and exchanging body rubs or pectoral pats. Specific instances were recorded between calves and other calves, between calves and more than one adult female, between adult males on meeting (in Set 4, two large adult males exchanged pectoral pats), between solitary juveniles encountering each other (rostrum touches, Set 4), between young adults, between large adult females and young adult females joining and parting, between large adult males briefly joining females in female-young groups, and between females and individual males in adult male subgroups. In summary, we saw greeting and affiliative behavior between every category of sex and age in the school except adult males and young adults and adult males and calves.

One unusual example of affiliative behavior occurred in Set 17, shortly before backdown. Both investigators were recording animals in adult male subgroups when a speedboat engine started up suddenly nearby, alarming the animals (the boat was attempting to drive animals away from a pocket in the net). Each group of males bunched together and speeded up, and each was simultaneously joined by females, calves, and young adults, crowded above, below, and beside the males. One group thus increased briefly from three males to ten individuals and the other, from eight males to seventeen individuals.

Conclusions

The sample size of this study is small—seventeen sets, of which only eleven were studied underwater. Nevertheless, we feel that subgroup identification was reliable, thanks to the focal animal sampling approach. The wealth of observational material allows some hypotheses on spotted dolphin organization.

Each autonomous aggregation of spotted dolphins in a set contained female-calf pairs, female-young subgroups, young adult subgroups, and adult male subgroups; some also contained juvenile subgroups. Because of the frequency and widespread nature of affiliative exchanges, we suspect that the spotted dolphin groups that we saw, or at least most of the animals in those groups, knew each other and had been in association for prolonged periods.

The adult male subgroups, in particular, exhibited unusually uniform behavior and constancy. In captivity, we have seen spotted dolphins (but not spinners) form fixed, long-term associations that excluded other individuals (Pryor 1975). We suspect that adult male subgroups will eventually prove to be neither temporary nor opportunistic but long-term associations of particular individuals.

What could be the function of such an alliance? Perhaps it facilitates breeding. In spotted dolphins, anatomical studies have demonstrated that fused-pattern males are the breeding animals (Hohn et al. 1985). About 50 percent of the females are reproductively mature in the mottled stage (and 4 percent in the speckled stage), but the testes size and function in many hundreds of samples show that mottled-pattern males are reproductively immature (Perrin et al. 1976; Perrin and Reilly 1984; Hohn et al. 1985). Therefore, what we identify as adult male subgroups are likely to be the principal reproductive males in their schools.

In examining specimens collected from four different schools, Perrin found several variations of color pattern between schools. For example, in one school, the darkish band from jaw to flipper, seen in all age groups, was narrow and simple, and in another, it was wide and complex; in one school, the adults had white rostrum tips, and in another, they did not (Perrin 1969). Behavioral evidence suggests to us that spotted dolphin schools could be stable enough to provide an environment favoring the perpetuation of genetic relatedness. Males within such a school might not need to compete with each other physically for dominance or access to mates. And, indeed, the amount of status conflict within adult male subgroups seemed to us to be very low. One can speculate, however, that defense of the group against other adult male subgroups, or maturing, younger males, would confer reproductive advantage. Again, our glimpses of social aggression by fused-pattern males and adult male subgroups are consonant with this premise.

There are many successful species of pelagic cetaceans. Spinners and spotted dolphins are among the most numerous. Their behavioral ecology is probably similar in some respects and different in others. For example, in the ETP in a year, spinners may travel no more than 300 miles, while spotted dolphins may move 1,000 miles or more, traveling 30 to 50 miles a day (Perrin et al. 1979). Possibly, a more rigid school structure is beneficial to the more nomadic spotted dolphins.

Several studies in progress at the Southwest Fisheries Center suggest that there are two kinds of spotted dolphin schools in the ETP. Some schools are breeding schools, generally numbering under 300 animals (Barlow and Hohn 1984, Myrick et al. 1986). These schools (which are the kind we saw in our research) are characterized by the presence of fewer males and more females and young than would be expected by chance and by a partly *missing* age class consisting of prepubescent animals (Hohn and Scott 1983). These juveniles are sometimes completely absent from the records of a given school, although young adults, the next age group, are present in expected proportions (Perrin and Myrick 1980).

The missing age class apparently forms the bulk of the second sort of spotted school: these are small groups, mostly male and mostly juvenile, which are often found in association with very large schools of spinners (Barlow and Hohn 1984, Hohn and Scott 1983). Since very large groups of spinners rarely carry tuna, they are seldom set on by tuna vessels; thus, fewer records exist of these animals that seem to have left the parental association. How and when (and if) these juvenile schools rejoin the breeding groups is at present only a matter for speculation.

REFERENCES

Altmann, Jeanne. 1974. Observational study of behavior: Sampling methods. Behaviour 49:3.

Armitage, Kenneth B. 1987. Social dynamics of mammals: Reproductive success, kinship and individual fitness. Trends in ecology and evolution 2:9.

Au, David W. K., Wayne L. Perryman, and William F. Perrin. 1979. Dolphin distribution and the relationship to environmental features in the eastern tropical Pacific. Southwest Fisheries Center Administrative Report no. LJ-79-43.

Barlow, J., and A. A. Hohn. 1984. Interpreting spotted dolphin age distributions. U.S. Department of Commerce NOAA Technical Memo NOAA-NMFS-SWFC-48.

Bateson, Gregory. 1974. Observations of a cetacean community. In: Mind in the waters, ed. Joan MacIntyre. New York: Scribner's.

Bertram, B. C. R. 1976. Kin selection in lions and in evolution. In: Growing points in ethology, ed. P. P. G. Bateson and R. A. Hinde. Cambridge: Cambridge University Press.

Brower, Kenneth, and William Curtsinger. 1981. Wake of the whale. New York: E. P. Dutton.

Defran, R. H., and Karen Pryor. 1980. The behavior and training of cetaceans in captivity. In: Cetacean behavior: Mechanisms and functions, ed. Louis Herman. New York: John Wiley-Interscience.

Douglas-Hamilton, Ian, and Oria Douglas-Hamilton. 1975. Among the elephants. New York: Viking Press.

Hohn, A. A., J. Barlow, and S. J. Chivers. 1985. Reproductive maturity and seasonality in male spotted dolphins in the eastern tropical Pacific. Marine Mammal Science 1(4):273–293.

Hohn, A. A., and Michael D. Scott. 1983. Segregation by age in schools of spotted dolphin in the eastern tropical Pacific. In: Abstracts: Fifth, Biennial Conference on the Biology of Marine Mammals. Marine Mammal Society.

Kasuya, T., N. Muazaki, and W. H. Dawbin. 1974. Growth and reproduction of Stenella attenuata in the Pacific coast of Japan. Sci. Rept. Whales Research Institute 26:157.

King-Thomas, Linda, and Bonnie Hacker (eds.). 1987. A therapist's guide to pediatric assessment. Boston: Little, Brown & Co.

Lorenz, Konrad. 1975. Foreword. In: Lads before the wind, Karen Pryor. New York: Harper & Row.

McArthur, R. H., and E. R. Pianka. 1966. On optimal use of a patchy environment. American Naturalist 100(916):603.

Myrick, A. C., Jr., A. A. Hohn, J. Barlow, and P. A. Sloan. 1986. Reproductive biology of female spotted dolphins, *Stenella attenuata,* from the eastern tropical Pacific. Fishery Bulletin 84:2.

Nollman, James. 1987. Animal dreaming. New York: Bantam Books.

Norris, Kenneth S., and T. P. Dohl. 1980. The structure and function of cetacean schools. In: Cetacean behavior: Mechanisms and functions, ed. L. M. Herman. New York: John Wiley and Sons.

Norris, Kenneth S., Warren E. Stuntz, and William Rogers. 1978. The behavior of porpoises and tuna in the eastern tropical Pacific yellowfin tuna fishery: Preliminary studies. National Technical Information Service PB-283 970.

Norris, Kenneth S., Bernd Würsig, Randall S. Wells, Melany Würsig, Shannon M. Brownlee, Christine Johnson, and Judy Solow. 1985. The behavior of the Hawaiian spinner dolphin, *Stenella longirostris.* Southwest Fisheries Center Administrative Bulletin LJ-85-06C.

Orbach, Michael K. 1977. Hunters, seamen, and entrepreneurs: The tuna seinermen of San Diego. Berkeley and Los Angeles: University of California Press.

Perrin, William F. 1969. Color pattern of the eastern Pacific spotted porpoise, *Stenella graffmani* Lonnberg. Zoologica 54:12.

———. 1975. Variation of spotted and spinner porpoise (Genus *Stenella*) in the eastern tropical Pacific and Hawaii. Bulletin of the Scripps Institution of Oceanography, vol. 21.

———. 1984. Patterns of geographical variation in small cetaceans. Acta Zoologica Fennica 172:137.

Perrin, William F., James M. Coe, and James R. Zweifel. 1976. Growth and reproduction of the spotted porpoise, *Stenella attenuata,* in the offshore eastern tropical Pacific. Fishery Bulletin 74:2.

Perrin, William F., W. E. Evans, and D. B. Holt. 1979. Movements of pelagic dolphins (*Stenella* spp.) in the eastern tropical Pacific as indicated by results of tagging, with summary of tagging operations, 1969–1976. NOAA Technical Report NMFS SSRF-737.

Perrin, William F., R. B. Miller, and P. A. Sloan. 1977. Reproductive parameters of the offshore spotted dolphin, a geographical form of *Stenella attenuata,* in the eastern tropical Pacific, 1973–1975. Fishery Bulletin 75:629.

Perrin, William F., and A. C. Myrick (eds.). 1980. Age determination of toothed whales and sirenians. Rept. Int. Whal. Comm. Special Issue 3.

Perrin, William F., and S. B. Reilly. 1984. Reproductive parameters of dolphins and small whales of the family Delphinidae. In: Reproduction in whales, dolphins, and porpoises, ed. W. F. Perrin, R. L.

Brownell, Jr., and D. P. DeMaster. Rept. Int. Whal. Comm. Special Issue 6.

Perrin, William F., Michael D. Scott, G. Jay Waler, and Virginia L. Cass. 1985. Review of geographical stocks of tropical dolphins (*Stenella* spp. and *Delphinus delphis*) in the eastern Pacific. NOAA Technical Report NMFS 28.

Perrin, William F., T. D. Smith, and G. T. Sakagawa. 1982. Status of populations of spotted dolphin (*Stenella attenuata*), and spinner dolphin (*S. longirostris*), in the eastern tropical Pacific. In: Mammals in the Seas, FAO Fisheries Series No. 5, Vol. IV.

Perrin, William F., R. R. Warner, C. H. Fiscus, and D. B. Holts. 1973. Stomach contents of porpoise, *Stenella* spp., and yellowfin tuna, *Thunnus albicares,* in mixed-species aggregations. Fishery Bulletin 71:4.

Pryor, Karen. 1973. Behavior and learning in porpoises and whales. Naturwissenschaften 60:412.

———. 1975. Lads before the wind. New York: Harper & Row.

———. 1981. Why porpoise trainers are not dolphin lovers: Real and false communication in the operant setting. Annals of the New York Academy of Sciences 364:137.

———. 1987. Reinforcement training as interspecies communication. In: Dolphin cognition and behavior: A comparative approach, ed. R. J. Schusterman, J. Thomas, and F. G. Wood. Hillsdale, N.J.: Lawrence Erlbaum Associates.

Pryor, Karen, and Ingrid Kang. 1980. Social behavior and school structure in pelagic porpoises (*Stenella attenuata* and *S. longirostris*) during purse seining for tuna. Southwest Fisheries Center Administrative Report LJ-80-11C.

Pryor, Karen, and K. S. Norris. 1978. The tuna/porpoise problem: behavioral aspects. Oceanus 21:2.

Stuntz, Warren E., and William F. Perrin. 1979. Learned evasive behavior by dolphins involved in the eastern tropical Pacific tuna purse seine fishery. Abstract, 3rd Biennial Conference of the Biology of Marine Mammals. Seattle, Wash.

Tayler, C. K., and G. S. Saayman. 1972. The social organization and behavior of dolphins (*Tursiops aduncus*) and baboons (*Papio ursinus*): Some comparisons and assessments. Annals of the Cape Provincial Museums 9:12.

Würsig, Bernd, and Melany Würsig. 1979. Day and night of the dolphin. Natural History 88:3.

Zar, J. H. 1984. Biostatistical analysis, 2d ed. 718 pp. Englewood Cliffs, N.J.: Prentice-Hall.

Randall Wells cradles a wild dolphin as colleagues and Earthwatch volunteers take measurements and use a hydrophone to record the animal's signature whistle. The dolphin, an adult male nicknamed Sparks, remains calm, accustomed to the handling. (Photo by Deb Williams.)

CHAPTER SIX

THE ROLE OF LONG-TERM STUDY IN UNDERSTANDING THE SOCIAL STRUCTURE OF A BOTTLENOSE DOLPHIN COMMUNITY

Randall S. Wells

INTRODUCTION

One of the many problems in studying dolphins is that they are long-lived, making an accurate understanding of their societies difficult to obtain from just one or two field seasons. The value of long-term studies has been clearly demonstrated for a variety of primate species (see Goodall

The Sarasota bottlenose dolphin research program has been the product of the efforts of many people between 1970 and 1988. Among the biologists who have contributed their time and expertise to collaborations with Blair Irvine, Michael Scott, and me on various field or laboratory aspects of the research are D. Black, J. Chamberlin-Lea, S. Chivers, D. Costa, D. Duffield, A. Hohn, V. Kirby, S. Kruse, A. Read, J. Sweeney, P. Tyack, and G. Worthy. We have Larry Fulford to thank for the safe and successful dolphin captures over the last five years. The program has benefited significantly from the efforts of R. Arden, C. Bell, J. De Metro, K. Fischer, B. Fortenberry, C. Gilligan, L. Mayall, K. Miller, P. Page, E. Patterson, H. Rhinehart, L. Sayigh, R. Spaulding, J. Zaias, and nearly 200 volunteers from Earthwatch, New College, and the Universities of Florida, South Florida, and California at Santa Cruz. Important logistical support has been provided by Cannon's Marina, Dick Wagner Realty, F. Wells, J. Wells, and F. Worl. Our research program has received support from a variety of sources over the years, including Mote Marine Laboratory, the U.S. Marine Mammal Commission, the National Marine Fisheries Service, the Inter-American Tropical Tuna Commission, the University of California at Santa Cruz, the Denver Wildlife Research Center, Earthwatch, the Office of Naval Research, Woods Hole

1986 and Smuts et al. 1987, for reviews), and such long-term research has become recognized as a viable approach for some odontocete cetacean species as well (chap. 1, this vol.; Bigg 1982; Wells, Scott, and Irvine 1987). One of the strengths of longitudinal studies is that the researcher may be able to recognize many of the animals under study as individuals and make repeated observations of social interactions. In time, it may also be possible to determine the subjects' sex, estimate their age, assess their reproductive condition, and learn their familial relationships. The compilation of case histories for individuals, and integration of life history and natural history data, can lead to a detailed understanding of the patterns underlying social associations.

Long-term studies provide at least two parallel approaches for testing hypotheses about the relative importance of societal features. Latitudinal tests can measure the persistence of structures through observations repeated over a number of seasons. For example, does a feature such as segregation by age and sex remain stable over time, in spite of such perturbations as unusual demographic or environmental fluctuations, that is, do the structural features stand the test of time? Longitudinal tests can address the same questions by following the fates of individuals over time. Do the basic components of the social structure remain intact during the course of a longitudinal study even though the individual members of the society may change their position within the structure? In other words, can similar patterns of social development be identified for different individuals?

In this chapter, I review the rationale, methods, and results of one longitudinal study of the social structure of bottlenose dolphins, *Tursiops truncatus*, along the central west coast of Florida, near Sarasota. With my colleagues, Blair Irvine and Michael Scott, I have been studying the behavior and ecology of a resident community of these dolphins since 1970. Using this long-term study as an example, I assess the value of applying parallel lines of investigation to hypotheses about the social structure of dolphins. In particular, I concentrate on patterns of social association between individuals. Our early research suggested that sex- and age-segregated groupings were important features of the dolphins' social system. Are patterns of association based on age and sex conserved over time? What are the consequences to the structure of the society as individual members pass into different phases in their life history? Do these bottlenose dolphins follow a regular pattern of social development?

Oceanographic Institution, and from donations of funds and equipment to Dolphin Biology Research Associates. The dolphin figures were prepared by E. Mathews. The manuscript was reviewed by B. Irvine, M. Scott, P. Tyack, K. Fischer, K. Pryor, and K. Norris.

Here I evaluate the approach and communicate the results of one particular long-term study of dolphin social behavior. A review of the status of our knowledge of bottlenose dolphin societies is beyond the scope of this chapter. For such reviews, with references to the fine work that has been done by other researchers, I refer the reader to Shane, Wells, and Würsig (1986) and Wells, Scott, and Irvine (1987). The data used in the descriptions presented below were drawn primarily from previously published sources (for 1970–71: Irvine and Wells 1972; for 1975–76: Wells 1978; Wells, Irvine, and Scott 1980; Irvine et al. 1981; for 1980–84: Wells et al. 1981; Duffield et al. 1985; Duffield and Wells 1986; Wells 1986; Wells, Scott, and Irvine 1987; Wells in press). More recent, previously unpublished data are presented in a few cases where appropriate.

When our research program began in 1970, it was not planned with the intention that it become a long-term study. The more we learned about the Sarasota dolphins, however, the more evident it became that this dolphin community afforded us excellent opportunities to ask increasingly refined behavioral questions. The program has progressed through five distinct projects during 1970–88 (see Scott, Wells, and Irvine 1990 for a review). The five projects have as a common focus efforts to understand the behavior of individually identifiable dolphins, and the projects differ in the kinds of information they have yielded about the individuals. The projects are (1) a pilot tagging study during 1970–71, (2) a tagging, radio tracking, and observational study during 1974–76, (3) censuses using photographic documentation of identifications from 1980 to the present, (4) focal animal behavioral observations during 1982–83 and from 1988 to the present, and (5) a capture, sample, mark, and release project from 1984 to the present.

Early on, we recognized that the validity of conclusions about the behavior of these animals depended on the quality of the background information that we collected for each member of the cast of characters. Strict reliance on real-time field identifications of natural markings may work for studies of terrestrial animals such as primates, but with bottlenose dolphins, these methods can result in mistaken identities and preclude subsequent confirmation of identifications or the tracking of changes in identifying characteristics (Scott, Wells, Irvine, and Mate 1990). Thus, a major thrust of the research program has been to refine our abilities to identify individuals positively and to obtain the most accurate information possible on sex, age, reproductive condition, and genetic relationships.

As our understanding of the dolphins improved, and as new technologies became available, new directions for the research were added to the basic task of identifying and following animals. At each stage, we evaluated whether the animals' long-term benefits from information gained in a new phase of research would outweigh any potential short-term impacts on their lives. The following description of the development of the research program presents the rationale for each project, relative to our studies of social structure. (For an overview of the results of other behavioral and ecological aspects of the Sarasota research program, see Scott, Wells, and Irvine 1990).

The pilot study initiated in 1970 came about because of an interest in the basic ranging patterns and daily activities of bottlenose dolphins. Little was known about the home ranges of bottlenose dolphins before 1970, and two extremes had been reported. Seasonal migrations had been described for bottlenose dolphins off Cape Hatteras (True 1891). In contrast, repeated, localized sightings of distinctive individuals had been reported from Cedar Key, Florida (Caldwell 1955), San Diego Bay (Norris and Prescott 1961), and Georgia-South Carolina (Essapian 1962). We began a tagging program to obtain information on ranging and social patterns of local dolphins.

During 1970–71, we accompanied a local commercial dolphin collector to tag those dolphins that were captured but were considered undesirable for captivity. Our study area was determined initially by the range of this collector's activities. Tagging efforts included the waters from southern Tampa Bay southward through Charlotte Harbor. We determined the sex of each animal and obtained length measurements. Each dolphin was freeze branded, and a numbered yellow plastic disk was attached to its dorsal fin (Evans et al. 1972). Of the thirty dolphins that were tagged, twelve were captured in the Sarasota area (fig. 6.1).

Most of our resighting effort during 1970–71 occurred from Sarasota northward to the southern edge of Tampa Bay. This effort was limited to opportunistic sightings during capture trips, occasional searches for tagged dolphins, and reports from the boating public. Resightings of tagged dolphins and several naturally marked individuals over a number of months indicated their apparent residency in the shallow inshore waters of Sarasota (Irvine and Wells 1972).

Following passage of the Marine Mammal Protection Act of 1972, it was incumbent on government management agencies to learn more about the biology of bottlenose dolphins so as to assess potential impacts on their populations from human activities. The apparent residency of dolphins near Sarasota and the relative ease with which they could be safely captured for tagging in the calm, shallow waters of the study area

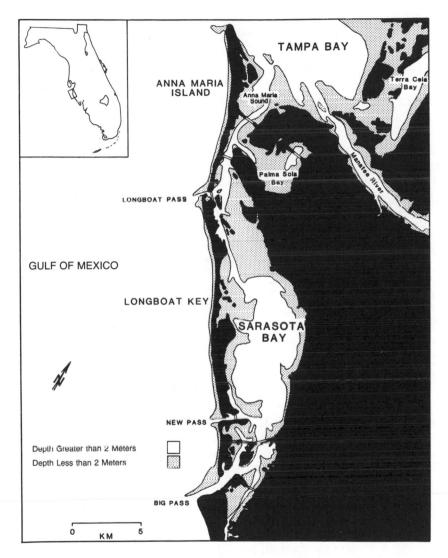

Fig. 6.1. The study area near Sarasota, Florida. The home range of the Sarasota bottlenose dolphin community includes the shallow inshore waters from Terra Ceia Bay, southward to near Big Pass, offshore to within several km of the barrier island chain.

were two factors leading to U.S. Marine Mammal Commission support for the second project, during 1974–76. This project involved more extensive tagging and observation efforts than the first. We were able to conduct dedicated captures and observations, as compared to our earlier more opportunistic efforts. From January 1975 through July 1976, we tagged forty-seven dolphins with freeze brands and a variety of attachment tags (Irvine, Wells, and Scott 1982). Acoustic recordings were obtained from each dolphin handled for a study of possible dialects (Graycar 1976).

Eleven of the forty-seven dolphins captured during 1975–76 had also been captured during 1970–71, suggesting that at least some dolphins were relatively permanent residents. We equipped ten dolphins with radio transmitters; these were tracked for up to three weeks. Tracking results and sightings of tagged and naturally marked dolphins indicated a resident community of about 100 individuals with a definable, year-round home range. This range extends southward from the southern edge of Tampa Bay to Siesta Key, off Sarasota, and includes all of the shallow waters from the mainland to several kilometers offshore of the barrier island chain (fig. 6.1; Irvine et al. 1981).

The availability of a large number of readily recognizable individual dolphins, of known sex and estimated age relative to maturity (from body length measurements compared to those of Sergeant, Caldwell, and Caldwell 1973), provided a unique opportunity to study social behavior. During 216 vessel surveys searching for tagged animals, we collected information on the composition of each dolphin school encountered. We used photography extensively to confirm field identifications.

Fieldwork during the late 1970s was limited to occasional surveys during which school composition was documented by photographic identification. In spite of the fact that most of the tags had been removed at the end of the second project, a high proportion of the previously tagged animals was still identifiable from photographs of natural marks, freeze brands, and tag scars. In April 1980, the continued availability of this by now well-known cast of characters led us to obtain field support, resulting in initiation of the third project.

The third project continues to date. In 1980, we initiated regular censuses documented by photographic identification to (1) assess the seasonal presence or absence of individuals, (2) document associations between mothers and calves for evaluation of birthrates and calving intervals, (3) examine long-term home range patterns, (4) document association patterns between individuals, (5) assess discreteness of Sarasota and adjacent population units, and (6) evaluate patterns of habitat use. These surveys included the study area of the 1970s, adjacent Gulf of

Mexico and Tampa Bay waters, and to a much lesser extent, Charlotte Harbor and Pine Island Sound to the south. Surveys conducted on 200 days from April 1980 through January 1984 brought the total number of dolphins in our photographic identification catalog to 466. Analyses of sighting data collected after January 1984 are in progress.

More than two-thirds of the dolphins tagged during the 1970s were recognized during the 1980s. Our concept of the home range for the Sarasota dolphin community remained essentially unchanged from the 1970s to the 1980s, except that we found Terra Ceia Bay (fig. 6.1) to be used regularly by Sarasota dolphins (surveys during 1975–76 rarely included this small bay). The estimated population size of about 100 dolphins, based on mark-recapture analyses, remained stable during both decades. Rates of immigration and emigration were estimated to be less than 3 percent per year (Wells et al. 1981, Wells 1986).

Adjacent, apparently resident communities have been identified in Tampa Bay and the Gulf of Mexico (Wells 1986, Weigle 1987). These adjacent communities share borders with the Sarasota community, but the other limits of their ranges have not been determined. Dolphins were identified repeatedly in adjacent waters, but some were not seen often enough to be classified as residents. Behavioral interactions between the Sarasota community and other dolphins were infrequent.

The fourth project (1982–83, 1988 to the present) involved focal animal behavioral observations (Altmann 1974). The Sarasota dolphins appeared to be habituated to the presence of small motorboats; thus, we were able to follow schools of dolphins from surfacing to surfacing for extended periods of time and to observe them without obviously disturbing them. Focal animal observations involved all classes of animals, but analyses to date have emphasized early development of calves (Fischer 1983). Concurrent underwater acoustic recordings were made in conjunction with many of the observations. Our focal animal observational efforts pointed to the need for more detailed background information on the community members. All too often, our known animals interacted with identifiable animals of unknown age or sex. This asymmetry in the data proved frustrating and was one of the primary reasons for initiating the next project. We resumed focal animal observations in 1988, taking advantage of a more thoroughly known cast of characters and more refined technology for recording and localizing underwater sounds while under way in our research vessel.

Beginning in 1984, and continuing to the present, we have conducted an annual or semiannual capture, sample, mark, and release program. The primary goals of these operations have been to learn the sex, age, reproductive condition, and genetic relationships of each member of the resident Sarasota dolphin community and to make certain that each individual could be identified unambiguously in preparation for continued focal animal observations. The capture operations provide opportunities to collect other data bearing directly on our understanding of the social system. The presence of dolphins of known age in our community has allowed us to calibrate age-determination techniques for use in our study as well as in other studies of bottlenose dolphins (Hohn et al. 1989). We are recording signature whistles from each animal for studies of the ontogeny and function of these sounds (Sayigh et al. 1990; chap. 11, this vol.). In addition, we are assessing the body condition of each animal, through measurements of weights, girths, and blubber thickness, for evaluation relative to social patterns as well as measuring seasonal variations for applications to improved husbandry for captive dolphins. Evaluations of the health of each dolphin garnered from physical examinations and blood sample analyses can provide insights into changes in social patterns.

We capture dolphins by encircling small schools (up to 10 dolphins) with a 300 m × 4 m seine net deployed from a commercial fishing boat as the animals swim over shallow seagrass meadows. Each dolphin is maneuvered by one or more members of our 36-person research team into a sling, weighed, and then placed in a 15-person raft. One hydrophone mounted on a suction cup is placed on the dolphin's melon, while another is either attached to another dolphin in the water nearby or is suspended in the water to record the acoustic emissions of other dolphins swimming in the net corral. Respiration rates are monitored, the dolphin's eyes are shaded from the sun, and its skin is kept moist while it is in the raft. The sex of the individual is determined, and then a suite of measurements is taken, including 25 standard lengths, 5 girths, and up to 30 ultrasonic readings of blubber layer thickness. Blood samples are drawn from the flukes for reproductive hormone measurements, genetic analyses, and standard blood chemistry and hematology assessments. A veterinarian applies a local anesthetic to the lower left jaw and gum tissue and removes one tooth (Ridgway, Green, and Sweeney 1975) for age determination (Hohn 1980). The dolphin is freeze branded, a series of photographs is taken of fins and scars, and it is returned to the water for release.

In addition to our own capture-release operations, we have obtained blood samples and have marked animals in nearby dolphin communities through the cooperation of a commercial dolphin collector. During 1984–85, we "piggybacked" on the operations of this collector in Pine Island Sound, Charlotte Harbor, and Tampa Bay. The samples obtained in this way were used in our genetic analyses of population discreteness.

The success of this project has been largely a function of four factors: (1) the tendency of the resident dolphins to use calm, shallow waters; (2) the dolphins' tendency to respond to the capture operation with greater calm with each subsequent capture; (3) the development of a number of simple, harmless, sampling techniques that can quickly provide a wealth of basic biological information; and (4) the willingness of a number of specialists to contribute their time and expertise to the collection and analysis of the samples. Among our collaborators for this phase of the program are Jay Sweeney (veterinary procedures), Peter Tyack and Laela Sayigh (Woods Hole Oceanographic Institution, acoustics), Debbie Duffield and Jan Chamberlin-Lea (Portland State University, genetics), Aleta Hohn (Southwest Fisheries Center, age determination), Vicky Kirby (University of California, Santa Cruz, reproductive hormones), Dan Costa and Graham Worthy (University of California, Santa Cruz, body condition), and Andy Read (University of Guelph, body condition and life history).

As of January 1988, 85 community members are readily recognizable, 65 have been handled since we began our sampling program in 1984, 74 are of known sex, 56 are of known age, and 53 are of known sex and age. Twenty-seven calves of known mothers are currently under observation.

SOCIAL STRUCTURE DESCRIPTION

The Sarasota dolphin community appears to be based on four main structural units: (1) mother-calf pairs, (2) mixed sex and single sex groups of subadults, (3) bands of females with their most recent offspring, and (4) adult males, as individuals or in strongly bonded pairs or trios. Our growing understanding of the significance of these four units is described in detail below. For each unit, I first present the state of our knowledge based strictly on our work during the 1970s. I then describe our understanding as a result of long-term latitudinal tests as well as longitudinal tests wherein I describe the individuals' transition through these stages over time, as their membership in each of the aforementioned structural units changes with age and social maturity.

Mother-Calf Pairs

Mother-calf pairs have been observed during all phases of the study. Females were associated with only one calf at a time during the 1970s. The strength of the mother-calf bond was indicated by the highly predictable associations of ten mothers with their calves and by observations of interactions between mothers and calves during capture situations. One mother-calf pair was observed for 15 months during 1975–76. At its first capture, the calf was the size of a 6- to 12-month-old, suggesting a minimum duration of the mother-calf bond of 21 months. No 1970–71 mother-calf pairs were observed during 1975–76.

Observations during the 1980s indicate that females with their young calves form among the most consistent associations within the Sarasota community. These associations are of longer duration than was indicated by our earlier work. Typically, calves remain with their mothers for 3 to 6 years and, in one case, 10 years. When they leave their mothers, young dolphins join groups of subadults.

The impetus for mother-calf separation is unclear. In various cases, separation has occurred while the mother was pregnant, soon after the birth of the mother's next calf, and apparently years before the next birth. In two cases, a female has been regularly accompanied simultaneously by her most recent calf and its next older sibling. In one case, the mother was with both calves for three years; in the other case, for only months until the younger calf was lost. Older offspring of both sexes are seen on occasion in brief association with the mother, especially following the birth of a new sibling.

By following the fates of calves through time, it should be possible to examine patterns of differential investment in offspring. Do mothers invest more in the raising of calves of one sex than in the other? What is the relationship between a mother's age and her investment in a given calf? Analyses are under way comparing calves relative to body length, weight, girths, and blubber thickness, as well as duration of association with their mothers and whether their separation was apparently brought about by the birth of a new sibling. The existing data set from eight calves (4 females, 4 males) that have already separated from their mothers is too small to be conclusive, but data from another thirteen calves of known age and sex (9 females, 4 males) still with their mothers after up to 7.5 years should provide the basis for examination of this question.

Subadult Groups

Subadult groups appear to be composed largely of dolphins born within the Sarasota community. On separation from their mothers at about 3 to 6 years of age, young dolphins join the subadult groups. One of the subadult males observed during 1975–76 was a small calf in the company of resident Sarasota females when he was first caught in 1970, and he has the same unique marker chromosome that has been found in some resident females (Duffield et al. 1985). Of the twenty-nine members of subadult groups that have been studied during the 1980s, ten are known to have been calves of Sarasota females, and at least one other has the unique marker chromosome. Many of the other dolphins that would be considered subadults now were born between 1977 and 1979, when few observations were made, so their maternal lineages are unknown.

Age-segregated schools of subadults were first noted during 1975–76. At this time, ages were estimated from body lengths. Repeated and consistent associations were observed among many of the subadults. These subadult groups were considered initially to be primarily bachelor groups because the four subadult females observed during 1975–76 were seen most frequently with adult females. The eight identifiable subadult males interacted with all other classes but rarely with adult males. Associations between several subadult males and "Killer," a female considered to be adult based on body length (Irvine et al. 1981), were exceptionally frequent, and we considered them unusual.

Continued observations and captures during the 1980s have provided a better understanding of subadult groups. We have learned that the subadult groups are often composed of both sexes, though male members have been more common than females. In contrast, subadult females tend to interact more frequently with adults than do the subadult males.

During 1975–76, two females were assigned to the wrong age classes through age estimates based on body length, thereby biasing our conclusions. Age determinations from teeth later indicated that subadult female 10 of Wells, Irvine, and Scott (1980) was actually a small adult, and Killer was actually a large subadult during 1975–76. As a subadult, Killer's frequent associations with subadult males during the 1970s were not anomalous; in fact, such associations are typical of our observations of subadults of both sexes during the 1980s.

The male-biased sex ratio of subadult groups appears to be a function of demography and of differences in life history patterns between the sexes. During 1975–76, we handled thirteen subadult or maturing dolphins (see Wells, Scott, and Irvine 1987, for age class designations), nine males and four females. During 1980–1987, we have handled or observed twenty-nine members of subadult groups, twenty males and nine females. This apparent discrepancy in sex ratios may be explained in part by two facts: (1) during the 1970s, there appeared to be more male than female calves born or surviving to an age when we could handle them (10 males : 2 females), and (2) females appear to be recruited into the breeding population at a younger age than males and therefore tend to belong to subadult groups for fewer years than males. Females typically give birth to their first calf at eight to twelve years of age. Males may continue to associate with subadult groups until they are roughly 10 to 15 years old. Thus, males were considered to be subadults for a longer period than were most females.

Subadult groups are the most active groups that we see in the Sarasota area. Leaping, chasing, and socializing are common within these groups. Socializing includes much physical contact, such as rubbing, stroking, pushing, and both heterosexual and homosexual interactions. More violent physical contact has also been noted, involving toothraking and tail-slaps. Perhaps these interactions are part of the process of development of dominance relationships, as reported for captives (Caldwell and Caldwell 1972).

By following individuals through time, we have learned that the term "subadult," as it refers to social groups, probably more accurately relates to social maturity than to sexual maturity. Females whose hormonal status indicates that they are ovulating continue to associate with subadult males. Similarly, males between 10 and 15 years of age, with testosterone concentrations indicative of adult breeding levels, continue to associate with younger males that have not yet reached sexual maturity.

Patterns of association independent of maternal associations become evident within these fluid groups of subadults. Strong bonds, as indicated by frequent associations, especially between pairs or trios of males, may be evident for individuals of similar age quite soon after entering the subadult groups. The subadult groups are often composed of several subgroups of these strongly bonded units, and it appears that bonds that are evident for young subadults may be maintained at least into early adulthood. As males get older, they tend to associate with fewer and fewer young males, and their associations during all months with adult females become more frequent.

Why do we find close associations between males within these subadult groups? Preliminary indications are that mortality rates are higher among subadult males than for all other age and sex classes. For example, three of the young males in figure 6.2 (#11, #38, #154) have died during 1984–87. Factors such as inexperience and small size may take their toll on this class during the first few years that these animals are away from the protection of the female bands. Perhaps the formation of close bonds improves the chances of survival for these young animals. Through continued observations, we may be able to quantify the survivorship of known subadults relative to their association patterns.

What are some of the characteristics of males who are found in close association with other males? Similar age appears to be one criterion. Of four pairs shown in figure 6.2 for which age data are available, all consisted of dolphins that were within less than three years of age of each other. Kinship may be another criterion. In two cases, pairs of males born

Subadult and Maturing

Fig. 6.2.   Subadult and maturing members of the Sarasota dolphin community in 1984. Pairs and trios indicate typical association patterns. Patterns of shading match those of their mothers in the Palma Sola band (fig. 6.3). A cross in the peduncle region indicates presence of an unusual marker chromosome. Sexes are indicated when known. Approximate ages are indicated by size. Dolphin identification numbers are those of Wells et al. (1987).

to females who associate with each other and share home range core areas have formed strong bonds as subadults (one pair consists of the younger brothers of the other pair). Preliminary analyses have shown that other female associates of these mothers are related to each other. If these two mothers are related to each other, then these males would be related at least along maternal lines. Genetic analyses of existing blood samples from these individuals, using chromosome banding patterns and DNA fingerprinting, are planned to assess the relatedness between the males and between their mothers. Because we have samples from both the mothers and the offspring, we plan to also conduct paternity exclusion tests and thereby evaluate the possibility of relationships along both paternal and maternal lines. Such a possibility would not be unreasonable if these animals live in a polygynous society.

Female Bands

During the 1970s, segregation by age and sex was found to be an important feature of the social system. Adult females were seen most often with other females, but school composition was fluid, and each female was seen in association with most of the available identifiable females at some point during observations. However, more persistent associations between some individuals were noted. Females were seen with their most recent offspring on a regular basis, and consistent combinations of mother-calf pairs, in nursery schools, were seen frequently. The home ranges of individual females were large, encompassing most of the community home range, but most females emphasized the northern half (fig. 6.1).

Continued observations into the 1980s clarified our understanding of females' position in the social structure. While associations between females on a day-to-day basis can be quite variable, certain groups of females tend to associate more with each other than with other females, and they share home range core areas. I apply the term "band" to the "female groups" described by Wells, Scott, and Irvine (1987) as a means of emphasizing the long-term relationships between these females. Band designations are based on statistical tendencies for associations and ranging patterns rather than absolute, permanent affiliations between individuals. Bands are not exclusive; members of different bands interact relatively frequently and may pass through each others' core areas. Some individuals are more closely affiliated with a particular band than are others, but the band as a unit appears to retain its structural integrity over time. Two of the individuals most frequently seen in one band are more than 40 years old.

Although many of the females were observed together frequently during the 1970s, female bands may not have been identified as structural features at that time because of the small size of the sample. The numbers of known animals and the numbers of observations of each animal were much smaller than during the 1980s. Furthermore, some of the dolphins from the 1970s were first recognized only a few weeks to a few months before the end of the 1976 field efforts. Because band designations are statistical rather than absolute determinations, the small sample size may have obscured the pattern during the 1970s.

Three female bands have been identified in the Sarasota community. The Palma Sola band is the largest of the three and the most thoroughly studied in terms of numbers of sightings and completeness of biological sampling. This band frequents the waters of Palma Sola Bay, and in 1984, it consisted of thirteen females and their most recent offspring (fig. 6.3). Two of these females were first captured in the same waters during 1970–71, and eleven were handled during 1975–76. All but one of the Palma Sola females have been handled for sampling during 1984–88.

The second-largest female band is the Anna Maria band. The seven females and their offspring that made up this band in 1984 (fig. 6.4) emphasize the waters around the north end of Anna Maria Island in their daily movements. Two of these females were first captured during 1970–71. Only three of the females of this band have been handled during 1984–88, but five were handled during 1975–76.

The third band, the Manatee River band, includes at least two females and their offspring as well as several other dolphins of unknown sex and relationships that frequent the waters of the Manatee River and Terra Ceia Bay, at the northeastern extent of the community home range (fig. 6.5). None of these dolphins have been captured to date.

Not all female members of the Sarasota community could be classified as members of particular bands. Five unclassified females are depicted in figure 6.6. Four of these and their three offspring have been handled for sampling during 1984–88, and one was captured in 1976.

Within the female bands, day-to-day associations appear to be based on reproductive condition. Females with young calves of similar age tend to swim together. The May-October peak in births results in well-defined cohorts of calves. Females without calves tend to swim together. When a calfless female gives birth, her association patterns may change and she may begin to swim with other mothers with newborns.

If bands of adult females are to be considered central features of the Sarasota social structure, then how do young females get recruited into these bands? Available evidence indicates that at least some young resident females join existing female bands, rather than creating a new unit or dispersing to other communities. Observations since 1980 have shown

Fig. 6.3. The Palma Sola female band in 1984. Paired dolphins indicate mothers with their calves. Identical patterns of shading indicate known maternal lines: full shading for the oldest generation, half shading for the second generation, and quarter shading for the third generation. A cross in the peduncle region indicates presence of an unusual marker chromosome. Sexes are indicated when known. Approximate ages are indicated by size. Dolphin identification numbers are those of Wells et al. (1987).

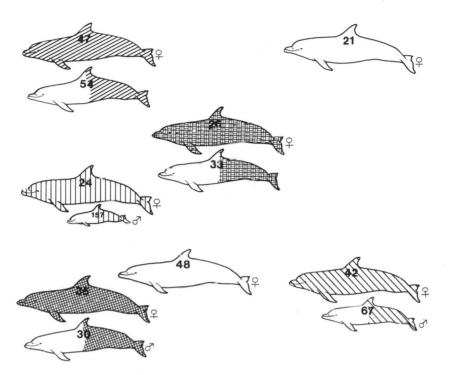

Fig. 6.4. The Anna Maria female band in 1984. Paired dolphins indicate mothers with their calves. Identical patterns of shading indicate known maternal lines: full shading for the oldest generation and half shading for the second generation. Sexes are indicated when known. Approximate ages are indicated by size. Dolphin identification numbers are those of Wells et al. (1987).

that of seven female calves or subadults identified in 1976, four have joined bands (#2, #4, #14, #28). Two of the remaining females were not reidentified after 1976, and the third disappeared in 1981. None of these females has been identified during photographic identification censuses of adjacent communities.

What factors draw young females into existing bands and maintain band structure over many years? Both long-term observations and preliminary genetic analyses indicate that at least some of the young females who join bands are in fact returning to their natal band. One female calf (#2) tagged in 1976 returned to her mother's band (Palma Sola) on the birth of her first calf and has had three calves as a member of this band. Another female (#15) was accompanied by her mother as she supported the remains of her (#15) first calf.

One subadult female (#14) tagged in 1976 who subsequently joined

Fig. 6.5. The Manatee River band in 1984. Paired dolphins indicate mothers with their calves. Identical patterns of shading indicate known maternal lines: full shading for the oldest generation and half shading for the second generation. Sexes are indicated when known. Approximate ages are indicated by size. Dolphin identification numbers are those of Wells et al. (1987).

the Palma Sola band has the same unusual marker chromosome that has been found in four other Palma Sola band members (Duffield et al. 1985). The uniqueness of this marker, its occurrence in at least one known mother-calf pair, and its occurrence in multiple generations of dolphins (ranging in age from 7 to 37 years) indicate its heritability and value as an indicator of close familial relationships. Investigations of the frequency of occurrence of this marker within the Palma Sola female band have not been completed; samples from four adults remain to be analyzed. The marker has not been found in four Sarasota females from outside of the Palma Sola band or eight females from other communities, but sampling and analyses are still incomplete. Thus, we feel the marker is a good indicator of the degree of relatedness within the Palma Sola female band.

The existence of a high percentage of apparently close familial relationships within a female band argues in favor of a kin selection explana-

Others

Fig. 6.6.　The recognizable Sarasota community members that did not fit clearly into any of the established groupings during 1984. Paired dolphins indicate mothers with their calves. Identical patterns of shading indicate known maternal lines: full shading for the oldest generation and half shading for the second generation. Sexes are indicated when known. Approximate ages are indicated by size. Dolphin identification numbers are those of Wells et al. (1987).

tion for altruistic acts (nepotism) between at least some bottlenose dolphins. Future analyses of samples from the Palma Sola and other bands should allow an assessment of the pervasiveness of close genetic relationships for females that associate with each other on a regular basis.

Adult Males

The earliest indications from the Sarasota community of segregation by sex and age and persistent associations between individuals came from observations of males of similar size that were captured and seen together frequently during 1970–71 and again during 1975–76. Observations from the 1970s suggested that adult males rarely associated with subadult males and that there was little overlap in the home ranges of these two classes.

The existence of two adult male social association patterns has been revealed by continued observations of known males during the 1980s. At least three adult males tended to be solitary, whereas other males can be found repeatedly with the same one or two adult male associates over a number of years (fig. 6.7). The single males tend to concentrate their activities in the northern half of the community home range, while the pairs and trios tend to range more extensively. Solitary males or pairs may occasionally disappear from the home range for days or weeks at a time.

The bonds between adult males can be quite consistent and longlasting. For example, members of one pair of large, heavily scarred males (#40 and #43) were seen together during 1975–76. These same dolphins, at least one of which is in his mid-thirties, were rarely seen separately. Another pair of adult males (#73 and #74), in their mid-twenties, has been together for at least five years. Another pair of large males, including one 29-year-old (#34) and one in his late thirties (#39), has been seen together for at least ten years.

The importance of strong bonds to some adult males has been indicated in several cases where individuals have been followed through time. In three cases, when one member of a pair died, the surviving male formed a close bond with another male (Wells, Scott, and Irvine 1987). One of these males (#27) has the marker chromosome and was first captured as a calf in 1970. Consistent and long-lasting pair bonds have developed between subadult males of similar age in four cases. Two of these pairs are still immature (#12 and #17, #36 and #44), but the other two pairs (#3 and #11, #31 and #132) have remained tightly bonded into early adulthood, over at least a six-year period. The bonds between adult males are apparently an outgrowth of relationships developed in subadult groups or earlier.

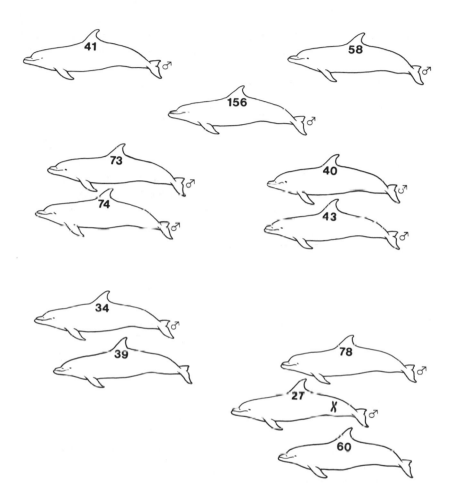

Adult Males and Associates

Fig. 6.7. Adult male members of the Sarasota dolphin community and their associates in 1984. Typical associations are indicated by pairings and the trio. A cross in the peduncle region indicates presence of an unusual marker chromosome. Sexes are indicated when known. Approximate ages are indicated by size. Dolphin identification numbers are those of Wells et al. (1987).

Why do we see the formation of tightly bonded adult male pairs and trios? These close associates appear to range more widely than do the single adult males, and their associations with resident females are more infrequent than those of the resident single males. Perhaps the close bonds of these males provide some of the same benefits of schooling that are presumably derived by single males during their more frequent interactions with female schools. If, for example, we assume that sensory integration (Norris and Dohl 1980) is important to these animals, then close, long-term associations among males could result in well-coordinated patterns of predator detection and defense as well as foraging and feeding. Perhaps a close association with at least one other individual allows one dolphin to rest safely while the other is vigilant.

Coordination or cooperation between adult males within tightly bonded bands may also provide benefits in such contexts as agonistic interactions with other dolphins or in securing mates. On several occasions, we have observed Sarasota male bands engaged in aggressive interactions with well-known, presumed males (large, heavily scarred individuals, seen over many years without calves) from adjacent communities. In each case, these interactions have occurred along the borders of the Sarasota community home range. The interactions have included such behaviors as tailslaps and violent leaps onto each other, resulting in bloodied fin edges and rostra. Perhaps interactions such as these explain some of the extensive scarring that we find on the fins of many older males. Are these cases of male-male competition for access to females? If so, then if males are related, the members of these pairs may increase their inclusive fitness by working together. Are they territorial conflicts that might help to explain the stability of the home range over so many years? The reasons behind the agonistic interactions remain unclear from the small sample of observations available.

Working as part of a closely knit team may increase mating opportunities for individual members of male bands. On several occasions, we have witnessed male pairs apparently separating single females from schools, which may involve chases, extensive socializing, and tailslaps. We have not yet seen these interactions culminate in mating.

The formation of strong bonds with other males may benefit the individuals as they travel outside the Sarasota community. The results of protein electrophoretic studies indicate that genetic exchange is occurring between communities (Duffield and Wells 1986). In light of the strong site fidelity demonstrated by adult females and the occasional disappearance of males, we hypothesize that the males are the primary vector for genetic exchange between communities. We are expanding our observational and capture studies into nearby Tampa Bay during 1988–92. It will be of great interest to see if we find Sarasota males in these adjacent waters and to observe their interactions with members of other communities. Similarly, we will be looking in our genetic analyses for indications of interbreeding between Sarasota males and females sampled from Tampa Bay.

SUMMARY AND CONCLUSIONS

Based on long-term study and the results from both latitudinal and longitudinal tests of hypotheses about the social structure of the Sarasota community, a consistent picture is beginning to emerge. The bottlenose dolphins that inhabit the shallow coastal waters near Sarasota, Florida, live in a relatively stable community of approximately one hundred individuals. This community shares home range borders with other resident communities. Annual rates of immigration and emigration appear to be low. The Sarasota community is composed of animals of both sexes and of ages up to at least the late forties. Dolphins born into the community tend to remain in the community range at least until they reach sexual maturity. Most observed social interactions are between members of the community, but there is evidence of genetic exchange with other communities. Social associations within the community appear to be based on sex, age, reproductive condition, and familial relationships.

Mothers and young calves form very strong bonds. Typically, mothers are accompanied by one calf at a time. Calves remain with their mothers for three to six years, well beyond the age at weaning of 18 to 20 months, as reported for captives. Even after separation, offspring may associate with their mothers from time to time.

On separation, young dolphins join groups of subadults. These groups are composed of both males and females, and while males tend to outnumber females, this may be a consequence of demographics and of earlier recruitment of females into the breeding population. Strong bonds may be developed between males of similar age, and these bonds may be maintained at least into early adulthood. As males reach 10 to 15 years of age, they tend to associate less with younger subadults and are found instead in smaller groups of strongly bonded young adults. Preliminary evidence suggests that some of these strong bonds may be between related males, but further analysis is required to determine the degree of relatedness. Young females primarily interact with other subadults until they produce their first calf at about 8 to 12 years of age, when they join bands of adult females and young.

Most of the adult females of the Sarasota community belong to one of three stable bands. Band members tend to emphasize the same core area of the home range in their daily activities. Evidence from observations of three generations and from preliminary genetic analyses indicates that female bands are composed of related females and that primiparous females may return to their natal band. Within these female bands, females tend to swim with others that have calves of the same age.

Adult males travel from one female school to another, either as individuals or as members of small, strongly bonded pairs or trios. Adult males may stay together over a period of at least fifteen years. Adult males tend to range farther than do adult females. Sometimes they are not seen for several months or years. Perhaps when they leave the community home range, they serve as a vector for genetic exchange between communities.

Our understanding of the Sarasota social system is by no means complete. Individual variation in dolphin behavior continues to challenge our abilities to identify general patterns. In time, we will likely find that parts of the description presented here are not entirely correct, but it offers the most accurate and parsimonious integration of our observations thus far.

Many questions remain. Of particular importance to our understanding of the social structure is a better knowledge of the mating system(s). Which males are siring calves during a given breeding season? Are breeding males from the Sarasota community, from other communities, or both? Males and females engage in very different reproductive strategies. Wells, Scott, and Irvine (1987) hypothesized the existence of a polygamous mating system for the Sarasota community, based on (1) the apparent brevity of associations between adult males and females, (2) a skewed adult sex ratio of more than two females to one male, (3) the inordinately large size of bottlenose dolphin testes relative to body size, (4) extremely high concentrations of sperm in bottlenose dolphin ejaculate, (5) a lack of marked sexual dimorphism, and (6) extremely low levels of male parental investment. The existence of sperm competition is suggested by several of these features.

The exact form (or forms) of polygamy that may occur in the Sarasota community is not clear. The brevity of associations between adult males and females and reports in the literature of apparently promiscuous matings in captivity argue against a polygynous system. But several facts suggest the possibility of polygyny: (1) some adult males are with receptive females much more frequently than are others; (2) adult males tend to be somewhat larger than adult females in length, weight, girth, and fluke span, features that might be considered factors in either inter- or intrasexual selection; and (3) adult males exhibit more extensive scarring from agonistic interactions with conspecifics than do females. Genetic analyses of existing blood samples from twenty-two known mother-calf pairs and a number of adult male members of the Sarasota and adjacent communities may provide insight into patterns of relative reproductive success of males. If likely sires can be identified, then by examining data on association patterns and body condition, it may be possible to describe several important features of the mating system.

Long-term study provides a powerful tool for developing an understanding of the social lives of long-lived animals such as bottlenose dolphins. The power of this tool can be optimized in several ways. A most important concern must be for the long-term well-being of the animals under study. Background data on the identities, sexes, ages, reproductive conditions, and genetic relationships of the cast of characters must be reliable. Continuity is crucial. Parallel lines of investigation are advisable as they provide a means for checking the internal consistency of social patterns. Are patterns that are apparent from repeated observations at intervals also apparent when tracking the fates of particular individuals through time and life history events? Collaborations among a variety of specialists can maximize the returns from precious field opportunities. Standardization of data collection techniques from one long-term study to another will greatly enhance the value of data collected from each by facilitating comparisons. Comparisons of social solutions from a variety of environments promise to provide important insights into the evolution of bottlenose dolphin social systems.

REFERENCES

Altmann, J. 1974. Observational studies of behavior: Sampling methods. Behaviour 49:227–267.

Bigg, M. A. 1982. An assessment of killer whale (*Orcinus orca*) stocks off Vancouver Island, British Columbia. Reports of the International Whaling Commission 32:655–666.

Caldwell, D. K. 1955. Evidence of home range of an Atlantic bottlenose dolphin. J. Mammal. 36:304–305.

Caldwell, D. K., and M. C. Caldwell. 1972. The world of the bottlenosed dolphin. New York: Lippincott.

Duffield, D. A., J. Chamberlin-Lea, R. S. Wells, and M. D. Scott. 1985. Inheritance of an extra chromosome in a resident female band of bottlenose dolphins in Sarasota, Florida. Sixth Biennial Conference on the Biology of Marine Mammals, Vancouver, B.C. (Abstract).

Duffield, D. A., and R. S. Wells. 1986. Population structure of bottlenose dolphins: Genetic studies of bottlenose dolphins along the central west coast of Florida. Contract Rept. to National Marine Fisheries Service, Southeast Fisheries Center. Contract No. 45-WCNF-5-00366. 16 pp.

Essapian, F. S. 1962. An albino bottle-nosed dolphin, *Tursiops truncatus*, captured in the U.S. Norsk-Hvalfangst-tid. 9:341–344.

Evans, W. E., J. D. Hall, A. B. Irvine, and J. S. Leatherwood. 1972. Methods for tagging small cetaceans. Fish. Bull. U.S. 70:61–65.

Fischer, K. 1983. Behavioral development in Atlantic bottlenosed dolphins, *Tursiops truncatus*—the first three months. Senior thesis (Honors), University of California, Santa Cruz.

Goodall, J. 1986. The chimpanzees of Gombe: Patterns of behavior. Cambridge: Harvard University Press.

Graycar, P. 1976. Whistle dialects of the Atlantic bottlenosed dolphin, *Tursiops truncatus*. Ph.D. dissertation, University of Florida. 90 pp.

Hohn, A. A. 1980. Age determination and age-related factors in the teeth of western north Atlantic bottlenose dolphins. Sci. Repts. Whales Res. Inst. 32:39–66.

Hohn, A. A., M. D. Scott, R. S. Wells, A. B. Irvine, and J. C. Sweeney. 1989. Growth layers in teeth from known-age, free-ranging bottlenose dolphins. Mar. Mam. Sci. 5(4):315–342.

Irvine, A. B., M. D. Scott, R. S. Wells, and J. H. Kaufmann. 1981. Movements and activities of the Atlantic bottlenose dolphin, *Tursiops truncatus*, near Sarasota, Florida. Fish. Bull. U.S. 79:671–688.

Irvine, B., and R. S. Wells. 1972. Results of attempts to tag Atlantic bottlenose dolphins, *Tursiops truncatus*. Cetology 13:1–5.

Irvine, A. B., R. S. Wells, and M. D. Scott. 1982. An evaluation of techniques for tagging small odontocete cetaceans. Fish. Bull. U.S. 80:135–143.

Norris, K. S., and T. P. Dohl. 1980. The structure and functions of cetacean schools. In: Cetacean behavior: Mechanisms and functions, ed. L. M. Herman. 211–261. New York: John Wiley and Sons.

Norris, K. S., and J. H. Prescott. 1961. Observations on Pacific cetaceans of Californian and Mexican waters. Univ. of Calif. Publ. Zool. 63:291–402.

Ridgway, S. H., R. F. Green, and J. C. Sweeney. 1975. Mandibular anesthesia and tooth extraction in the bottlenosed dolphin. J. Wildl. Dis. 11:415–418.

Sayigh, L. S., P. L. Tyack, R. S. Wells, and M. D. Scott. 1990. Signature whistles of free-ranging bottlenose dolphins (*Tursiops truncatus*): Stability and mother-offspring comparisons. Behav. Ecol. and Sociobiol. 26:247–260.

Scott, M. D., R. S. Wells, and A. B. Irvine. 1990. A long-term study of bottlenose dolphins on the west coast of Florida. In: The bottlenose dolphin, ed. S. Leatherwood and R. R. Reeves. Orlando: Academic Press. 235–244.

Scott, M. D., R. S. Wells, A. B. Irvine, and B. R. Mate. 1990. Tagging and marking studies on small cetaceans. In: The bottlenose dolphin, ed. S. Leatherwood and R. R. Reeves. Orlando: Academic Press. 489–514.

Sergeant, D. E., D. K. Caldwell, and M. C. Caldwell. 1973. Age, growth, and maturity of bottlenosed dolphins (*Tursiops truncatus*) from northeast Florida. J. Fish. Res. Bd. Canada 30:1009–1011.

Shane, S. H., R. S. Wells, and B. Würsig. 1986. Ecology, behavior and social organization of the bottlenose dolphin: A review. Mar. Mam. Sci. 2:34–63.

Smuts, B. B., D. L. Cheney, R. M. Seyfarth, R. W. Wrangham, and T. T. Struhsaker (eds.). 1987. Primate societies. Chicago: Univ. of Chicago Press.

True, F. W. 1891. Observations on the life history of the bottlenose porpoise. Proc. Nat. Mus. 13:197–203.

Weigle, B. L. 1987. Abundance, distribution, and movements of bottlenose dolphins, *Tursiops truncatus*, in lower Tampa Bay, Florida. M.S. thesis, University of South Florida. 47 pp.

Wells, R. S. 1978. Home range characteristics and group composition of Atlantic bottlenose dolphins, *Tursiops truncatus*, on the west coast of Florida. M.S. thesis, University of Florida. 91 pp.

———. 1986. Population structure of bottlenose dolphins: Behavioral studies along the central west coast of Florida. Contract Rept. to National Marine Fisheries Service, Southeast Fisheries Center. Contract No. 45-WCNF-5-00366. 58 pp.

———. In press. The marine mammals of Sarasota Bay: Present status and patterns of habitat use. Proc. Sarasota Bay Area Scientific Information Symposium, 24–26 April, 1987, Sarasota, Fla.

Wells, R. S., A. B. Irvine, and M. D. Scott. 1980. The social ecology of inshore odontocetes. In: Cetacean behavior: Mechanisms and functions, ed. L. M. Herman. New York: John Wiley and Sons. 263–317.

Wells, R. S., M. D. Scott, and A. B. Irvine. 1987. The social structure of free-ranging bottlenose dolphins. In: Current mammalogy, vol. 1, ed. H. Genoways. New York: Plenum Press. 247–305.

Wells, R. S., M. D. Scott, A. B. Irvine, and P. T. Page. 1981. Observations during 1980 of bottlenose dolphins, *Tursiops truncatus*, marked during 1970–1976, on the west coast of Florida. Contract Rept. to National Fisheries Service, Southeast Fisheries Center. Contract No. NA-80-GA-A-196. 29 pp.

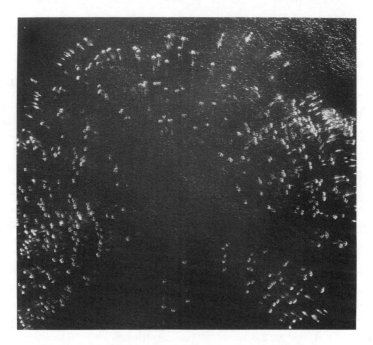

*Frightened by the aerial photographer's noisy airplane, a
school of Pacific spotted dolphin explodes in all directions.
(Photo courtesy of IATTC.)*

CHAPTER SEVEN

USING AERIAL PHOTOGRAMMETRY TO STUDY DOLPHIN SCHOOL STRUCTURE

Michael D. Scott and Wayne L. Perryman

INTRODUCTION

The chapters in this volume document the numerous advances made in the understanding of cetacean behavior in coastal environments. These gains were made through long and often painstaking observations of animals from cliffs or small coastal vessels. Unfortunately, logistics have severely hampered the study of pelagic dolphin schools. The difficulty in observing very mobile and very large schools, combined with the high operation costs of ships from which long-term behavioral observations can be made, has led us to take another approach to the study of dolphin behavior. Instead of looking at a few animals over a long period of time, we have tried to study dolphin school structure by looking at hundreds or thousands of animals at one instant in time through the use of aerial photography.

Our work grew out of management concerns for the stocks of dolphins

The authors gratefully acknowledge the following individuals who participated in the survey as pilots or observers: R. DeRosa, J. Laake, J. Niehaus, F. Ralston, and J. Rutledge. We thank the members of the EROS staff at Bay St. Louis, Mississippi, for their help and the use of their Variscan projectors. Funding for the study was provided by the IATTC; camera equipment was provided by the NMFS. We also thank R. Allen, P. Hammond, and M. Hall for their advice and support during this project. G. Schnell provided an enthusiastic introduction to the Mantel test. W. Bayliff, D. DeMaster, M. Hall, and P. Moore reviewed the manuscript.

impacted by the tuna purse seine fishery in the eastern tropical Pacific. A major problem for abundance estimation of these stocks is how to determine accurately the numbers of dolphins in the large pelagic schools that are commonly encountered. Since researchers in a wide range of wildlife management fields have demonstrated that observer estimates of the number of animals in large aggregations can be greatly in error, we began an aerial photographic program to determine dolphin school size accurately. We used large-format cameras to take vertical aerial photographs of dolphin schools from which estimates of school size could be made (Scott, Perryman, and Clark 1985). One fortuitous spinoff from taking vertical photographs is that the lengths of individuals and interanimal spacing can be calculated. Knowing the length distribution of a school can provide data on reproductive parameters (Hammond, Scott, and Perryman 1986), and it is our belief that much can be revealed about school structure from the manner in which animals space themselves in relation to conspecifics. We hope that a combination of data on school size, animal lengths, and interanimal distances will help us to interpret the structure of pelagic dolphin schools. Our study is far from complete, but here we will demonstrate the techniques we are using and pose many of the questions that can be tested with these methods.

FIELD METHODS

Our photographs were collected during a joint experiment conducted by the Inter-American Tropical Tuna Commission (IATTC) and the U.S. National Marine Fisheries Service (NMFS). We used a twin-engine Beechcraft AT-11, a World War II-vintage bombing trainer, with a large glass nose bubble and a long-range capability that made it an excellent platform for offshore photographic missions. We mounted an array of camera equipment in the floor of the aircraft, but our primary camera system was a 5-inch format camera (KA-62A) that we obtained from the U.S. Navy. This camera has a 3-inch lens, providing coverage of a large area, and automatic exposure control. The cameras were triggered remotely by an observer in the bow bubble, and a series of frames were taken until the observer turned off the cameras.

We evaluated several films in the course of our aerial photographic studies in the eastern tropical Pacific and achieved the best results with Kodak Aerochrome MS 2448 color transparency film. We found that black-and-white films did not allow recognition of subtle differences in shading and color pattern by which different species can be distinguished (mixed-species schools are prevalent in the eastern tropical Pacific). We compared color transparency and color negative films and found that the resolution of the original transparencies was better than that of prints made from negatives.

One major advantage of the system was its image motion compensation (IMC) system. This allows the camera to advance the film while the shutter is open at the same speed that the camera "sees" the ground passing beneath the aircraft. Otherwise, the advance of the aircraft during the fraction of a second that the lens is open blurs the image, particularly at low altitudes. This system also provided a significant advantage in safety because the pilot could maintain a safe air speed without losing photographic quality.

The survey flights were flown out of airports along the west coast of Mexico. Flight tracks covered areas of predicted high dolphin school concentrations up to 200 nautical miles (nm) offshore. We generally chose to fly during the morning when the breezes were light and the sun angle was high enough to provide sufficient lighting, yet not directly overhead so that glare did not cover the water directly below. Wind speed is also critical; once the sea state reaches Beaufort 3 (12 to 15 knots wind speed), the reflection of the sun on the choppy water creates glitter that obscures animals lying beneath the surface. In general, we photographed under nearly ideal conditions—light winds, calm seas, and extremely clear water.

Photographic passes were made at altitudes between 240 and 275 m at a ground speed of about 100 knots. These altitudes were low enough to obtain measurable images, yet high enough to easily get the whole school in the frame. If animals react to the plane by diving (as *Delphinus* were seen to do in our study), then higher altitudes may be necessary to avoid disturbing them. The camera systems were leveled to the floor of the aircraft, and the pilots trimmed the aircraft to ensure that the cameras were as level as possible during each photographic pass. Four observers were aboard the aircraft to search for dolphins, record data, and operate the cameras. The following data were recorded for each sighting: date, time, position, sea state, altitude, the roll and frame numbers of the film, camera shutter speed, and the estimates of the size and species composition of the school.

Large, tightly bunched schools of spotted dolphins (*Stenella attenuata*) and spinner dolphins (*S. longirostris*) were favored for photography since the more scattered schools would not be included in a single image and because we needed a large sample size to test the feasibility of our techniques. One must be aware of potential biases with this approach, because schools of different sizes may be comprised of different age classes (Chivers and Hohn 1985; Hohn and Scott 1983). Clearly, a comprehensive effort should be made to photograph schools so as to minimize potential sources of bias.

ANALYTICAL METHODS

To demonstrate the techniques we are using to examine school structure, a sample school of the Central American stock of spinner dolphins (*S. longirostris centroamericana*) will be used throughout this chapter. The estimated school size, using the methods described below, is 2,314 dolphins. Of these, 1,454 could be measured with photogrammetric methods.

Methods for Estimating School Size

Determining the number of animals in a school is usually the first step in studying its structure. This can be a deceptively difficult task for pelagic dolphins that may occur in schools of hundreds or thousands of animals. The problem in accurately estimating school sizes has led to the development of aerial photography for counting animals and using these counts to estimate school size (Scott, Perryman, and Clark 1985).

To obtain these estimates, photographs were first examined to select only those schools that were entirely within the frame. Schools were rejected if the field notes indicated that large numbers of dolphins were observed to be diving during the pass. The photographs were then magnified twelve times with a Variscan rear projector onto an acetate-covered screen. Usually, the school was contained on more than one frame per pass; this overlap allowed questionable images to be reexamined and allowed the counting of animals visible on one frame but not on others. The images were marked on the acetate and counted by several readers. In addition, the readers made a school size estimate based on these counts by adding a small percentage (usually less than 6%) to account for dolphins that were not seen. This factor was determined by considering the quality of the photograph, the degree of sun glare or splashing that may obscure animals, and the number of diving animals that were noted either in examination of the photographs or in the field notes (see Scott et al. 1985, for detailed methodology). We generally viewed the counts as underestimates of the true school size, particularly as detectability of the dolphins decreased.

How valid are these counts? How many dolphins are too deep to be counted? The accuracy of these counts was tested aboard the tuna purse seiner M/V *Gina Anne* during a research cruise in the eastern tropical Pacific (Allen et al. 1980). Dolphin schools were photographed from the ship's helicopter, encircled with the purse seine, and then counted by several observers as the dolphins were backed out (released) from the net. When compared with the backdown counts, the photographic estimates gave no indication of bias (Scott et al. 1985). Replicate estimates by different readers indicate the method is quite precise as well.

This study was conducted over waters clear enough to count deep-swimming dolphins (to a depth of about 10 m). Photographs of dolphins in more turbid waters may not provide accurate school size estimates. The best school size estimates of the sample school made by two readers (2,268, 2,361), for example, averaged 2,314 dolphins. We used the median of the readers' estimates (instead of the largest estimate) because, while the counts are likely to be underestimates, the estimates based on these counts may overestimate school size due to reader variability.

Methods for Calculating Dolphin Lengths

The methods for determining the lengths of objects on the ground from aerial photographs are well established (Slama, Theurer, and Henriksen 1980). While measuring dolphins from aerial photographs has proved to be more complex than measuring objects that do not dive, splash, flex, or twist, the basic equations are quite simple:

(1)
$$S = \frac{f}{H}$$

where   S = scale of the photograph
               f = focal length of the camera lens
             H = altitude of the camera

and

(2)
$$a = \frac{i}{S}$$

where   a = actual length of the object
          i = image length on the photograph.

Usually the focal length, altitude, and image length are known, from which the actual length and scale can be calculated. Since we wanted the actual length, a, to be calculated in centimeters, the units of f, H, and i are converted to centimeters as well. The height of the camera above the water (H) was considered to be the distance to the animals because, although the dolphin could be counted several meters below the surface, they could be measured only at or near the surface. If an animal swimming at a depth of 2.5 m could have been measured, however, the length would have been underestimated by less than one percent.

Equations 1 and 2 are applicable when the camera is perpendicular to the earth's surface, but frequently the images are distorted because the camera is tilted. This requires that the angle of tilt be measured or calculated and the image length be adjusted according to the degree of tilt and the position of the image on the photograph:

$$(3) \qquad\qquad S = \frac{f - y \sin(t)}{H}$$

where

$y = $ distance from the isocenter[1] of the photograph to the image in the direction of tilt, and

$t = $ angle of tilt (in degrees).

The angle of tilt should be measured at the time the camera is fired, although Hammond, Scott, and Perryman (1986) have developed a method for estimating the angle of tilt based on the differences in lengths of dolphins photographed on overlapping frames. The same individual photographed on successive frames will occupy different positions on each frame (as a result of the movement of the aircraft over the school) and will have different image lengths for each frame (as a result of camera tilt). Since the differences in image length and position can be calculated for several dolphins, the angle of tilt can be estimated.

Because of the large sample of dolphins to be measured, a mono-comparator operated by experienced technicians (Applied Photogrammetric Surveys, Inc., Riverside, Calif.) was used to digitize the head and tail positions of the images. The coordinates of these points were stored on a computer, allowing not only the measurement of the images but also the production of a scaled plot of the school.

Hammond, Scott, and Perryman (1986) have examined several sources of variance and bias in photogrammetric measurements of dolphins. They made replicate measurements of images of varying photographic quality and found that these measurements proved to be relatively precise, particularly for larger animals. The accuracy was influenced by two sources of bias. First, measurements from aerial photographs were greater than standard measurements (Norris 1961) because fluke notches were not visible in photographs taken from an altitude of 275 m. Our measurements were made from the tip of the rostrum to a line connecting the fluke tips. We made measurements on twenty-five dolphin carcasses to adjust for this bias. Adjustments of 3.2 percent (large dolphins > 150 cm)

---

[1] The isocenter is the point where the bisector of angle r intersects the photograph (see Figure 2-3 in Slama, Theurer, and Hendriksen 1980). For the method described above, however, the exact center (or "principal point") is used. The center is determined by the intersection of four marks ("fiducial marks") on the borders of the photograph.

to 4.4 percent (small dolphins) were made in each measurement. The second bias, flexure by the dolphins, tended to produce underestimates of the length. By examining and editing the photographs, we could delete many measurements of dolphins that were flexed or twisted. Typically, a moderate amount of flexure caused the flukes or rostrum to disappear from our overhead view and could be easily discerned and deleted. A small amount of bias remains, however, because of the presence of slightly flexed dolphins. To estimate the maximum extent of this bias, we measured dolphin carcasses bent to various degrees. Hammond, Scott, and Perryman considered that the maximum measurement error due to slight arching by the dolphin was 7.1 percent of standard length. The maximum extent of the first bias was estimated by measuring dolphin carcasses bent to various degrees; the second was estimated by measuring the distance from the fluke notch to a line connecting both fluke tips.

In general, aerial photogrammetric lengths tend to be more variable than standard lengths taken with conventional techniques. Figure 7.1 presents the length distribution of the sample school corrected to standard length measurements. While the sample of Central American spin-

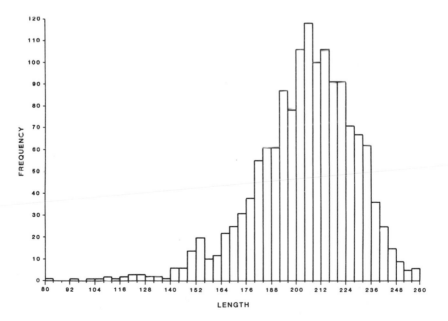

Fig. 7.1.   Length distribution (in cm) of a school of Central American spinner dolphins calculated from aerial photogrammetric measurements. The number of dolphins measured was 1,454; the estimated school size based on photographic counts was 2,294.

ner dolphins collected from the tuna fishery is too small and too sex biased for statistical comparison ($N = 15$ adult specimens, 12 females : 3 males; Perrin et al. 1985), the mean of the largest mode of the photogrammetric length distribution (208 cm) is similar to the mean of the fishery sample (203 cm).

Methods for Examining School Structure

Distribution of Size Classes within the School   As mentioned above, computer-generated plots are produced to scale from the coordinate data (fig. 7.2). Each dolphin can be numbered on the plot and color coded by

Fig. 7.2. Computer-generated plot from a photograph of a school of Central American spinner dolphins. The small darkened figures represent calves (< 156 cm), and the large darkened figures represent the largest adults (> 245 cm). The small squares indicate the head position of dolphins that could be counted but not measured; the cross marks the center of the frame. The lines divide the school into three sectors containing equal numbers of dolphins.

length. The computer plots of the schools permitted visual examination of the spatial relationships of various size classes. Hypotheses then can be posed about the distribution of size classes in the schools.

If we are interested in how a school protects its most vulnerable members, for example, we can test the hypothesis that calves are equally likely to occur in various sections of the school. This can be done in various ways, but first we must ask, what is a calf? If life history data are available on lengths at age, weaning, or independence, we can use such data to define a "maximum calf length." In the present example, however, scant life history data are available for the Central American stock of spinner dolphins (Perrin et al. 1985), and we were compelled to employ indirect methods that used our length and nearest-neighbor data. We first selected an arbitrary (and very low) maximum calf length; we then determined the length of the first nearest neighbor for each calf. The percentage of these nearest neighbors that were also calves was calculated. We assumed that very small calves virtually always swim next to an adult (presumably their mother) and that there would be no "calf-calf" neighbors when the maximum calf length was low. When this procedure is iterated, however, using successively greater maximum calf lengths, this percentage eventually will begin to increase as the young animals become more independent of their mothers and begin to associate with other juveniles (as suggested by studies on spinner dolphins by Pryor and Kang [1980 and chapter 5, this volume] and by Chivers and Hohn [1985]). We used the length at which the first non-zero percentage occurred as an estimate of the length at which calves begin to display independence from their mothers. As seen in figure 7.3, the estimated length of independence is 156 cm in the sample school.

Once a range of lengths has been defined for calves, the numbers of these individuals can be counted in different sectors of the school. The sample school was divided into three sectors containing equal numbers of dolphins in the middle, front, and rear of the school (fig. 7.2). The divisions are somewhat arbitrary, but the general procedure involved dividing the school in half to separate the front and rear sectors and counting from the perimeter of the school inward until each of the sectors contained one-third of the total school. An effort was made to keep identifiable subgroups together in the same sector. This particular school showed no significant differences among sectors using a chi-square test.

Norris et al. (1985) have suggested that when Hawaiian spinner dolphins are harassed, calves become enveloped into the center of the school. By examining successive passes over the same school, we can ask whether repeated passes cause this pattern to change. Our preliminary studies suggest that there is no statistical difference in the location of calves in schools photographed on early passes versus those photographed on later

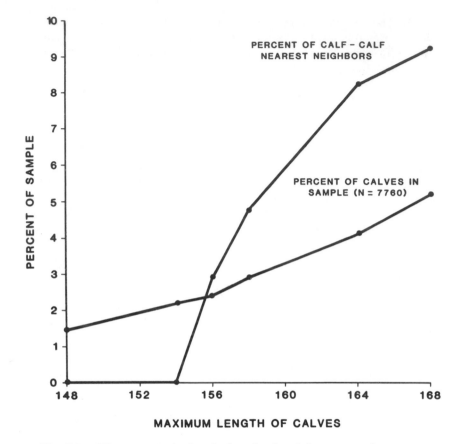

Fig. 7.3.   The percent of calves in the school and the percent of nearest-neighbors pairs in which both members are calves plotted as a function of increasing the maximum length (in cm) that defined a calf.

passes, suggesting that either the calves are not found in any one particular sector when harassed or that the photographic passes did not constitute a sufficient annoyance to cause this response.

Similarly, we can test for differences in the distribution of other classes within the school. For example, examining the distribution of the very largest dolphins would provide an interesting glimpse of dolphin social organization. These animals are likely to be adult males due to sexually dimorphic size differences in spinner dolphins (Perrin 1975). Their distribution within the school could be used to examine such questions as, do large males provide protection to other school members by positioning themselves on the periphery of the school? Are they found in the center of the school, serving as focal animals for the rest of the school? We found no significant differences, however, in the distribution of large animals among the sectors.

This method can also be used to examine how different species interact in mixed-species schools. Schools comprised of both spinner and spotted dolphins are common in the eastern tropical Pacific, and the spatial relationships between the two species may tell us much about the social structure of both species.

Spacing of Individuals   Many questions about the spacing of individuals within the schools can be answered with photogrammetric techniques. By examination of the plots, one discovers that the density of dolphins within the school is not constant. There are two classical approaches to this type of problem: (1) a grid method in which the school is sectioned into quadrats and the number of dolphins in each quadrat is counted and compared with other quadrats, or (2) distance methods in which interindividual distances to the nearest neighbor of all plotted dolphins are compared among different areas of the school. Because of fewer problems with arbitrary grid sizes and dealing with grids containing the edge of the school, we have used the distance method. This method, however, is prone to problems with "edge effects," that is, the nearest neighbor to an individual may be outside the sampled area, resulting in biased interanimal distances. In this study, the entire school is sampled and only those photographs with entire schools with no animals near the edge of the frame were selected, so this bias is negligible. To demonstrate this method, we tested whether dolphin densities differed in the front, middle, and rear of the school, using the sectors illustrated in figure 7.2. We measured the interanimal distances from the rostrum of one dolphin to the rostrum of its neighbor. In the sample school, differences in the distances between first nearest neighbors were not significant among sectors (Wilcoxon rank sum test), but differences were highly significant among distances between second and third nearest neighbors ($P < 0.001$). These differences are a result of the relative dispersion of dolphins in the front of the school and relative crowding in the other sectors. It is interesting to note that this is not reflected in the first nearest-neighbor distances, suggesting that, even when relatively dispersed, social bonds between pairs of animals are important.

Nearest-neighbor analyses can also be useful for analyses of other social spacing questions. One can compare, for instance, the differences in nearest-neighbor distances among schools engaged in different activities or arrayed in different formations or the differences among different spe-

Table 7.1.   *Average Distances between Nearest Neighbors of Size Classes of Dolphins*

Size Class [a]	Sample Size	Average Nearest-Neighbor Distances (in m)		
		1ST	2ND	3RD
Calves	50	0.92 (0.36)	1.92 (0.77)	2.58 (0.86)
Medium	897	1.51 (0.86)	2.43 (1.20)	3.11 (1.39)
Large	507	1.79 (0.96)	2.69 (1.47)	3.34 (1.69)

[a] Calves: less than 156 cm; medium-sized dolphins: 156–219 cm; large adults: greater than 220 cm. Standard deviations are included in parentheses.

cies in separate schools or in mixed-species schools. We have used, as an example, a comparison of nearest-neighbor distances of dolphins of various size classes (table 7.1). These results show that the average nearest-neighbor distance increases with increasing size. This would be expected for two reasons. First, physical reasons would dictate that as animals grew in both length and girth, the interanimal distances would be likely to increase. Second, interanimal distances may increase for social reasons, such as the dependence of calves on their mothers and increased aggressiveness of large males. Norris, Stuntz, and Rogers (1978) found a similar trend for pelagic spotted dolphins captured and photographed underwater in a tuna purse seine in the eastern tropical Pacific.

Many standard statistical methods exist for analyzing these types of data. For example, we are currently assessing one such method, the Mantel test (Mantel 1967), which has been adapted for analyses of animal behavior and social structure. Schnell, Watt, and Douglas (1985) have demonstrated its use in statistical analyses of the spatial patterns of social individuals. The Mantel test compares the corresponding elements of two different matrices to evaluate associations. Each matrix is composed of pairwise comparisons between all individuals in the sample (e.g., the length differences between each pair of animals) and then compared with another similar matrix (made up, for example, of the length differences that would be expected under a particular hypothesis, or another interanimal variable such as distances between individuals). Many questions about dolphin school structure can be tested with statistical techniques: Do large dolphins occur more often near the center of the school or the periphery? Are dolphins that are found close together likely to be of similar size? How does interanimal distance vary with school size?

School Formations The shape of a school can be readily observed and measured in photographs, and thus aerial photogrammetry can be a good tool for detailed description and analysis of school geometry. The temptation, of course, is to correlate school formation with ongoing behavior. During this study, behavior could be observed only during the relatively brief passes over the school, and only broad categories of behavior could be identified. The school in figure 7.2, for example, was swimming rapidly during the pass after apparently being startled by the plane's shadow; the formation matches that of Norris and Dohl's (1980) description of fleeing schools as being broader than long and tightly packed.

The internal geometry of subunits within the school can be readily described, as well. Swimming in echelon formations by subgroups within the school is apparent in figure 7.2 and has been described previously in free-ranging dolphins (e.g., Norris and Dohl 1980). We believe that using many of the methods described above to study these subgroups may prove to be most fruitful. For example, Norris, Stuntz, and Rogers (1978) found relative interanimal distances to decrease with increasing subgroup size in captured spotted dolphins; this hypothesis can be readily tested on free-ranging dolphin schools as well.

Methods for Studying Dolphin Locomotion One photograph (see frontispiece) taken during this study was used to analyze the "porpoising" behavior of dolphins (Au, Scott, and Perryman 1988). From this aerial photograph of rapidly moving spotted dolphins, it was possible to measure the distance between the characteristic exit and entry splashes the dolphins make when they leap. The average leap distances were calculated to be 6.8 m. The minimum speed required to leap this far is 15.9 knots. If dolphins "porpoise" to lessen water-induced drag, then one might expect that at least as much time would be spent airborne as swimming. The average distance the dolphins swam between leaps, however, was 12.7 m, almost twice the distance of the leaps. The puzzle of "porpoising" behavior awaits further explanation.

CONCLUSIONS

We believe that aerial photogrammetry will prove to be particularly valuable for studying coastal marine mammals. Such studies would be relatively inexpensive and easy to conduct and would provide information that could neatly dovetail with data from shore-based studies. Aerial photographs provide an intensive look at school structure that can be best appreciated by having the broader background data possible from long-term observations. One example of such an approach is currently being conducted; the NMFS Southwest Fisheries Center is applying aerial photographic techniques to the spatial patterns of hauled-out pinnipeds in California to complement the numerous and extensive ground-based studies conducted on these populations.

Studying the school structure of pelagic dolphins, however, is much more difficult and expensive. Aerial photogrammetry is one of the limited number of tools available to the field biologist. Studying pelagic schools in this manner offers the unique opportunity to examine the structure of very large schools. This can alleviate the bane of many behavioral studies, the search for an adequate sample size (e.g., we are currently examining 15 schools of spinner dolphins, ranging in size from 150 to over 6,000 animals, containing over 20,000 plotted animals). We think we should sound a note of warning about such large schools, however. While it took us a matter of minutes to photograph a school of dolphins, we have spent years poking and prodding this large data set, and we still have not explored all the possible avenues of research.

REFERENCES

Allen, R. L., D. A. Bratten, J. L. Laake, J. F. Lambert, W. L. Perryman, and M. D. Scott. 1980. Report on estimating the size of dolphin schools, based on data obtained during a charter cruise of the M/V *Gina Anne*, October 11–November 25, 1979. Inter-Amer. Trop. Tuna Comm. Data Rept. No. 6:1–28.

Au, D. W., M. D. Scott, and W. L. Perryman 1988. Leap-swim behavior of "porpoising" dolphins. Cetus 8(1):7–10.

Chivers, S. J., and A. A. Hohn. 1985. Segregation based on maturity state and sex in schools of spinner dolphins in the eastern Pacific. Abstract: Sixth Biennial Conf. Biol. Marine Mammals. Vancouver, B.C. Nov. 22–26, 1985.

Hammond, P. S., M. D. Scott, and W. L. Perryman. 1986. Obtaining dolphin lengths from aerial photographs. Unpub. ms.

Hohn, A. A., and M. D. Scott. 1983. Segregation by age in schools of spotted dolphin in the eastern tropical Pacific. Abstract: Fifth Biennial Conf. Biol. Marine Mammals. Vancouver, B.C. Nov. 27–Dec. 1, 1983.

Mantel, N. 1967. The detection of disease clustering and a generalized regression approach. Cancer Res. 27:209–220.

Norris, K. S. 1961. Standardized methods for measuring and recording data on the smaller cetaceans. J. Mamm. 42(4):471–476.

Norris, K. S., and T. P. Dohl. 1980. The structure and functions of cetacean schools. In: Cetacean behavior: Mechanisms and functions, ed. L. M. Herman. John Wiley and Sons. 211–261.

Norris, K. S., W. E. Stuntz, and W. Rogers. 1978. The behavior of porpoises and tuna in the eastern tropical Pacific yellowfin tuna fishery: Preliminary studies. Marine Mammal Comm. Final Rept. NTIS Pub. No. PB283-970. 86 pp.

Norris, K. S., B. Würsig, R. S. Wells, M. Würsig, S. M. Brownlee, C. Johnson, and J. Solow. 1985. The behavior of the Hawaiian spinner dolphin, *Stenella longirostris*. Southwest Fisheries Center Admin. Rept. No. LJ-85-06C. 213 pp.

Perrin, W. F. 1975. Distribution and differentiation of populations of dolphins of the genus *Stenella* in the eastern tropical Pacific. J. Fish. Res. Bd. Canada 32(7): 1059–1072.

Perrin, W. F., M. D. Scott, G. J. Walker, and V. L. Cass. 1985. Review of geographical stocks of tropical dolphins (*Stenella* spp. and *Delphinus delphis*) in the eastern Pacific. NOAA Tech. Rept. NMFS 28. 28 pp.

Pryor, K., and I. Kang. 1980. Social behavior and school structure in pelagic porpoises (*Stenella attenuata* and *S. longirostris*) during purse seining for tuna. Southwest Fisheries Center Admin. Rept. No. LJ-80-11C. 113 pp.

Schnell, G. D., D. J. Watt, and M. E. Douglas. 1985. Statistical comparison of proximity matrices: Applications in animal behaviour. Animal Beh. 33: 239–253.

Scott, M. D., W. L. Perryman, and W. G. Clark. 1985. The use of aerial photographs for estimating school sizes of cetaceans. Inter-Amer. Trop. Tuna Comm. Bull. 18(5): 381–419. [In English and Spanish.]

Slama, C. C., C. Theurer, and S. W. Hendriksen. 1980. Manual of photogrammetry, 4th ed. Amer. Soc. Photogrammetry. 1056 pp.

PART II

Laboratory Studies

Essay

MORTAL REMAINS: STUDYING DEAD ANIMALS

Karen Pryor

Queen Elizabeth I of England loved to eat dolphin meat. I can't think why; to me it tastes like steak fried in cod liver oil.[1] Elizabeth decreed that any dolphin or porpoise corpse that washed up on shore was "The Queen's Fishe" and had to be sent to the palace cooks at once.

The queen passed on, but the decree remained; in due course, the British Museum became the repository for dolphin remains, whether from the British Isles or elsewhere, thereby acquiring one of the world's greatest collections of cetacean skeletons. From this trove, a series of British Museum scientists defined our basic understanding of the species and distributions of small cetaceans.

Meanwhile, the whaling industry of the eighteenth and nineteenth centuries provided data about large cetaceans—where they were caught, what kinds were caught, their measurements. And until very recently, about all we knew about most cetaceans was the location of death and the measurements of their carcasses and bones.

This told us very little about behavior. For example, in 1965, Sea

---

[1] Elizabeth's choice should not be confused with mahimahi, or dolphin-fish, commonly served in restaurants, which is not a mammal but a white-fleshed, saltwater fish, *Coryphaenus* spp., which tastes like a snapper or a cod.

Life Park was the first oceanarium to capture a pigmy killer whale (*Feresa attenuata*), a strange little black animal with white lips, like a clown. It was such a rare specimen that Ken Norris flew out from California just to measure it. Museum data could tell us what species it was and where it had been seen before; but nothing warned us that in behavior it was more like a wolf than a normal dolphin, that it was going to growl and snap like a canid and would not hesitate to attack people and other cetaceans.

And yet, behavioral information can be deduced from dead data, from bones and teeth, ovaries and stomach contents. It is only necessary to ask the right questions. Helene Marsh and Toshio Kasuya, whose chapter follows, took advantage of the Japanese onshore fishery for small cetaceans to study pilot whales. Being able to look at a whole school simultaneously made it possible to age and sex every animal and thus see the structure of the school; looking at ovaries showed the reproductive status of each female. But it seems to me it took a leap of the imagination to perform the simple test for milk sugar on the stomach contents of *all* animals and thus find out that females may nurse their young up to seventeen *years* and that lactation may have unusual bonding functions in pilot whale society.

One of the most impressive collections of dolphin remains in the world occupies the basement of the Southwest Fisheries Center in La Jolla, California. Here, for over two decades, the National Marine Fisheries Service (NMFS, pronounced "nymphs") has been amassing specimens collected from tuna vessels engaged in purse seining for tuna in association with dolphins in the eastern tropical Pacific: the so-called tuna-porpoise fishery. The procedure involves locating and encircling dolphin schools that are often accompanied by tuna and then releasing the dolphins.

Dolphin mortality in some years was enormous. Scientists at the Southwest Fisheries Center and elsewhere expended major efforts both to mitigate the mortality and to understand the dolphin populations and the effect of the fishing. These efforts were, on the whole, successful; cooperative research with the fishing industry resulted in improvements that reduced dolphin mortality from over 100,000 a year in the early 1970s to under 14,000 in 1986. (Unfortunately, improvements in the U.S. fleet have recently been offset by mortality caused by new and increasing foreign fleets.)

During the course of this research, government scientists and observers on hundreds of ship voyages, across a million square miles of ocean, collected data and samples of anatomical material from specimens incidentally killed in the nets. In the basement of the Southwest Fisheries Center, the library stacks holding the resulting collections reach in shadowed corridors from floor to ceiling, wall to wall, and one end of the building to the other. The shelves are crowded with jars of uteri and stomachs; jars of teeth and jawbones; books full of stained slides of cross sections of testes; racks and racks of bones and skulls. A conservative estimate is that the collection includes samples of over 22,000 spinner and spotted dolphins.

Every specimen is precisely labeled—sex, size, when and where it was taken, and much more. The intangibles that cannot be pickled in formalin, such as color patterns (a clue to population distributions), are stored in computer data bases and also on paper, in carefully organized library binders, in other rooms upstairs. Here one can locate the original, grimy, water-stained work sheet, filled out on shipboard by one of hundreds of data collectors, for every dolphin. And the Southwest Fisheries Center houses yet another enormous collection of data resulting from this fishery: the computer data on thousands of sets of the net by boats in the U.S. fleet, accumulated by NMFS and by the Inter-American Tropical Tuna Commission (IATTC), which for many years has overseen the conservation and management of the tuna themselves.

A primary organizer of this enormous collection has been NMFS biologist William Perrin. The methodical studies by Perrin and his associates of these thousands of anatomical specimens have themselves provided a vast quantity of data for other scientists to work on (it is informative to glance at the references in this book to see how often Perrin's work is cited).

The NMFS and IATTC scientists have done much to define the geographic distribution of spinner and spotted dolphins, to depict the apparent division of these populations of many millions of animals into separate stocks, and to identify their growth rates and reproductive cycles, all crucial information for making management decisions and protecting the population's survival. Now other researchers are beginning to tease behavioral information out of this mass of samples and data. For example, Southwest Fisheries Center scientist Aleta Hohn and her associates performed extensive statistical analyses of computerized data on the sizes of encircled schools, compared to the sex and maturity of animals taken in the same sets. They have been able to show, among other things, that spotted dolphins have two different kinds of schools: breeding schools, such as those described by myself and Shallenberger in this volume (chap. 5), and small schools of immature males that spend most of their time living among large aggregations of spinners. This arrangement, which I find amazing, would have been very hard to pinpoint by underwater observation, since the fishermen do not encircle large aggregations of spinners because they are seldom accompanied by tuna; but it immediately made sense of occasional anecdotal reports, from fishermen and government observers, of "a few little spotters" among big spinner schools.

Two chapters in this book describe behavioral information gleaned from the tuna fishery sampling and data collections: in Part I, Michael Scott and Wayne Perryman describe what they have learned about social behavior from the aerial photographs taken in population survey studies. In Part II, Albert Myrick tells us how he and others unraveled some interesting mysteries from those jars of teeth down in the basement. To me, these contributions demonstrate the ultimate value of what sometimes seems like the dustiest and most useless part of biology—museum collections, the warehousing of specimens, the laborious filling out of labels, the minute measurements of taxonomy. The dolphin teeth Myrick used had been collected over the years not in the certainty but only in the hope that someone might be able to use them all. The Southwest Fisheries Center's donations of material to other museums and its own present collection of interrelated specimens and data derives from a depressing loss of life, to be sure, but it constitutes an unequaled research tool. We can only guess at what questions will be asked of it in the future and what new understandings may be acquired, which in turn can be used, in fact, to save the dolphins.

Albert C. Myrick, Jr., discusses his work with students touring the Southwest Fisheries Center in La Jolla.

CHAPTER EIGHT

SOME NEW AND POTENTIAL USES OF DENTAL LAYERS IN STUDYING DELPHINID POPULATIONS

Albert C. Myrick, Jr.

INTRODUCTION

Ever since the discovery by Nishiwaki and Yagi (1953) that it is possible to estimate the age of a dolphin by analyzing the layers in a single tooth, scientists have used tooth ages in conjunction with other data to conduct their biological investigations of dolphins (see Scheffer and Myrick 1980). By studying enough tooth samples from animals taken in the field, it is possible to identify important age-related characteristics of whole dolphin populations, such as the age structure of the population, the average age at sexual maturity (determined from reproductive organs and tooth layer counts), and the average reproductive longevity. This information can then be used, along with data derived from pregnancy and mortality rates, to assess the current condition of a given population and predict its future growth rates. Such assessments become particularly valuable for populations jeopardized by direct hunting, by alteration of their habitat, or by inadvertent kill through entanglement in or entrapment by fishing gear.

I thank A. Dizon, D. DeMaster, and the several anonymous reviewers whose useful comments helped improve the manuscript. I am grateful to W. Walker, who provided specimens and data regarding parturition in captive bottlenose dolphins. R. Allen assisted in preparing some of the figures.

Thousands of dolphins are killed incidentally each year in the eastern tropical Pacific (ETP) when purse seines are set on dolphin schools to capture yellowfin tuna that associate with them (Perrin 1969; Smith 1983; Hammond and Tsai 1983). In 1972, concern over the biological consequences of this considerable annual mortality prompted the U.S. National Marine Fisheries Service to mount a long-term monitoring and research program to study the problem.

In 1978, I began work on the problem of trying to estimate ages from burgeoning samples of teeth collected from the kill by federal technicians aboard selected ETP seiners. To extract age or other biological information from layers in dolphin teeth requires a certain basic understanding of the teeth, such as which tooth to use, when the tooth tissues are formed, how and where layers are accumulated, what factors may affect layer deposition, how they appear when viewed microscopically, where in the layered tissue to begin making a tooth age reading, and what the layers represent in terms of time.

THE UNIQUE TEETH OF DOLPHINS

Teeth of dolphins are different from those in any other mammal group in a number of important respects. First, except for the rodents and lagomorphs (pikas, rabbits, and hares), in which teeth are highly modified for gnawing and grinding, and the edentates and allies (pangolins, some anteaters, sloths, and armadillos), in which teeth are strongly reduced or absent, almost all toothed mammals produce baby or milk teeth and adult or permanent teeth (Scott and Symons 1964; Peyer 1968; Walker 1968). Dolphins produce only a permanent set of teeth: in a dolphin, the teeth form before birth, erupt within a few months after birth, and remain in place throughout its life.

Second, with the exception of the Amazon River dolphin (*Inia geoffrensis*), whose rear teeth are almost molarlike, and a few species in which teeth have undergone retrogression (*Phocoenoides* sp.) or reduction (narwhals and beaked whales), virtually all dolphins have homodont dentition, that is, all teeth of an individual have one shape—usually simple cones or pegs. Homodonty is common in fish, amphibians, and reptiles, but among the mammals, it is limited only to toothed whales and dolphins. Most mammals have heterodont dentition, with teeth differentiated into incisors, canines, premolars, and molars, or some combination of two or more types (Peyer 1968).

Another feature peculiar to dolphin teeth is that they are far more numerous in most delphinid species than in any other mammals with nor-

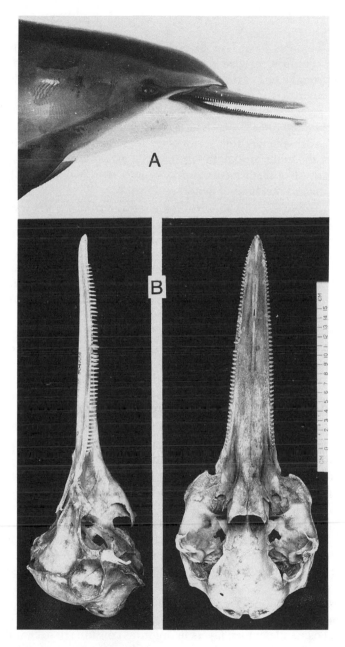

Fig. 8.1. Reiterative homodont teeth in dolphins. A. Head of spinner dolphin (*Stenella longirostris*) showing numerous teeth in upper and lower jaws. B. Right lateral and ventral aspects of a spinner dolphin skull, showing teeth. (Photo from Perrin 1972.)

mal teeth. Humans have 7 or 8 teeth on each side of both upper and lower jaws. Primitive placental mammals have 11 on a side, with a total of 44. But (to take one of the extreme examples) among spinner dolphins (*Stenella longirostris*), individuals may have a total of between 180 and 220 teeth, arranged in rows of at least 45 teeth on each side of each jaw (fig. 8.1). Among other living mammals, only the African dog (*Otocyon megalotis*), with a heterodont series of up to 50 rather weak teeth (Peyer 1968), and the giant armadillo (*Priodontes giganteus*), with up to 100 intermittently shed teeth (Scott and Symons 1964, Walker 1968), show any trend toward this kind of dental reiteration.

SOME DENTAL ANATOMY

A dolphin tooth is a natural recording device, somewhat like a trunk of a tree that registers changes through its own life by the characteristics of its accumulating rings. The tooth differs from trees, however, in having three layering tissues instead of one. Each tissue, enamel, dentine, and cementum, is deposited in incremental layers, apparently with clockworklike regularity (fig. 8.2). In small delphinid species, deposition of enamel, which forms the apical mantle of the tooth, probably commences in the fetus three or four months after conception and is completed shortly before the dolphin's birth (Myrick 1980). The layers of enamel appear to represent daily increments (ibid., Boyde 1980; fig. 8.3).

The body of the tooth is composed of layers of dentine formed by continuous application of the tissue to the inner surface of the hollow dental cone. Dentine deposition begins concurrently with the onset of enamel formation, and thus, like enamel, prenatal dentine represents a record of a large part of the fetal life. However, unlike enamel, which forms prominent daily layers, prenatal dentinal layers show what seems to be a monthly depositional pattern (Myrick 1984b, Myrick et al. 1984).

Dentine deposited after birth, that is, postnatal dentine, is more conspicuously layered than prenatal dentine. As in most other mammals, postnatal dentinal deposition in dolphins begins at birth with the formation of a distinctive hypomineralized *neonatal layer* on the inner surface of the prenatal dentine (fig. 8.2). Formation of dentine usually slows down and gradually stops in most mammals but probably not in most dolphins. Postnatal dentinal layers in dolphins continue to accumulate internally until death or until the pulp cavity fills with dentine sufficiently to close off the supply of nutrients to the dentine-producing cells. Unless factors alter a dolphin's calcium or protein physiology sufficiently to interfere with the layering process, dentinal layers, from the neonatal layer to the pulp cavity margin, represent the dolphin's complete postnatal depo-

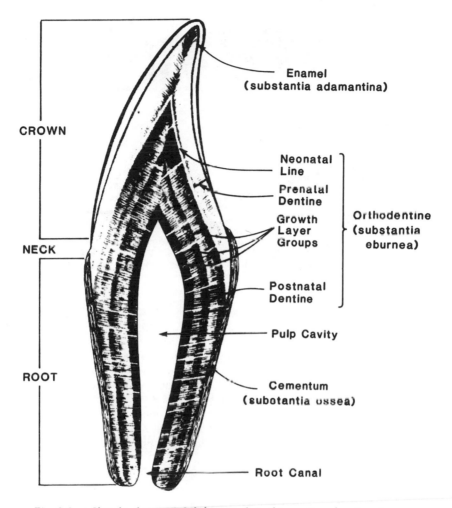

Enamel
(substantia adamantina)

CROWN

Neonatal
Line

Prenatal
Dentine

Growth
Layer
Groups

Orthodentine
(substantia
eburnea)

NECK

Postnatal
Dentine

Pulp Cavity

ROOT

Cementum
(subotantia ussea)

Root Canal

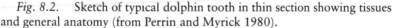

Fig. 8.2. Sketch of typical dolphin tooth in thin section showing tissues and general anatomy (from Perrin and Myrick 1980).

sitional record. Of course, the dentinal record would be truncated if the dolphin lived beyond the time of pulp-cavity occlusion.

Formation of cemental layers begins shortly after birth. Succeedingly younger layers of this tissue are applied to the external surface of the tooth's basal half. Because deposition occurs externally in the relatively unconfined space of the tooth socket, or alveolus, accumulation of cemental layers is thought to continue until death. This tissue should represent an uninterrupted layered recording of the entire postnatal life, presuming again that there is no sustained physiological interference.

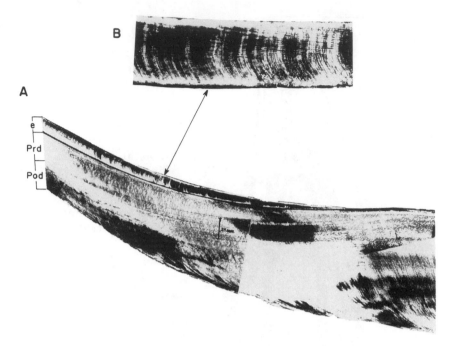

Fig. 8.3.   Daily (?) layering in enamel in a tooth of a spotted dolphin (*Stenella attenuata*). A. Half of a tooth section viewed with polarized light and showing layering in the enamel (arrow). B. Highly magnified section of enamel as shown in A showing layers as indicated by marks. Because of the oblique (off-lapping) arrangement of layers, many transects at various intervals from apex to neck may be necessary to obtain full counts. A total of 223 enamel layers was seen in tooth depicted. Abbreviations: e, enamel; Prd, prenatal dentine; Pod, postnatal dentine. (From Myrick 1980.)

As a corollary of the dolphin's homodont and reiterative dentition, deposition of the layered record in one tooth occurs at the same time, in the same sequence, and with virtually identical patterns as in almost every other tooth in a dolphin's jaws (Myrick et al. 1984). This means that the number and distinctness of dentinal layers identified in one tooth will be the same as those in adjacent teeth. More important, it indicates that all of the teeth are under complete systemic control. The subtle uniqueness of the layering pattern is so well replicated in each tooth of an individual that by comparing the layering patterns of two teeth, even if they have the same number of layers, I have usually been able to determine whether or not they were taken from the same specimen (fig. 8.4).

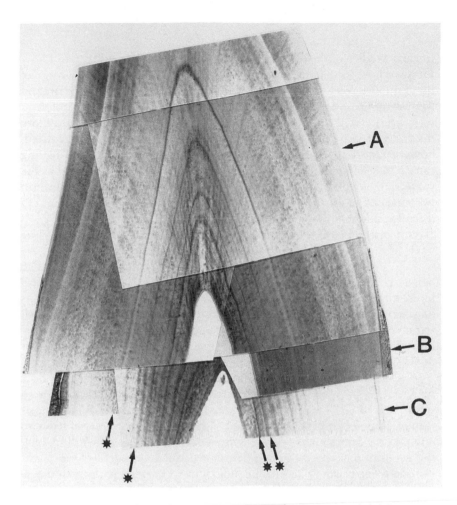

Fig. 8.4. Dentinal layering patterns of three individual spotted dolphins compared. Photographs of teeth from three specimens (A, B, C) are superimposed to show that a pattern common to all exists. B is the youngest, A the next youngest, and C the oldest. The three specimens were collected in different areas of the eastern central Pacific in different years. Although the patterns resemble one another closely, it is not difficult to locate individual differences (probably reflecting physiological differences), some of which have to do with intensity of some of the incremental layers (starred).

CALIBRATION

In early studies of the teeth of wild dolphins, there was no way to know for sure how long it took to lay down each similarly layered group of tissues, more recently termed growth layer groups, or GLGs (Perrin and Myrick 1980). Because of this initial uncertainty, tooth ages of dolphins were often stated in layer or GLG units, not in months or years. Before the full importance of thoroughly calibrating dental layers was realized, however, more than a few studies treated layer units as though layering rates had been measured. The lack of thorough calibration, as well as the subjectivity involved in selecting an "annual" layer or GLG, may account for some of the disparity in age-specific parameters given by different authors for different dolphin populations (Perrin and Reilly 1984).

One method used to calibrate the GLG deposition rate is to study the teeth of captive-born dolphins, because the ages are already known exactly. A count of GLGs in a single tooth from a dolphin of known age helps to verify the elapsed postnatal time they represent. Another method in use compares GLGs in teeth pulled from an animal in different seasons or years. A third approach is to mark the teeth of captive dolphins with tetracycline. Later, after the animals eventually die or when a tooth is removed from each live animal, elapsed time is compared with the number of GLGs formed after the tetracycline mark to estimate deposition rate.

A tooth from a captive-born dolphin of known age may show how many GLGs have been deposited since birth, but it gives no clue as to when a given GLG was formed or how long the process took: a deposition rate cannot be firmly established. A series of teeth extracted from an animal over time should provide a number of time periods within which the deposition of specific annual GLGs is defined with greater precision. But many teeth would have to be pulled over extended periods to determine deposition rates—a drawback to subject and experimenter alike.

Tetracycline marking, however, can be carried out repeatedly on a single animal to produce labels that are deliberately introduced weeks, months, or years apart. This permits examination and verification of deposition rates at almost any scale of detail.

Four features make tetracycline the GLG calibration tool of choice: (1) it combines immediately with the calcium incorporated into a newly forming incremental layer of dentine or cementum, (2) unabsorbed amounts of the drug are excreted within a few days after entry into the circulatory system, (3) when combined with the calcium in a layer viewed microscopically in ultraviolet light, tetracycline shows up as a fluorescent label among the layers in a tooth thin section, and (4) because it is an antibiotic, tetracycline causes no ill effects when introduced intramuscularly into a dolphin in low or moderate concentrations (30 to 75 mg/kg body weight, scaled to resting metabolic rate).

Of the three methods in use, multiple introductions of tetracycline over several years would seem most useful for determining deposition rates. Of course, the use of all methods together would be superior to any one of the three alone.

In 1979, I began a three-and-a-half-year experiment, assisted by staff members of Sea World, Inc., and Hubbs Marine Research Institute, to monitor the dentinal and cemental deposition rates in dolphin teeth. We used twelve bottlenose dolphins (*Tursiops truncatus*), including three captive-born animals and two control animals, maintained for public display by Sea World, Inc. Two of the wild-captured dolphins were from the California Pacific Coast and the remainder were from a population occupying coastal waters near Florida.

At the beginning of the experiment, we injected each animal with tetracycline and we extracted a tooth from each. (We used an extractor and an elevator after temporarily deadening the interalveolar nerve with a local anesthetic. The procedure is simple and apparently causes the animal little discomfort, and dolphins have plenty of teeth to spare.) About every three months thereafter we gave tetracycline to all but the control animals, and at approximately six-month intervals, we removed a tooth from all but the controls. In addition, sham events, in which animals were handled but did not undergo treatment or tooth removal, were conducted on the project animals at arbitrary intervals. At the end of the project (fall 1982), a tooth was taken from all animals (Myrick and Cornell 1990).

During the experiment, we kept weekly records of water temperature and salinity measurements, and we recorded weekly averages of types and amounts of food consumed. We also noted any changes in behavior, activity, and apparent health and any episode that might be stressful, such as unusual amounts of handling of an animal or transporting of any animals between Sea World parks (in Cleveland, Ohio; Orlando, Florida; and San Diego, California). In addition, we recorded body weight and length measurements at each labeling or extraction session.

In 1980, with the cooperation of colleagues at Sea Life Park, Waimanalo, Hawaii, I also conducted a GLG calibration study of captive Hawaiian spinner dolphins (*Stenella longirostris*), in two stages: (1) a retrospective phase, and (2) a direct monitoring phase. For the retrospective phase, we used teeth of four carcasses (including a captive-born specimen) and three live animals. All seven specimens had received numerous therapeutic doses of tetracycline at various times over periods of years during their captivity, and I thought it highly likely that each treatment had inadvertently labeled teeth of the recipient. (Fortunately, the labels

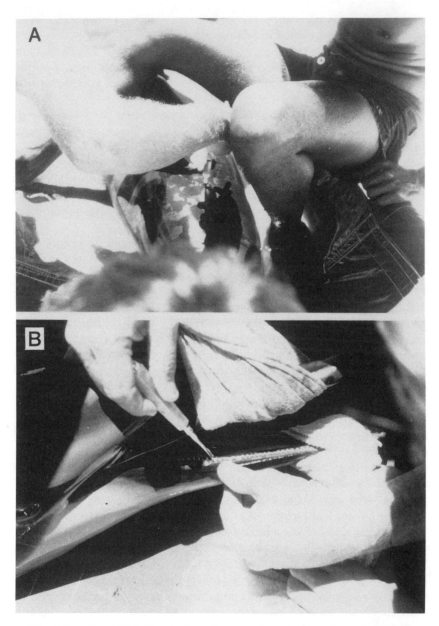

Fig. 8.5. Dental labeling and tooth extraction. A. Hawaiian spinner dolphin (*Stenella longirostris*) receiving tetracycline injection to introduce a label into the dental tissues for layer calibration studies. B. Removing a tooth from the anesthetized lower jaw.

had been introduced.) With labels identified to dates of past therapeutic treatments taken from records provided by Sea Life Park, we were able to document depositional rates from many years earlier.

In the direct monitoring phase of the Sea Life Park study, we extracted three teeth and gave three additional tetracycline injections, over a one-year period, to each of the three live captive animals to monitor dental tissue accumulation directly (Myrick et al. 1984; fig. 8.5). During the year, no environmental or behavioral records were systematically made.

Appropriately prepared, multiply labeled tooth thin sections from both the Sea World and Sea Life Park experiments were examined separately under plain and ultraviolet light. By tracing labels and layers on plastic overlays of ultraviolet- and plain-light photographs of each specimen, I was able to determine the position of each label within the layering pattern of each tooth. I matched almost every label to a treatment date by comparing the spacing, brightness, and thickness of each label in a tissue with the relative elapsed time between treatments and the length and dose strength of each treatment. With labels identified and series of units of layered tissue bracketed by dates, it was a simple step to identify, measure (in μm), and characterize repeating annual and subannual depositional patterns (fig. 8.6).

What made the two studies successful was the scrupulous manner in which the extensive medical records were maintained for each animal by Sea Life Park and Sea World staff members. Although the records were complete and virtually error-free, a certain amount of detective work had to be done to try to identify labels in a few specimens for which no corresponding dates of treatment were found. In the case of one of the captive-born bottlenose dolphins in the Sea World project, a narrow label occurred in the postnatal dentine a scant distance from the neonatal layer, but the animal had no corresponding treatment record. Since it was not usual for very young calves to be treated with antibiotics or to take solid food, I guessed that the dam who suckled the calf had been given tetracycline, which combined with her blood calcium and was imparted to the calf through her milk. An examination of the dam's records revealed that about one month after the calf was born, the dam was being treated with tetracycline (Myrick and Cornell 1990).

A more complicated mystery was encountered in the Hawaiian spinner study (Myrick et al. 1984). Teeth of the (deceased) captive-born dolphin contained numerous labels, but its records indicated that it had been treated with tetracycline on only three occasions during the three and three-fourths years that it lived (fig. 8.7). I found two faint closely spaced labels within the dentine formed in the first year of postnatal life (fig. 8.7A), which corresponded to records showing that the calf's dam was

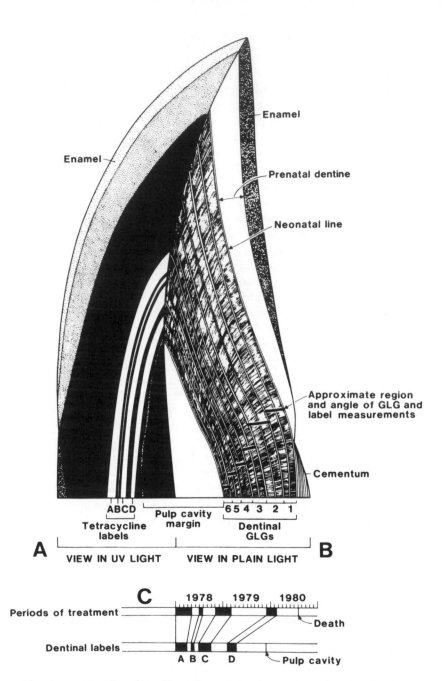

Enamel

Enamel

Prenatal dentine

Neonatal line

Approximate region
and angle of GLG and
label measurements

Cementum

A B C D   Pulp cavity   6 5 4 3 2 1
            margin
Tetracycline   Dentinal
labels   GLGs

A   **B**

VIEW IN UV LIGHT   VIEW IN PLAIN LIGHT

C   1978   1979   1980

Periods of treatment

Death

Dentinal labels

A B C   D

Pulp cavity

Fig. 8.6. Line drawing of hypothetical dolphin tooth in thin section showing appearance of tetracycline labels, A, B, C, D, under UV light (A, left-hand
side) and dentinal growth layer group (GLG) layering patterns under plain
transmitted light (B, right-hand side). C illustrates method of identifying labels
by comparing relative thickness and spacing of labels to duration of and elapsed
time between treatments. (From Myrick et al. 1984.)

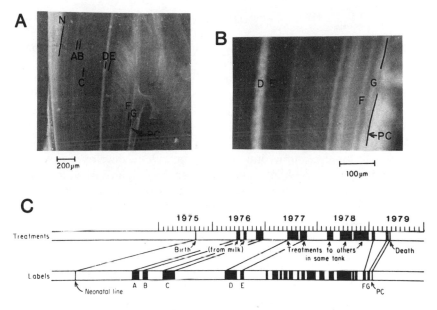

Fig. 8.7.   A, B. Ultraviolet-light view of tetracycline labels in dentine of known-age Hawaiian spinner dolphin. C. Illustration of how labels were identified with intentional and unintentional introductions of tetracycline fluorophores. Labels A and B apparently were introduced when calf's dam was treated with tetracycline for two periods while the calf nursed. Labels received in 1977 and 1978 were introduced when the youngster stole and ate tetracycline-dosed smelt intended as treatment for other dolphins in the tank. (From Myrick et al. 1984.)

treated with tetracycline during two periods while she suckled the calf. (We proposed that a method of treatment might be developed using nursing as a vehicle to transmit certain medicines to sick calves in captivity without disturbing the calves or separating them from their dams (Myrick et al. 1984).

The mystery of the other undocumented labels in the teeth of the captive-born Hawaiian spinner dolphin solved itself when I matched the records of other dolphins that, at various times, had shared tanks with this young animal. Dolphins are usually treated with tetracycline by slipping the pills inside fish that are then given as food to the patients. As I discovered, many of this young animal's tank companions were being treated with tetracycline-dosed smelt. Interviews with a staff member confirmed that the young dolphin was very adept at stealing and eating fish intended for others and had been observed doing so with some frequency (fig. 8.7B,C).

MONTHLY LAYERS

Multiple labels introduced into the dentine only weeks or months apart showed repeatedly, in various specimens, that annual GLGs consist of thin layers that are deposited with lunar monthly regularity (Myrick et al. 1984). These results support observations made earlier suggesting that incremental layers of annual dentinal GLGs represent monthly records (Laws 1962; Kasuya 1977; Myrick 1979, 1980; Hohn 1980; fig. 8.8).

With the assumption that the dentinal recording device is keeping some sort of lunar monthly time, we may be able to develop a new tool that could furnish us with information about the dolphins in which lunar monthly layers are distinctly visible. A back-count of monthly layers might yield the year and month of birth of an individual if the date of its death is known.

Yearly peaks or other patterns in reproduction, including year-to-year changes in reproductive activity, should be detectable by monthly layer determinations of birth dates of specimens from a population sample

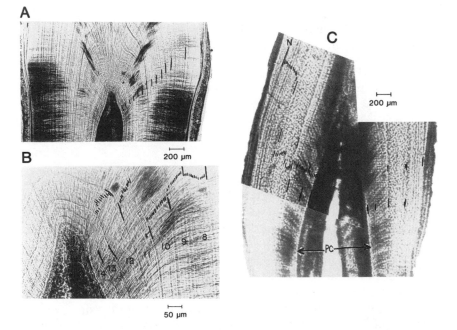

Fig. 8.8.   Dentine of Hawaiian spinner dolphins showing apparent lunar monthly layers (fine marks) within the GLGs (heavy marks). A, B. Two magnifications of same specimen. C. Tooth of known-age dolphin. (From Myrick et al. 1984.)

(Myrick 1980, 1984*a*). If, for example, monthly layer counts for most of the specimens sampled from one population produced a significant birth peak in March and the gestation period was twelve months long, then serious perturbations of the population in the early spring might interfere with courting, breeding, and calving sufficiently to alter reproductive rates. Suppose another sample of the same hypothetical population is taken ten years after the initial sample and seven years after three successive years of strong El Niño conditions in the spring. Monthly layer counts from the new sample produce a broad birth peak at May-June-July. From this, a hypothesis might be formulated that the protracted change in the environment in the spring was connected with the shift in and broadening of the reproductive season.

Inferences might be made of the reproductive compatibility or isolation of two or more geographically contiguous populations from reproduction patterns generated from the counts (Myrick 1984*a*). If counts for the sample from one population produce a significant peak for spring and counts for the sample from another population produce a peak only for fall, a case for limited interbreeding between the two populations might be considered.

If the reproductive season of a population is limited to a short period, monthly layer counts might be used to estimate the approximate months of death in analyses of "die-offs" suspected of being connected with environmental changes (Myrick 1984*b*). Consider a hypothetical case in which several months after a huge oil spill, remains of many dolphins become washed up on nearby beaches. Morphometrics indicate that the specimens are from an offshore population whose calving season is in October and November. The oil spill is suspected of being connected with the large die-off of dolphins. Counts of monthly layers (from neonatal layer to pulp cavity) in teeth of the beach-cast specimens would give the number of lunar months from the beginning of November (the middle of the two-month peak calving season) until they died. In other words, counts would indicate the approximate month in which the animals died and whether deaths occurred at about the same time as the oil spill.

GENETICALLY BASED PATTERNS

Our calibration experiment with the bottlenose dolphins included detailed monitoring of food consumption, weekly average water temperature and salinity, and notation of handling, activity, and behavior. I examined dentinal and cemental layers that were deposited during periods of sudden fluctuations in some of these monitored conditions, for ex-

ample, abrupt drops in water temperature, expecting to find corresponding alterations or interruptions in the regular layering patterns. Instead,
comparisons showed that monitored exogenous changes had no percep-
tible effect on dentine deposition or layering pattern.

When I compared patterns of all the Sea World project specimens captured from the Florida coast, I found them to be very similar, especially in
the annual thickness of dentine deposited at a specific age. I then compared the Florida pattern with the pattern exhibited in common by the
two Pacific Coast captives in the project sample. In this comparison, I
noted a sharp difference in age-specific GLG thicknesses between Florida
and Pacific Coast specimens, even though they belonged to the same
nominal species (*T. truncatus*). For example, Florida coast captives deposited dentinal GLGs in the first, second, and third years of life which
were at least half again the thickness of those deposited by project specimens representing the Pacific Coast population. Such results are suggestive of a genetically determined dentinal depositional pattern, but the
possibility exists that patterns and their differences could be an artifact,
somehow, of the captive situation.

To consider the idea of genetically based patterns in greater detail,
some colleagues and I compared annual GLG depositional patterns of
Sea World captives from Florida waters with patterns in teeth taken from
thirteen wild Florida coastal bottlenose dolphins captured and quickly released back into the wild (Myrick et al. 1985). Because they were from a
population that had been the subject of extensive field studies for many
years, the wild animals were individually identifiable and had been monitored from year to year, so that they were either of known age or known
minimum age.

In the captive/wild pattern study, we used the age-specific dentinal
GLG pattern (developed from studies of Sea World Florida coast captives) to estimate the tooth ages of the wild dolphins. The ages of the wild
dolphins were already documented by the field studies, but they were unknown to me at the time I made the tooth age estimates. We wanted to
determine whether wild and captive animals from the same population
had similar depositional rates and patterns. An answer to this question
would also resolve whether or not the wild pattern was measurably altered by conditions of captivity. Results showed that the age estimates
made with the captive pattern of GLG thicknesses as a guide were very
close to the known ages of the wild dolphins (Myrick et al. 1985). These
results and the results of an earlier study in which I found layering and
thickness patterns to be nearly identical in spotted dolphins and Hawaiian spinner dolphins (*Stenella attenuata* and *S. longirostris;* fig. 8.9)
are evidence that at least some of the pattern is genetically determined
(Myrick et al. 1983).

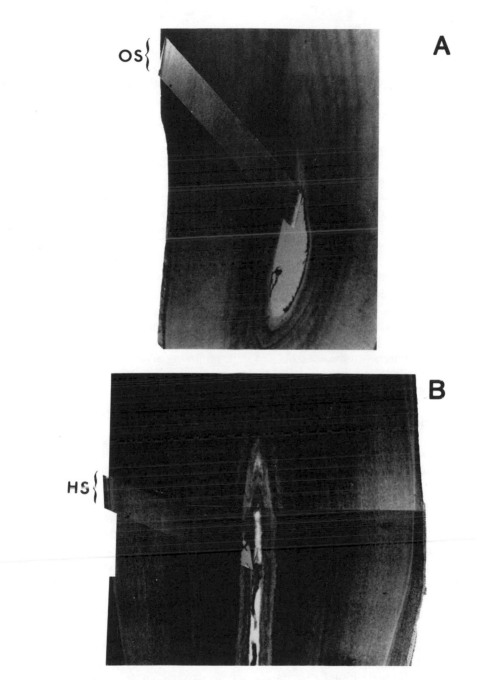

Fig. 8.9. Swatches of dentine from a specimen of one species of dolphin imposed on the dentinal pattern of a specimen of a different species to show pattern similarities and dissimilarities. A. Offshore spotted dolphin (OS), *Stenella attenuata*, swatch on Hawaiian spinner dolphin, *Stenella longirostris*, tooth. B. Hawaiian spinner dolphin swatch (HS) on offshore spotted dolphin tooth. (From Myrick et al. 1983.)

If adequate criteria can be established to define unique genus-, species-, or subspecies-specific patterns, future studies might use patterns to determine genetic closeness (fig. 8.4), perhaps even the degree of interbreeding, of contiguous or adjacent dolphin populations. The degree of physical and genetic separation of stocks belonging to the same species is of vital interest in fisheries management. In the case of the ETP purse seine fishery, for example, it is sometimes difficult to distinguish between two stocks of spinner dolphins (*Stenella longirostris*) affected by fishing operations because of the rather high overlap in external identifying characteristics (Perrin et al. 1985, Anon. 1986).

INDIVIDUAL VARIABILITY

It is not difficult to detect individual variability within what I suspect are species-specific patterns (fig. 8.4). However, based on calibration studies previously described, it must be concluded that exogenous factors that we monitored probably do not cause measurable pattern variability. I found no pattern variation associated with removing an animal from the water, changing the diet from pure fish to mixed fish and squid, or environmental changes, such as seasonal water temperature fluctuations or transporting animals to locations differing in latitude by up to 13 degrees (Orlando to Cleveland) (Myrick and Cornell 1990). An animal in one of the calibration studies accidentally leaped out of its tank. It remained out of the water, on the ground, perhaps for hours before it was discovered and returned to the tank, apparently uninjured. This unpredictable accident provided us with an excellent opportunity to test whether an "instantaneous," presumably stressful situation causes dentinal pattern variability. Later, when I examined the dentine formed during the period encompassing the accident, I could discern no changes in the normal pattern.

Variation within dentinal layering patterns is more conspicuous and more frequent in female dolphins than in males. In decalcified and stained thin sections of teeth, much of the variability exists as thin, intensely stained layers that are intermittently distributed in the dentine and that stand out rather starkly against the comparatively muted normal pattern of differentially stained layers of the GLGs.

In a study of the dentinal patterns in *Stenella* spp. (Klevezal' and Myrick 1984), we found that most of these strongly contrasting layers (SCLs) occur in females in dentine formed after attainment of sexual maturity. It seemed likely to us that SCLs could represent dentinal calcium fluctuations connected with reproductive cycles of the females—possibly calving events. To test this idea, Klevezal' and I first looked at thin sections of tetracycline-labeled teeth from a captive female Hawaiian spinner dolphin that had given birth to a calf while in captivity. This animal was one of the seven specimens used in the study to calibrate dental layers in teeth of Hawaiian spinner dolphins (Myrick et al. 1984). We then studied tooth thin sections of seventy-five sexually mature female spotted dolphins selected from a larger, random sample used for age-specific reproduction studies (Myrick et al. 1986). Sexual maturity of a female was determined by the presence of at least one ruptured Graafian follicle (corpus luteum) or a scar from such a rupture (corpus albicans) on either ovary, indicating that the animal had ovulated at least once.

The Hawaiian spinner dolphin had been in captivity for about ten months when she delivered a calf; the calf survived for only three days. One month after parturition, the dam was started on a therapeutic treatment of tetracycline that lasted more than one month. This treatment produced a dentinal label identified as Label B in the teeth of this animal (Myrick et al. 1984; fig. 8.10). To locate Label B for the SCL study, we had to use untreated thin sections (because treating teeth involves decalcification, which destroys the labels). To look for any SCLs near the position of Label B, we had to use decalcified and stained thin sections (because SCLs are narrow, dark-stained layers). After only a little searching of the decalcified and stained dentine, we identified an SCL in about the same position as Label B in the untreated thin sections (fig. 8.11). This reinforced the idea that SCLs were connected with some part of the female reproduction cycle. But, what part? Did SCLs represent pregnancy, parturition, early lactation, or all or none of the above?

To limit the number of possibilities, we began the other part of the study using teeth from the seventy-five female spotted dolphins. This exercise was different from the search for the SCL in the Hawaiian spinner dolphin tooth, in that we independently identified, counted, and noted the position of SCLs for each specimen without knowing anything about the reproductive condition of the specimen in advance (except that each female was sexually mature).

We erected five hypotheses directed at the question of which condition (if any)—pregnancy, ovulation, parturition, lactation, or parturition and lactation—was most frequently associated with the presence of an SCL. The following reasoning was used in reaching our conclusions:

1. If the layers have no specific connection with reproduction, then they are just as likely to occur in reproductively inactive females as in reproductively active females. [They did not.]

2. If the layers represent ovulations, then they should be exactly equal in number to the total count of ovarian corpora (because corpora, once formed, probably never disappear) (Perrin and Donovan 1984).

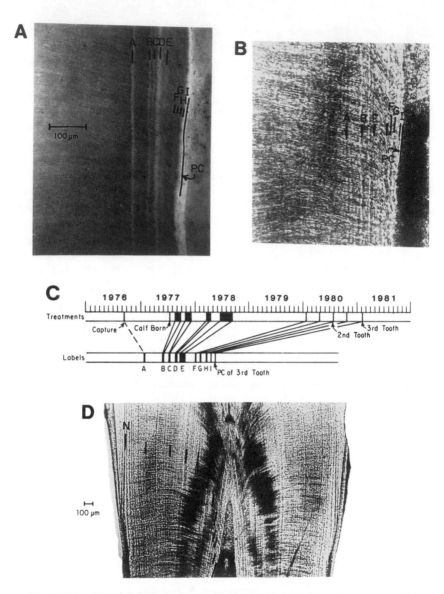

Fig. 8.10.   Tooth labels, layers, and matching of labels with treatment dates of female Hawaiian spinner dolphin that gave birth to a calf in captivity. A. Labels are lettered A-I. Label B was introduced one month after the female gave birth. PC = pulp-cavity margin. B. Labels are located in layered tissues by marking plastic overlays of UV photos and plain-light photos. C. Labels matched to treatment dates using label spacing and treatment duration. D. Gross view of layering pattern across entire tooth. (From Myrick et al. 1984.)

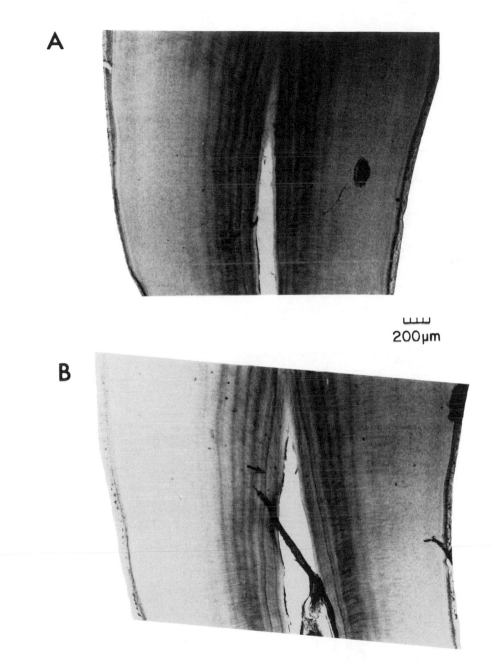

A

B

200μm

Fig. 8.11. A, B. Photographs of two decalcified and stained thin-sectioned teeth from female Hawaiian spinner dolphin that gave birth in captivity. A strongly contrasting layer (SCL), indicated by arrow, is near the position of a tetracycline label introduced one month after the female's calf was born. (From Klevezal' and Myrick 1984.)

Table 8.1. *Tooth Layering and Reproductive Condition in Female Spotted Dolphins*

Hypotheses— Formation of SCL Associated with:	Tests of Coincidence	Total Number of Specimens	Coincidence		Noncoincidence		Result of Test
			NUMBER OF SPECIMENS	PERCENT	NUMBER OF SPECIMENS	PERCENT	
1. Ovulation	1a. SCL count = CAs + CLs	75	10	13	65	87	Negative
a. all ovulations	1b. SCL count = CAs + CLs excluding CLs of pregnancy	75	17	22	58	78	Negative
b. all but most recent fertilized ovulation							
2. Pregnancy	2a. In all pregnant females, SCL in last GLG	30	1	3	28	97	Negative
3. Parturition or lactation	3a. SCL count ≤ CAs + CLs excluding any CLs of pregnancy	75	70	93	5	7	Positive
	3b. In all pregnant females, SCL count < CAs + CLs of pregnancy	75	70	93	5	7	Positive
4. Lactation only	4a. In all lactating females, SCL in last GLG without a space between SCL and pulp cavity	40	0	0	40	100	Negative
5. Parturition only	5a. In all lactating females, at least one SCL within the last two GLGs	40	35	88	5	12	Positive

Note: This table compares the percent coincidence tests of five hypotheses of relationships between SCLs and reproductive condition in female spotted dolphins (*Stenella attenuata*). SCL: strongly contrasting layers; GLG: growth layer group; CA: corpus albicans; CL: corpus lutem. (Modified from Kevezal' and Myrick 1984.)

3. If the layers represent pregnancies, then in nonpregnant females, they should be equal to or less than the total count of ovarian corpora (because not all ovulations may result in pregnancy). Furthermore, in all pregnant females, an SCL should be present within the last-formed dentinal GLG (because the female was pregnant at death).

4. If the layers represent lactation, then SCLs should be equal to or less than the total count of ovarian corpora (because lactation is preceded by a pregnancy and not all ovulations may result in pregnancies). Furthermore, in all lactating females, an SCL should be present within the last-formed dentinal GLG without a space between it and the edge of the pulp cavity (because the female was lactating when killed).

5. If the layers represent parturition only, then the first point in #4 should be true. Furthermore, in all lactating females, an SCL should be present within the two last-formed GLGs (because nursing may continue for more than a year after parturition) (Perrin and Reilly 1984).

The percent coincidence tests we used showed a consistently strong (83–93%) association of layers with parturition exclusive of lactation (table 8.1). The tests excluded the high likelihood of SCLs as indicators of any other condition that we considered.

Since that study, I have had the opportunity to examine the teeth of three long-lived, captive female bottlenose dolphins with extensive reproductive and calving histories. Two had each given birth to three calves; the other, to one calf. I found a parturition layer in the dentine formed during the date recorded for each calving event (fig. 8.12). I also noted similar spacing of anomalous layers in the cementum of these specimens.

Reliable criteria for detecting SCLs that indicate parturition would be extremely useful in managing dolphin populations. In the case of the dolphins associated with the yellowfin tuna purse seine fishery, evaluation of the fishery's historical impact on the size and health of the dolphin populations has been a major problem. The dolphins became directly affected by the fishery in 1959 when the fishing method changed from bait fishing to purse seining. From that time through the 1960s and into the early 1970s, dolphin fishing mortality is estimated to have reached between 100,000 to 400,000 animals annually (Smith 1983). The impact assessment problem exists because (1) there are no firm estimates of pre-1959 population levels of dolphins (numbers of individuals), (2) the dolphin populations are presumed to have undergone changes in reproductive rates as a density-dependent response to the substantial fishing mortalities, and (3) almost no mortality samples were collected before 1972 (long after the populations are presumed to have responded to the effects

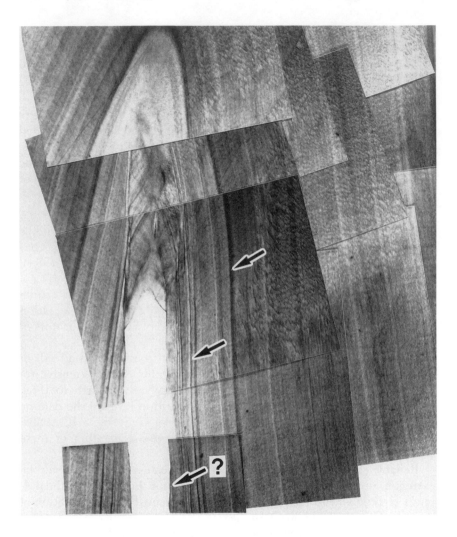

Fig. 8.12.   Highly magnified section of decalcified and stained tooth from a female bottlenose dolphin that gave birth to three calves in captivity. Two strongly contrasting layers, SCLs (arrows), occur in the regularly layered dentine, a third may be next to the pulp-cavity margin (the female died not long after giving birth to the third calf).

of the high kills). In short, virtually no information exists concerning the dolphins before the 1970s.

If SCL definitions can be refined for ready identification, possibly they could be used to interpret the reproductive histories, including per-capita calf production and calving frequency, of sexually mature females from their dentinal and cemental patterns. The years of heavy kill did not take place before 1959 (13 years before sampling started), and the average age at sexual maturity for females is between 6 and 11 years old (depending on the population). These two facts indicate that tooth samples of females between 20 and 35 years old collected in 1973, 1974, and 1975 should contain SCLs that were formed well before the introduction of purse seining. If such specimens exist in sufficient number, they could be useful in the estimation of reproductive rates that occurred before the impact of purse seine fishing. In addition, counts and spacing of SCLs in those and younger specimens might help fill the data gap concerning any changing rates during the heavy-kill period (1960–1975). This possibility has not yet been examined in any detail.

SUMMARY

Results from research that I initially conducted to develop more accurate age determination methods show that layers in dental tissues have a much greater potential as tools for biological inquiry than merely servicing the need for age estimates. Now that layering patterns in various species are being calibrated with time, it soon may be possible to use monthly layers to estimate peak reproductive seasons for a population and, thereby, to measure the degree of reproductive compatibility between adjacent populations.

Several of our studies have shown that local exogenous fluctuations do not perceptibly alter the genetically based dentinal layering patterns. This permits consideration of the use of dentinal patterns to estimate genetic closeness of sampled populations at the genus, species, and even perhaps at the subspecies level.

Recognition of the existence of genetic patterns has facilitated the identification of individual variation within the patterns. Because external factors do not seem to affect the patterns, variation must reflect, to a large extent, certain individual physiological changes. Parturition layers in sexually mature females seem to be a likely example of this. These layers are a prime candidate for development as a tool for population management because they may provide a record of the complete calving history of a female. In addition, other variations in the dental patterns are known to exist in both male and female delphinids. There is little doubt that they reflect other physiological events or biological cycles yet to be identified. Extensive study and experimentation will be needed to determine if such potential tools can be fully developed for practical use.

Despite the very difficult task of trying to study such secretive animals as dolphins, techniques are being developed and improved to address some of our most important questions concerning delphinid biology. In the 1950s, researchers were developing a tool for estimating a dolphin's age. Since then, the method has been used to help provide answers to questions about the reproductive health and future growth of dolphin populations. Results of the calibration studies that I have described should add greater firmness to some of those answers, but it is becoming increasingly clear that dental recording devices of dolphins have been underutilized as potential sources of biological information.

When we realize that the recordings themselves are chronological histories written in the biochemical language of the animals' own physiology, we are prompted to search for a Rosetta stone that will allow us to translate these histories. By learning what physiological conditions cause the recorded features, we can then use the features to interpret past conditions, as we are beginning to do with parturition layers. Perhaps, eventual interpretation of large parts of an individual's physiological history may be possible (see, for example, Myrick 1988).

After decades of work to understand the natural and human-caused environmental effects on tree growth, dendrologists are now able to determine much of the historical growth physiology of trees from details of their annuli. With dolphins, the problem differs only in our limited ability to monitor their wild environment and their physiological responses to it closely enough. Until that ability is achieved, we must settle for remote and retrospective monitoring to answer our questions about the dolphin populations. No doubt, some of these interim answers will come from layers in the teeth.

REFERENCES

Anon. 1986. A guide to shipboard identification of spinner and spotted dolphin schools (genus *Stenella*) in the eastern and central tropical Pacific Ocean. NMFS, SWFC Amin. Rpt. LJ-86-30. 23 pp.

Boyde, A. 1980 (1981). Histological studies of dental tissues of odontocetes. In: Age determination of toothed whales and sirenians, ed. W. F. Perrin and A. C. Myrick, Jr. Rep. Int. Whal. Comm. Spec. Issue 3. 65–87.

Hammond, P. S., and K. T. Tsai. 1983. Dolphin mortality incidental to purse-seining for tunas in the eastern Pacific Ocean, 1979–1981. Rep. Int. Whal. Comm. 33:589–597.

Hohn, A. A. 1980 (1981). Analysis of growth layers in teeth of *Tursiops truncatus,* using light microscopy, microradiography, and SEM. In:

Age determination of toothed whales and sirenians, ed. W. F. Perrin and A. C. Myrick, Jr. Rep. Int. Whal. Comm. Spec. Issue 3. 155–160.

Kasuya, T. 1977. Age determination and growth of Baird's beaked whale with a comment on the fetal growth rate. Sci. Rept. Whales Res. Inst. Tokyo 29:1–20.

Klevezal', G. A., and A. C. Myrick, Jr. 1984. Marks in tooth dentine of female dolphins (genus *Stenella*) as indicators of parturition. J. Mamm. 65(1):103–110.

Laws, R. M. 1962. Age determination of pinnipeds with special reference to growth layer in the teeth. Saugetier Mitt. 27(3).129–146.

Myrick, A. C., Jr. 1979. Variation, taphonomy, and adaptation of the Rhabdosteidae (=Eurhinodelphidae) (Odontoceti, Mammalia) from the Calvert formation of Maryland and Virginia. Ph.D. dissertation, Univ. of California, Los Angeles. 411 pp. (Avail. Univ. Microfilms Intl.)

———. 1980 (1981). Examination of layered tissues of odontocetes for age determination using polarized light microscopy. In: Age determination of toothed whales and sirenians, ed. W. F. Perrin and A. C. Myrick, Jr. Rept. Int. Whal. Comm. Spec. Issue 3. 95–97.

———. 1984a. Reproductive seasonality in multi-herd groups of northern offshore spotted dolphins, *Stenella attenuata*, and age-related reproductive parameters for females. (Abstract) In: Reproduction in whales, dolphins and porpoises, ed. W. F. Perrin, R. L. Brownell, Jr. and D. P. DeMaster. Rep. Int. Whal. Comm. Spec. Issue 6. 480.

———. 1984b. Time significance of layering in some mammalian hard tissues and its application in population studies. Acta. Zool. Fennica 171:217–220.

———. 1988. Is tissue resorption and replacement in permanent teeth of mammals caused by stress-induced hypocalcemia? In: The biological mechanisms of tooth eruption and root resorption, ed. Z. Davidovitch. EBSCO Media, Birmingham. 379–389.

Myrick, A. C., Jr., and L. H. Cornell. 1990. Calibrating dental layers in captive bottlenose dolphins, *Tursiops truncatus*, from serial tetracycline labels and tooth extractions. In: The bottlenose dolphin, *Tursiops* spp., ed. S. Leatherwood and R. Reeves. Orlando: Academic Press.

Myrick, A. C., Jr., A. A. Hohn, J. Barlow, and P. A. Sloan. 1986. Reproductive biology of female spotted dolphins, *Stenella attenuata*, from the eastern tropical Pacific. Fish. Bull. 84(2):247–259.

Myrick, A. C., Jr., A. A. Hohn, R. S. Wells, M. D. Scott, and A. B. Irvine. 1985. Dentinal growth layer similarities in captive and wild known-age bottlenose dolphins, *Tursiops truncatus.* Proc. Abstracts, Fourth Intl. Theriological Congress, Edmonton, Alberta. Aug. 13–20, 1985.

Myrick, A. C., Jr., A. A. Hohn, P. A. Sloan, M. Kimura, and D. P. Stanley. 1983. Estimating age of spotted and spinner dolphins (*Stenella attenuata* and *Stenella longirostris*) from teeth. NOAA-TM-NMFS-SWFC-30. 17 pp.

Myrick, A. C., Jr., E. W. Shallenberger, I. Kang, and D. B. MacKay. 1984. Calibration of dental layers in seven captive Hawaiian spinner dolphins, *Stenella longirostris,* based on tetracycline labeling. Fish. Bull. 82(1):207–225.

Nishiwaki, M., and T. Yagi. 1953. On the age and the growth of teeth in a dolphin, (*Prodelphinus caeruleo-albus*). I. Sci. Rept. Whales Res. Inst. Tokyo 8:133–146.

Perrin, W. F. 1969. Using porpoise to catch tuna. World Fishing 18(6):42–45.

———. 1972. Color patterns of spinner porpoises (*Stenella* cf. *S. longirostris*) of the eastern Pacific and Hawaii, with comments on delphinid pigmentation. Fish. Bull. U.S. 70(3):983–1003.

———. 1975. Variation of spotted and spinner porpoise (genus *Stenella*) in the eastern tropical Pacific and Hawaii. Bull. Scripps Inst. Oceanog. (Univ. Calif. Press), vol. 21. 206 pp.

Perrin, W. F., and G. P. Donovan (eds.). 1984. Report of the workshop. In: Reproduction in whales, dolphins, and porpoises, ed. W. F. Perrin, R. L. Brownell, Jr., and D. P. DeMaster. Rept. Int. Whal. Comm. Spec. Issue 6. 1–24.

Perrin, W. F., and A. C. Myrick, Jr. (eds.). 1980 (1981). Age determination of toothed whales and sirenians. Rept. Int. Whal. Comm. Spec. Issue 3. 229 pp.

Perrin, W. F., and S. B. Reilly. 1984. Reproductive parameters of dolphins and small whales of the family Delphinidae. In: Reproduction in whales, dolphins, and porpoises, ed. W. F. Perrin, R. L. Brownell, Jr., and D. P. DeMaster. Rept. Int. Whal. Comm. Spec. Issue 6. 97–133.

Perrin, W. F., M. D. Scott, G. J. Walker, and V. L. Cass. 1985. Review of geographical stocks of tropical dolphins (*Stenella* spp. and *Delphinus delphis*) in the eastern tropical Pacific. NOAA Tech. Memo. NMFS-SWFC 28, 28 pp.

Peyer, B. 1968. Comparative odontology. Trans. and ed. R. Zangerl. Chicago: Univ. of Chicago Press. 347 pp. + 96 pls.

Scheffer, V. B., and A. C. Myrick, Jr. 1980 (1981). A review of studies to 1970 of growth layers in the teeth of marine mammals. In: Age determination of toothed whales and sirenians, ed. W. F. Perrin and A. C. Myrick, Jr. Rept. Int. Whal. Comm. Spec. Issue 3. 51–63.

Scott, J. H., and N. B. B. Symons. 1964. Introduction to dental anatomy. 4th ed. Edinburgh: E. & S. Livingstone Ltd., 406 pp.

Smith, T. 1983. Changes in size of three dolphin populations (*Stenella* spp.) in the eastern tropical Pacific. Fish. Bull. U.S. 81(1):1–13.

Walker, E. P. 1968. Mammals of the world. 2d ed., revis. J. Paradiso. Baltimore: Johns Hopkins Univ. Press.

A school of pilot whale females and young in Hawaiian waters. (Photo by William Curtsinger.)

CHAPTER NINE

AN OVERVIEW OF THE CHANGES IN THE ROLE
OF A FEMALE PILOT WHALE WITH AGE

Helene Marsh and Toshio Kasuya

The short-finned pilot whale (*G. macrorhynchus*) has been hunted by traditional Japanese whaling teams since the early seventeenth century. To study the life history and reproductive biology of this species, data and specimens were collected during the flensing of 483 females and 234 males from twenty schools caught in a driving fishery between 1975 and 1984, inclusive. This fishery is unselective in that it captures entire schools of whales; however, there is evidence of some geographic segregation of schools in the hunting season. Schools with a high proportion of pregnant females apparently congregate in the whaling area, which occurs in coastal waters at the northern limit of the range (Kasuya and Marsh 1984).

In the absence of known-age material, age was estimated by counting dentinal and/or cemental growth layers as outlined in Kasuya and Matsui (1984), the growth layer deposition rate being deduced from the seasonal pattern of layer deposition. The reproductive organs were examined using standard macroscopic and histological procedures (Kasuya and Marsh 1984, Marsh and Kasuya 1984). The age determination and reproductive studies were done independently.

The apparent pregnancy rate (proportion of mature females pregnant) declined markedly with increasing maternal age. Even though the ages of the females examined ranged up to 63 years, none of the 92 females over 36 years old was pregnant. The two oldest pregnant whales were aged 34.5 years and 35.5 years, respectively. Using the fetal growth equation developed by Kasuya and Marsh (1984), we calculated that these animals were scheduled to give birth before age 36. Another female subsequently estimated to have been 35.5 years old was observed to give birth the day before being killed. Using the life table data in Kasuya and Marsh (1984), we calculated that the mean life expectancy of a female pilot whale aged 36 years is fourteen years.

There was a parallel age-related decline in the ovulation rate. All females over 40 years old had apparently ceased to ovulate, as corpora lutea and young corpora albicantia were absent from their ovaries (Marsh and Kasuya 1984). Ovulation was less likely to be followed by pregnancy in older whales, the proportion of well-developed corpora lutea accompanied by a confirmed pregnancy being significantly lower in females older than 20 years than in younger whales. The infertility of the older females was associated with ovarian changes. Macroscopic (> 1 mm in diameter) follicles occurred in only six of the forty-nine females over 40 years old studied by Marsh and Kasuya (1984). When examined histologically, such follicles were invariably atretic and thus incapable of ovulation. Semiquantitative histological study of the ovarian cortex of thirty whales spanning the age range of 4 to 63 indicated that the oocyte stock of older females was severely depleted (ibid.). Old pilot whale ovaries also exhibited other features characteristic of postmenopausal human ovaries, including a general decrease in cortical volume, thickening and sclerosis of the arterial walls, in-growth of the surface epithelium, and increased pathology (Benirschke and Marsh 1984, Marsh and Kasuya 1984). We conclude that female pilot whales in this population become reproductively senescent by age 36 to 40 and that about 25 percent of the mature females sampled were probably postreproductive.

Investment in rearing as opposed to bearing young apparently increases with maternal age in *G. macrorhynchus*. Although pilot whales start taking solid food when less than one year old (Kasuya and Marsh 1984), the duration of lactation seems to be highly variable. The number of females that are lactating in relation to the number that are pregnant is significantly greater for whales older than 30 than for those in the age range of 15 to 30. Our estimates of the lactation period for pilot whales in various age groups (corrected for the geographic bias in the sample discussed above as detailed in Kasuya and Marsh 1984) indicate an increase from 2.8 years for the 15 to 20 age group to 6.4 years for both the age groups above 30. Older female pilot whales clearly have fewer calves but lactate longer.

The ages of the lactating *G. macrorhynchus* ranged up to 51, at least eleven years after ovulation had presumably ceased. Fostering and/or communal nursing have been documented in several terrestrial mammals

(Gubernick 1981). However, these phenomena have not been confirmed in any wild cetacean (although Whitehead [1984] recently observed two calves apparently suckling from one sperm whale). While not denying the possibility that postreproductive, lactating *G. macrorhynchus* may be exhibiting these behaviors, our analysis of the schools in which such females occur (see Kasuya and Marsh 1984) suggests that extended lactation by older females will, in most cases, be attributable to the prolonged nursing of their own calves. Indeed, a female's last calf may occasionally be suckled until puberty, that is, up to about 8 years (female calves) and up to about 13 years (males).

The majority (80/102) of the older (>age 35) female pilot whales in our sample were not lactating; their role within the pilot whale school is unknown. Recent examination of mucus from the uteri of pilot whales of a range of ages, however, indicates that mating is not restricted to estrous females in this species. Small quantities of spermatozoa were found in the uteri of lactating and resting females, including one postreproductive female, suggesting that mating has more than a purely reproductive function in this species.

Schools of *G. macrorhynchus* are frequently involved in mass-stranding incidents and are believed to be highly cohesive (Geraci and St. Aubin 1979). Our analysis with respect to age, sex, and reproductive status of thirteen schools sampled completely (Kasuya and Marsh 1984) provides insight into the social structure of this species. With an average membership of twenty-five whales, most schools are breeding units composed of adult male(s), adult females spanning the entire range of age and reproductive status, and immature animals of both sexes. The marked excess of reproductive females over mature males (which weigh twice as much as mature females) suggests polygyny (see Ralls 1977). These schools appear to be essentially matrilineal kinship groups like pods of killer whales (Bigg 1982).

The oldest male pilot whale sampled was aged 46, seventeen years younger than the oldest female. This sex difference in longevity (which is characteristic of polygynous species in which males are larger than females; Ralls et al. 1980) means that most of the older whales in a breeding school are female. However, postreproductive resting females and/or old (>35 years) lactating females do not occur in every school (see Kasuya and Marsh 1984).

Rather than focusing on the significance of the postreproductive phase in the life cycle of female *G. macrorhynchus*, it is more meaningful to consider it as a consequence of the trade-off between bearing and rearing in a species that has the following characteristics: (1) a low lifetime reproductive output; on average, female pilot whales produce only four or five calves (Kasuya and Marsh 1984); (2) maternal care that is so lengthy that the care of successive offspring overlaps; (3) a female adult mortality rate that increases with increasing age (the usual mammalian pattern; see Caughley 1966, Ralls et al. 1980). The mortality rate of female pilot whales older than 45 is significantly higher than in younger adults (Kasuya and Marsh 1984).

In such a species, the death of an older mother could substantially reduce the survival chances of several dependent young and perhaps other close relatives as well. In these circumstances, selection would tend to favor females that phase out their investment in bearing young as the probability of living long enough to rear them declines. This would be particularly important if reproduction in old age substantially increases the risk of mortality. It is significant that the expected length of post-reproductive life in short-finned pilot whale females is close to the maximum time taken for the young to reach puberty.

REFERENCES

Benirschke, K., and H. Marsh. 1984. Anatomic and pathologic observations of female reproductive organs in the short-finned pilot whale, *Globicephala macrorhynchus*. Rept. Int. Whal. Comm. Spec. Issue 6. 451–455.

Bigg, M. 1982. An assessment of killer whale (*Orcinus orca*) stocks off Vancouver Island, British Columbia. Rept. Int. Whal. Comm. 32:655–670.

Caughley, G. 1966. Mortality patterns in mammals. Ecology 47:906–918.

Geraci J. R., and D. J. St. Aubin. 1979. Biology of marine mammals: Insights through strandings. U.S. Dept. of Commerce NTIS Report no. MMC-77/13.

Gubernick, D. J. 1981. Parent and infant attachment in mammals. In: Parental care in mammals, ed. D. J. Gubernick and P. H. Klopfer. New York: Plenum Press. 243–305.

Kasuya, T., and H. Marsh. 1984. Life history and reproductive biology of the short-finned pilot whale, *Globicephala macrorhynchus,* off the Pacific coast of Japan. Rept. Int. Whal. Comm. Spec. Issue 6. 259–310.

Kasuya, T., and S. Matsui. 1984. Age determination and growth of the short-finned pilot whale off the Pacific coast of Japan. Sci. Rept. Whales Res. Inst. (Tokyo) 35:57–91.

Marsh, H., and T. Kasuya. 1984. Changes in the ovaries of the short-finned pilot whale, *Globicephala macrorhynchus,* with age and reproductive activity. Rept. Int. Whal. Comm. Spec. Issue 6. 311–335.

Ralls, K. 1977. Sexual dimorphism in mammals: Avian models and unanswered questions. Am. Nat. 111:917–938.

Ralls, K., R. L. Brownell, Jr., and J. Ballou. 1980. Differential mortality by sex and age in mammals, with special reference to the sperm whale. Rept. Int. Whal. Comm. Spec. Issue 2. 233–243.

Whitehead, H. 1984. The unknown giants: Sperm and blue whales. Nat. Geog. Mag. 166:774–789.

SOME THOUGHTS ON GRANDMOTHERS

Kenneth S. Norris and Karen Pryor

I

In recent years, it has emerged that the social patterns of the various whales and dolphins are not all alike. They range from the very fluid schools of spinner dolphins, in which school membership changes substantially from day to day, drawing from a population of as many as one thousand dolphins, to the very rigid pods of killer whales that seem to consist of exactly the same members, apparently groups of mothers and offspring, for years on end.

Now comes the news, from the work of Toshio Kasuya and Helene Marsh, that pilot whale schools, which include family units, are different in another way. They contain large cadres of postreproductive females, including some that are more than a decade older than the age of last reproduction for the species. And some of these females are still lactating. Why?

Marsh and Kasuya suggest that a postreproductive female might continue to let her last calf nurse for many years, in some cases until it reaches adulthood. The fact that some old females are still nursing, fourteen years past their last pregnancy, could thus be an artifact of failure to wean. Still, it seems anomalous that only the last calf of a whale should receive such preferential treatment.

And why so many postreproductive females? Normally, each member of a biological population has important duties to perform in the society of which it is a part. I do not know of another animal that commits a substantial percentage to an age class with nothing to do (except our own). Does the postreproductive pilot whale female serve some important social function, one that could be supported by natural selection and one that is not carried out by the other whales in a school?

One possibility might be that postreproductive females nurse the calves of others. Pilot whales are diving animals that apparently feed at considerable depths (navy scientists have been able to train pilot whales to dive on command deeper than two thousand feet). One wonders if such a species requires a cadre of school members who can tend nursing young while others, including nursing mothers, are feeding deep beneath the surface. If so, postreproductive females capable of providing "milk rewards" to calves incapable of the dives of adults might be a crucial part of pilot whale society.

Marsh and Kasuya point out that there is little evidence for fostering (nursing offspring other than your own) in cetaceans but ample evidence of alloparenting (providing other types of care for young other than your own). It is suggestive that the only other odontocete known to have significant numbers of postreproductive females in its societies is the deep-diving sperm whale.

If this scenario proves to be true, it might be found more widely among diving species, such as the various beaked whales. No one yet knows the answers to the intriguing questions raised by this work.

—*K.S.N.*

II

So the only use for postreproductive females is babysitting? I suggest that the role of postreproductive females in pilot whale schools may be as a repository of cultural information, such as the whereabouts of feeding grounds.

Pilot whales and people are not alone in their production of old ladies. Postreproductive females, often surviving to a grand old age, are commonplace in several terrestrial mammals, including horses and cattle (as many farmers can testify) and elephants. In some of these species, mature males join the female groups only temporarily, for breeding. In all of them, a dominant male may be replaced by a rival at any time. If long-term experience is necessary, for example to learn and remember the terrain beneath your feet or fins, then continuity of individuals in the group is vital: and older females can provide it.

The current evidence suggests to me that pilot whales, at least in New England and California waters, do not cruise at random, finding food where they can, but migrate along the coasts feeding on squid by diving in particular canyons along the edge of the continental shelf (and perhaps on seamounts, as well). Random 2,000-foot dives in the wrong spot would be very expensive, energetically. Locating food, year-round, might be at least partly a function of long-term memory of the oldest animals, comparable to the need of wild horses, cattle, and elephants to remember migratory routes and the location of water holes.

While extended lactation certainly plays some role, not yet clear, in pilot whale society, not all the old females lactate; I suggest their principal biological contribution might be to learn, remember, and transmit what pilot whales need to know.

—K.P.

PART III

Captive Studies:
The Key to Understanding Wild Dolphins

Essay

LOOKING AT CAPTIVE DOLPHINS

Kenneth S. Norris

I was a graduate student at the University of California about the time
we first began to learn about the lives of captive dolphins. That is very
recent history, considering the present complexity of what we know.
The following paragraphs are a somewhat idiosyncratic account of the
events I experienced or knew about in the decades that followed. I did
look up some things in an attempt to get the record straight, but other
people at other places would have seen things somewhat differently, I
am sure.

The history of our learning started out with a few very simple observations and tests. Today, dolphin research has grown into a quite
sophisticated branch of natural science often involving subtle and complicated collaborations between scientist and animal. It has also been
the beneficiary of the political competition of the superpowers, as both
the Russians and the Americans made sure that neither learned things
from these animals that might help its military mission that the other
did not learn, too.

The scientists have generally worked wherever the dolphins are already being kept. Few have braved the expense and difficulty of keeping
dolphins for their own research purposes. So, as dolphin exhibits have
spread around the world, the number of researchers doing dolphin

work and the number of countries involved have both increased substantially. The pace of discovery has not slackened in this history. The new things we find out today about these intriguing animals continue to be as fresh and challenging as they were when we knew nothing at all.

Dolphins, just like other wild animals, had been kept in zoos for a long time, but no one paid very much attention to their singular attributes until a small colony was developed at Saint Augustine, Florida, in the late 1930s. The establishment where they came to live was an unlikely one to spur an entire lineage of scientific endeavor into being. It was, in fact, a big movie studio.

A group of movie makers had succeeded in filming African veldt animals such as antelopes by enclosing them in a pen so large a cameraman could photograph the animals as they ran by without also photographing the fence. Why not do the same thing with marine animals such as sharks and manta rays, they reasoned. So, on cheap palmetto swampland located far down a winding road in back of the beach, they built a curious establishment called Marine Studios (later to be renamed Marineland, Florida), which consisted of a series of steel tanks ringed with big windows for photography. Word of their venture got around, and more than 14,000 people showed up on opening day. The movie makers found, to their surprise, that they were in the public exhibition business. (They were also in the business of science and public education.)

As an afterthought, and without any visions of the trained shows we see today, they sent their collection crew to bring in some of the locally available bottlenose dolphins. Soon, they accumulated a little society of dolphins, caught in the inland channelways in back of the tank complex. There was no show routine at first; the dolphins were simply regarded as exhibits, like the fish. The dolphins, however, regarded themselves as a society and began to do many of the things they had done at sea. They courted, mated, gave birth, helped each other, caught fish, quarreled, and generally behaved like the good mammals they were.

None of this went unnoticed, thanks to the fact that the filmmakers had hired a young biologist-psychologist named Arthur McBride to care for the exhibit animals. In the 1940s, he began to record many things about the marine life that paraded before him. More and more, his attention turned to the bottlenose dolphins as he noted their complex behavior. He and his colleague, Frank Essapian, and visiting scientists

Henry Kritzler, Margaret Tavolga, and D. O. Hebb wrote a series of landmark scientific papers about these captives. The animal that emerged from this scrutiny was a rather typical large, social mammal, but it also proved to be a species with an unusual degree of cooperative behavior in its makeup.

This dolphin colony attracted a small group of biologists to Marine Studios, and a diverse collection it was. It brought acoustic pioneers William Schevill and his wife, Barbara. It brought neurologist John Lilly and a group of associates from the National Institutes of Health. Along came psychologist Winthrop Kellogg, whose fame rested on raising his child and a chimpanzee together to compare their developmental rates. And in 1953, I ventured down the winding palmetto road because I had taken the job as curator of a new and affiliated oceanarium at the Palos Verdes Peninsula of California, Marineland of the Pacific.

Bill and Barbara Schevill (née Lawrence) came to see if they could learn how high pitched a sound a dolphin could hear. There had been hints from the dolphin's behavior that it could hear sounds very much above our own upper hearing threshold. And the dolphins had been heard to emit long trains of clicks that contained very high frequencies. In the process, the Schevills performed the first conditioned response experiment on a dolphin.

The accommodating curator, who by this time was Forrest Wood, assigned the Schevills a bottlenose dolphin that lived in an earthen pool in back of the main display tanks. They taught the dolphin to return to a feeding station each time they played a sound into the water, in return for a fish reward. The dolphin soon learned this simple task, allowing the experimenters to raise the stimulus tone higher and higher; so high, in fact, that the Schevills began to think the animal was using a curious means of hearing called "bone conduction." The animal began to falter when it was played sounds about ten times the upper limit of the average human adult, or about 150 kilohertz (kHz). (This level, after corroborating experiments by other workers, is still thought to be the approximately correct upper limit for the species.) The important thing for us here is that these workers had contrived to ask a dolphin a specific question and to receive a definite answer in terms of its behavior.

The investigators were puzzled, though, about how their animal had been able to make its decisions. Because the dirt pond was a haven for snapping shrimp, a species that makes an intolerable racket by snapping one of its claws closed, they had heard no dolphin clicks above the noise. So they asked Curator Wood if he could let them take a dolphin to faraway Woods Hole Oceanographic Institute in Massachusetts for further tests. After much preparation, the animal was taken by air to a newly dug pond near the Cape Cod beach. There, the Schevills performed their second experiment, and it too was a cornerstone in what we later came to know of dolphins.

This time there were no snapping shrimp. They asked their dolphin a slightly more complex question: "Can you detect a fish presented below the water from a distance too far to see it?" The dolphin was required to make its choice beyond a projecting net that partly divided the tank in two. It had to decide at a distance of several meters, and in murky water, whether to swim on one side of the net or the other to receive its reward. Murky water and all, it chose easily; at the same time, they could hear it emit those rusty hinge noises. This, to me, is the first demonstration that dolphins navigate by the echoes of their own clicked sounds. This work led to the studies of dozens of investigators from around the world who have explored the remarkable dimensions of the dolphin's acoustic capability.

Kellogg had been interested in this sensory system too. He had watched the fright reactions of dolphins to sounds of various frequencies played into the Marine Studio's tanks. Up to about 60 kHz he could see the dolphins react and concluded that that was their upper hearing limit. He had also recorded their rusty hinge noises at sea, in the first use of high frequency listening gear by a dolphin scientist. He too decided to import dolphins to his own laboratory. Another big tank was dug in the marl near a Tallahassee beach, and dolphins were installed. There, Kellogg carried out a series of tests of their echolocation capability. He blocked the tank with a central barrier in which two gates were placed. The dolphins had to choose the open gate from one blocked with clear Plexiglas, a task they performed unerringly.

Kellogg had erected a cable scaffolding over the tank which allowed his students to creep out above the water. When objects, even tiny lead shot, were dropped into the tank and began to sink, the dolphin could be heard tracking them with clicks. Kellogg asked his student to drop a spoonful of water instead of a solid object. It hit the water with an audible splash; the dolphin turned toward it, emitting clicks, and then, detecting nothing, turned away. Later, Kellogg carried out a controlled test of a dolphin's discrimination of two species of fish—the most involved psychophysical experiment to that date. The work was not conclusive in the sense that it was difficult to tell what, exactly, dolphin had been discriminating; that is, between the two species of fish using unknown cues or simply between the relative sizes of the two targets. But the test was important because it showed that a dolphin could be made a partner in a rather involved and rigid training task and that many aspects of its cognition were thus available to the patient experimenter.

Another pair of visitors, later to take up permanent residence in Saint Augustine, were David and Melba Caldwell. For many years, they traveled wherever dolphins were to be found and recorded many interesting features of dolphin behavior. For instance, they found that each captive dolphin appeared to use, almost exclusively, a single whistle, which they called a *signature whistle*. This signal is now thought to identify individuals as they swim in schools large enough so that they cannot always see each other.

John Lilly, M.D., had arrived at Marine Studios as part of a team of neurophysiologists who attempted to anesthetize dolphins so they would be available for studies of brain function. After many frustrating failures, they found that the respiratory apparatus of a dolphin is unlike that of a terrestrial mammal. The dolphin larynx, instead of emptying into the pharynx and mouth, is shunted across the esophagus by means of an elongate extension. This tubelike structure is held by a powerful sphincter muscle in the base of the skull below the openings of the two bony nasal passages. In this way, dolphins and their allies are able to separate food and air completely and are also able to use the same air over and over to make their many sounds during long dives. Once Lilly and his associates understood this, they were able to reach into a dolphin's mouth, detach the laryngeal spout from its sphincter, and insert a tube into it from the respirator.

Finally, using a respirator that mimicked the long breath holds of a diving dolphin, they achieved successful anesthesia. A decade later, using similar techniques, physiologists such as T. H. Bullock, A. D. Grinnell, K. Katsuki, and Sam Ridgway, and later a group of Russian workers, learned all we know about the functioning of the dolphin brain.

In the early and middle 1960s, Lilly performed a number of experiments of interest. He showed that dolphins could make two kinds of sounds at once: whistles and clicks. He showed that two animals, given an acoustic link, would engage in an exchange of signals. He showed that they could make approximate mimics of a variety of sounds presented to them. These useful, early contributions were followed by a series of books in which Lilly spun out scenarios related not to scientific reality but instead to his experimentation with altered consciousness states and imagination. These extended his real findings into claims that dolphins possessed a language and that some, such as the sperm whale, possessed an intelligence whose complexity far exceeded our own. They extended the hope of interspecies communication between humans and dolphins.

These imaginings, which are without scientific foundation, caught the public fancy and for the serious cetologist, created two quite different results. On one hand, we scientists soon found ourselves working with an animal that was in the public eye. Unless we escaped to some lonely atoll to study our animals in isolation, nothing we did would ever be private again. We found ourselves to be the conservatives, constantly countering the imagined super-animal found in the public mind. On the other hand, existence of this fanciful public view sometimes made relations difficult with our scientific colleagues, who wondered if we had accepted Lilly's imaginings. On occasion, we still face the question, Can you be a serious scientist if you work with dolphins?

But the same public interest created support for research and for conservation efforts wherever whales (which a good fraction of the public hardly separates from dolphins) and dolphins are found.

Another party with early interest in dolphins was the U.S. Navy. Military goals led to research support for a number of nonmilitary scientists. The focus was on what the navy called "bionics," or exploration of biological systems to find design criteria that might be applied to the mission of the navy. Were dolphins swimming faster than a comparable ship could go, and if so, how did they do it? Were their echonavigation systems better than comparable navy sonar systems? I suppose navy officials wanted to know anything they could find out about a high-order animal that shared the ocean with them and that seemed to do things their ships could not do. I remember, in particular, when William Evans and Bill Powell demonstrated that dolphins could use their sounds to inspect the composition of targets; for example, they could be trained to tell aluminum from iron. We heard that some navy officials worried that a dolphin, using this capability, could look into the innards of objects such as surface ships and submarines.

Two civilian navy officials, Dr. Sidney Galler and Dr. William McLean, had much to do with the progress of experimental cetology in the 1950s and 1960s. Galler assembled around him a group of young scientists looking at various aspects of dolphin biology. He arranged meetings, publications, and support. Recognizing that his biologists were mostly innocent of the physical sciences, he brought workers from these disparate disciplines together. At the time, scientists studying bat echolocation were far ahead of the dolphin workers in the study of animal acoustics, so a mix of these two kinds of scientists appeared at meetings. McLean, the director of the Naval Ordnance Test Station in the California desert, developed a program within the navy establishment itself where in time much of the finest work on the acoustic system and mind of the dolphin would be done.

I was not even aware of all this until, while I was curator of Marineland of the Pacific, my colleague John Prescott and I developed the first dolphin blindfold and thereby showed once and for all that dolphins could navigate by echoes without sight. The last uncertainty of Schevills' and Kellogg's tests, performed by dolphins in murky water but with unimpaired vision (could they somehow see better than we could?), was dispelled. I still can see our peppery young dolphin, Kathy, with her ridiculous lumpy hand-formed suction cup blindfolds stuck over her eyes, swim unerringly through a complex maze, locate a tiny fragment of a fish, and delicately take it as it sank within inches of a wall. She truly could "see with sound." In that momentary vignette, a considerable new scientific vista opened before us as we comprehended that Kathy was demonstrating a sophisticated system whose dimensions we did not understand.

On the basis of some very simple observations, I also concluded that it was likely that Kathy beamed these clicked sounds from her forehead and upper snout and received the echoes through the general area of her lower jaws. My functional speculations had been strongly influenced by two sets of elegant anatomical studies of the dolphin head and ear, one from the British Museum and Drs. F. C. Fraser and P. E. Purves and the other from Harvard University performed primarily by Barbara Lawrence. Also important in my thinking had been the functional explanation of dolphin hearing presented by a Dutch otolaryngologist,

F. Reysenbach de Haan. Not long afterward, I joined the faculty at the
University of California, Los Angeles, and decided to explore those insights, a difficult thing to do on an inland campus, I found. First, I built
a complex of plastic swimming pools, pumps, and diatomaceous earth
filters by the student dorms. There I installed a geriatric lady bottlenose
dolphin named Alice. Then, a young Skinnerian psychologist, Ron Turner, and I set to work. Later, Bill Evans joined us to help with the recording and analysis of her sounds.

We decided to ask Alice to tell us just how fine her biosonar discrimination was. We used pairs of ball bearings of different sizes which we
hung in the water. We asked her to direct her sounds at the sphere,
decide which was the larger, and then to indicate her choice by pressing
one of two levers. Alice could discriminate ball bearing sizes that we
could barely tell were different by eye, using about one second's worth
of echolocation. Most puzzling of all, she refused to answer when we
presented her with identical pairs of ball bearings. She somehow knew
an impossible problem from a merely difficult one. We concluded that
she "rang" the ball bearings with her clicks and listened to the different
tones produced by the different sizes of spheres. (That difference, if we
could have heard it, was well within the capability of the human ear to
discriminate.) Of course, if the spheres were identical, the returning
tones would be, too. Alice, therefore, could tell instantly that we had
presented her with identical spheres.

The time was the mid-1960s. Within a few years, important experimental contributions were coming from several places in the world.
From the Laboratories d'Acoustique in France, where René Guy Busnel,
Albin Dziedzic, and their colleagues told us many things about harbor
porpoise and dolphin echolocation; from Bertel Møhl and Soren Andersen of Denmark, who also worked with harbor porpoises; and from a
burgeoning Russian dolphin research establishment. A trip to the USSR
at the end of the 1960s showed me how extensive this new Russian
effort was. Full-scale programs, involving many scientists, were under
way on bioacoustics, physiology, anatomy, radiotelemetry, and dolphin
locomotion; and a research facility with experimental dolphins had
been set up in an old hotel by the Black Sea.

In the late 1960s, I decided to see if we could reverse the usual sequence of capturing a wild animal and then studying it in captivity.
Could we, I wondered, train a captive animal and return it to sea, for
studies impossible to perform in the confined space of a seawater tank?

For example, could we train a dolphin, take it to sea, and ask it to dive as far down as it could go? I decided to start by taking a trained dolphin to sea for tests of its true top speed. Karen Pryor, hydrodynamicist Tom Lang, my wife Phyl, our daughter Susie, and I trained a bouncy young bottlenose dolphin to swim laps, and then we took him to sea to run a measured racecourse. He was to sleep at night in a cage moored offshore. At essentially the same time, a similar U.S. Navy effort was going on. In time, the Navy program led to the remarkable work of Ridgway and John Kanwisher, who trained a bottlenose dolphin to dive as deep as 1,000 feet and then breathe on return into an inverted funnel so that the respiratory gas could be caught and analyzed. Similar navy efforts have since induced a pilot whale to dive deeper than 2,000 feet on command.

In more recent years, studies of captive animals have emanated from so many places and so many workers that I will only sketch in a few highlights. As marine exhibits proliferated in several countries, so did reports on the behavior, physiology, psychology, and veterinary medicine of dolphins. Russian work had gone on apace. By 1975, one bibliography listed 420 works by Russian scientists such as Ayrapetyants, Bel'kovich, Burdin, Dubrovskiy, Giro, Krushinskiy, Ladygina, Markov, Morozov, Mukhametov, Romanenko, Saprykin, Sokolov, Solntseva, Sukhoruckenko, Titov, Tomilin, Voronin, and Yablokov. Included in this trove are explorations of brain function, sleep, sound processing, hearing, and social behavior, among many other topics. Evidence that dolphins can listen to individual clicks down to a separation of about 300 microseconds is there (our brains merge sounds at about 30 per second). Russian sleep work suggests that dolphins may have very separate functions on the two sides of their brains. Perhaps they may even sleep one hemisphere at a time. Ridgway, the U.S. Navy's senior dolphin physiologist, has recently presented corroborative evidence for this novel capacity, first suggested by Lilly.

Ridgway, who began his studies in the early 1960s, has sketched out whole realms of knowledge using captive dolphins: how they breathe, dive, and sleep, how their brains function, how their kidneys work, and a great deal more. He, more than any other Western scientist, has told us about their health problems, which in aggregate strikingly resemble our own, good mammals that we both are. Many others are now involved in these studies, enough to form a Marine Mammal Veterinary Association.

In recent years, civilian U.S. Navy scientists such as Whitlow Au, Robert Floyd, Patrick Moore, Paul Nachtigall, Earl Murchison, Wayne Turl, and Ralph Penner, through a series of elegant experiments using trained dolphins, have told us many intriguing things about dolphin sound systems. They have described both the emitted sound beam of dolphins and the cone of received sound. They have defined the ranges over which dolphin sonar can work (e.g., a dolphin can detect a target about the size of a tangerine more than 100 m away, a remarkable feat for us to perform by vision). They have shown that dolphins can make quite detailed assessments of surface texture of targets and of their internal composition.

Explorations of the dolphin mind were only hinted at by the early work I have described. Karen Pryor extended our understanding while training dolphins for show purposes at Sea Life Park oceanarium in Hawaii. Instead of leading a dolphin laboriously through a desired behavior, she wondered if she could simply ask the animal to do "something new." Using the very flexible-minded, rough-tooth dolphin (*Steno bredanensis*), she required the animal to innovate a new behavior in each training session. At first, these new behaviors occurred only by chance, but soon the dolphin "got the idea" and began to pour out new patterns in profusion. It had made the quite remarkable intuitive leap that it must offer behavior that had not been rewarded before in order to receive a reward. It had perceived the context of her requests, displaying a cognitive function of a high order called second order learning.

On the other side of Oahu, at the University of Hawaii, Louis Herman began similar cognitive studies about 1970 working with a pair of captive bottlenose dolphins. Since that time, Herman has collaborated with a number of students and professional colleagues to define many of the dimensions of the dolphin mind. He has studied their memory, their ability to transfer tasks from one modality to another (for example, a task given first in acoustic terms which might then be performed by vision), their air and water vision, and their ability to respond to complex commands given in the form of syntactical sentences. Now his dolphins undoubtedly perform the most complex trained behavior ever achieved for cetaceans including responding to questions, classifying objects, and remembering absent articles.

So advances in knowledge about dolphins have come from many sites where captives are kept. We frequently learn of new work from

Japan, the Netherlands, Canada, Switzerland, or Australia in these days in which the opportunities to learn from captives are spread around the globe.

The chapters that follow are a sampling of new work, from our newest colleagues and from those who have taken a part in the older story I have sketched here.

Schooner and Bayou perform with other animals in the Killer Whale and Dolphin Show at Marine World Africa USA, in Vallejo, California. (Photo by Marine World Africa USA.)

CHAPTER TEN

CHANGES IN AGGRESSIVE AND SEXUAL BEHAVIOR BETWEEN TWO MALE BOTTLENOSE DOLPHINS (*TURSIOPS TRUNCATUS*) IN A CAPTIVE COLONY

Jan Östman

INTRODUCTION

Studies of several species of dolphins and porpoises, both in the wild and in captivity, indicate that sexual behavior is prevalent in their social interactions (Brown and Norris 1956; Caldwell and Caldwell 1967; Tayler and Saayman 1972). These interactions can be seen any time of the year between and within all age groups, including calves as young as six weeks

I wish to express my deepest thanks to Hal Markowitz for supporting me during the course of this study and for commenting on previous drafts. I also wish to thank Amy Samuels, Bernd Würsig, Adrienne Zihlman, Stuart Mackay, and John Kaufmann for reading and commenting on previous drafts of this chapter. My sincere thanks are extended to Patrick Corcoran, whose help and extensive knowledge of computers was invaluable during data analysis. I am very grateful for all the support and valuable suggestions provided by Diana Reiss. I am also indebted to Mrs. Demetrios and Marine World Africa USA for giving me the opportunity to study their animals. Thanks are also due to the trainers, Jim Mullen, Deborah Merrin, and Alison Seacat, who endured numerous questions and gave me insight into the interactions among the dolphins during training sessions and shows. Veterinarian Laurie Gauge provided information on the animals. Marshal Sylvan at the University of California, Santa Cruz, gave helpful suggestions on useful statistical methods. Financial support was provided during the study by Thanks to Scandinavia, Inc., in cooperation with the Swe n-America Foundation, and by the Adolf Lindgren Foundation, Örebro, Sweden.

copulating with their mothers or other older animals, and heterosexually as well as homosexually among both sexes (McBride and Kritzler 1951; Caldwell and Caldwell 1967, 1972; Bateson 1974; Dudok van Heel and Mettivier 1974; Saayman and Tayler 1977). It is therefore likely that these interactions have other functions besides proliferation of the species.

The term "sexual behavior" implies that we know more about these behavior patterns than we really do. Generally, the stated or implied definition of sexual behavior is that the genital area of one or several of the interacting animals is involved; however, the apparent sexuality may be assumed by the observer (Gaskin 1982, 144).

With a few exceptions, for example, Puente and Dewsbury (1976) and Wells (1984), information on sexual behavior in dolphins has been presented in anecdotal form. Quantitative data analyzed statistically are needed to support subjective impressions if we are to ascertain the meaning of these behavior patterns. In the study presented here, I looked at the homosexual interactions between two male bottlenose dolphins. I related changes over time in their homosexual interactions to changes in their aggressive interactions. The data suggest that the sexual interactions between the two males were a component of their dominance relationship.

METHODS

This study surveys the behavior of a group of dolphins in a public show tank at Marine World Africa USA, in Redwood City, California. One hundred fifty-six observation sessions were made from October 21, 1982, through May 11, 1983, with six additional sessions in August 1983. All observations were made from underwater viewing windows through which most of the tank could be seen.

The tank was 14.6 m in diameter by 5.5 m deep. There were five animals in the tank at the beginning of the study: four bottlenose dolphins (*Tursiops truncatus*) and a pilot whale (*Globicephala macrorhynchus*). Shilo, a 13- to 14-year-old female bottlenose dolphin, was an adult (she had given birth to a stillborn calf the year before). Based on length and estimated ages, the two males, Bayou and Schooner, were probably still subadults at the time of the study. They were both about 2.2 m long and 7 to 8 years old. Stormy, the remaining female dolphin, may or may not have been sexually mature. She was about 2.4 m long and 7 to 8 years old. The pilot whale was an adult male, about 18 years old.

I used focal animal observation techniques (Altmann 1974) and tape-recorded notes. Four 20-minute observation sessions were conducted two days per week, in the mornings; some afternoon sessions also took place in February, March, and August. The order in which the animals were observed was rotated in such a way that over a two-week period, each dolphin was observed an equal number of times during each observation period over the course of a day. All dolphins within 2 m of each other were considered to be in a subgroup. I noted all behavior patterns in any subgroup containing the focal animal. I also noted which animal was the "actor," or initiator of the behavior. If the behavior was obviously directed toward one or more specific recipients, I noted that as well. Observed behavior patterns were listed sequentially in 20 one-minute segments each session. Any one-minute section during which a training session or show was in progress, or divers were in the water cleaning the tank, was discarded during analysis. The composition of subgroups was sampled three times during each observation session to enable me to estimate how much time each dolphin spent in the same subgroup with each of the other dolphins.

For purposes of analysis, behaviors were clumped into four categories: attack, threat, erection, and sexual behavior (any behavior involving the genital area of at least the actor). Behavior patterns were assigned to categories as follows.

Attack	Hit, bite, mouth
Threat	Jaw clap, head jerk, head down, flukes down, head and flukes down, arch/hump (a very pronounced head and flukes down)
Erection	Penis display
Sexual behavior	Penis rub, intromission attempt (penis rub around the genital area), intromission

The recipient of attacks, and of actions listed in the category "sexual behavior" can easily be identified during observation, since these behavior patterns involve contact. The intended recipient of threats, or of displays of erections, may or may not be obvious, however. If the recipient could not be specifically identified, the behavior was scored as being directed toward all the dolphins in the actor's subgroup.

Chi-square statistics, using the Yates correction for continuity, have been used unless otherwise specified. All statistical analyses were done on the original frequencies, although the data displayed in the figures have been adjusted to six focal animal observations per animal and time period.

RESULTS

Subgroup Structure

During fall 1982, Bayou spent more time in subgroups than any other dolphin (68% of 72 subgroup samples). Schooner, Shilo, and Stormy swam with other dolphins in 49 percent, 51 percent, and 58 percent of

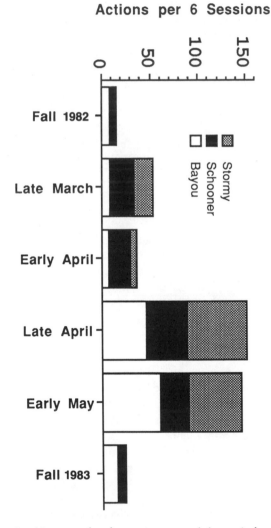

Fig. 10.1.   Looking at or head scanning toward the genital area of Shilo.

the subgroup samples, respectively. Bayou and Shilo were seen together in 39 percent of the subgroup samples. Schooner and Shilo, however, were together in only 5 percent of the samples during the fall; they were seen only twice in the same triad and were never sampled as a pair.

During the course of the study, the time spent in various groups changed for all four dolphins; the largest change occurred for Shilo in the spring. In the last two weeks of April, she spent 87 percent of 46 subgroup samples with other dolphins. Both males spent much more time with her; Schooner was with her in 65 percent of the samples and Bayou was with her in 52 percent of the samples. This change, and the interest the other three dolphins showed in Shilo's genital area (fig. 10.1), suggest that she might have come into estrus at this time.

Aggressive Interactions

The relationship between the two males, Bayou and Schooner, seemed to be fairly stable during fall 1982, with comparatively few aggressive interactions observed between them (fig. 10.2). Bayou was the more aggressive of the two, attacking Schooner more than Schooner attacked him. In spring 1983, Shilo was removed from the tank temporarily because of illness. When Shilo was gone, during most of February and early March, Schooner threatened Bayou much more frequently than in the previous fall. Bayou responded by attacking Schooner more. In late March, after Shilo had been returned, Schooner's threats escalated into overt attacks. By early April, the overall rate of attack had decreased between the two males, but Schooner still attacked Bayou more. In fact, there was a steady increase in the proportion of attacks made by Schooner from fall 1982 through the first half of April ($r = 0.98$).

In late April, Bayou again was the more aggressive of the two. By August, the relationship between the two males had returned to that of the previous fall; Bayou once again was more aggressive toward Schooner than vice versa. Both males, however, were now using threats more than open attacks. For Schooner, threats had increased from 47 percent of his aggressive behavior in fall 1982 to 67 percent in August 1983; and for Bayou, threats went from 34 percent of aggressive acts to 62 percent.

Sexual Interactions

The sexual interactions between Bayou and Schooner went through a similar change. During fall 1982, Bayou displayed more erections than Schooner did. Bayou was always the actor during homosexual inter-

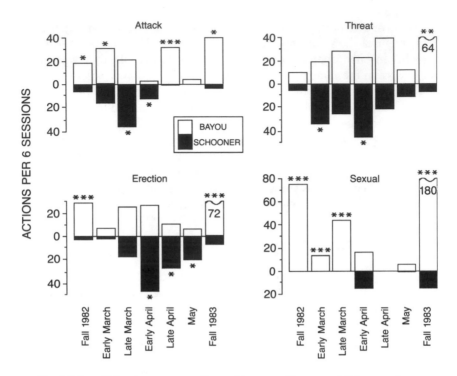

Fig. 10.2.   Behavior patterns directed between Bayou and Schooner irre-
spective of subgroup composition. Significant differences between number of behavior patterns originated by each male are indicated: * P < 0.05; ** P < 0.01;
*** P < 0.001.

actions (figure 10.2). In early March, when Shilo was gone and the
aggression level between the males was high, homosexual interactions
decreased drastically; in fact, it was the low level of sexual interactions
between the males in this period that first alerted me to the change in
their relationship. By late March, Schooner began displaying erections
more frequently than Bayou, and in early April, Schooner, for the first
time, was observed to be the actor during homosexual interactions and
made intromissions. The proportion of erections displayed by Schooner
increased through the first six time periods (r = 0.98), just as did the proportion of attacks by Schooner on Bayou. By August, observations indicate that Bayou again was the actor during most homosexual encounters
and displayed more erections than Schooner; Schooner, however, was occasionally the actor in homosexual interactions, a change from the previous fall.

Long-term changes in the relationship between the two males seemed to be triggered by the removal of Shilo. Even when Shilo was in the tank, however, the interactions between the two males were affected by her presence or absence in their subgroup. It is instructive to compare the interactions between the two males during three different parts of the study when Shilo was present in the tank but restricting the comparison to those times when Shilo was not swimming in the same subgroup as the males (fig. 10.3). During fall 1982, Bayou was the aggressor and the actor

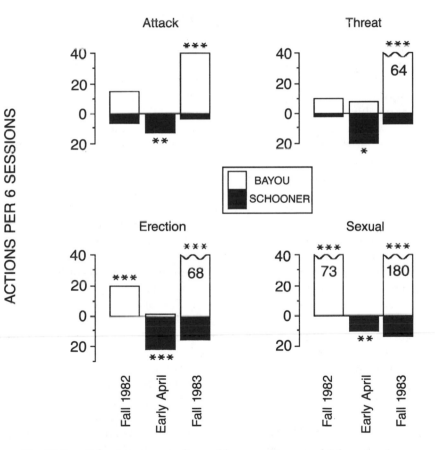

Fig. 10.3. Behavior patterns directed between Bayou and Schooner when Shilo was in the tank but not in the male subgroup during the three parts of the study, highlighting the changes in the male-male relationship. Significant differences between number of behavior patterns originated by each male are indicated: * P < 0.05; ** P < 0.01; *** P < 0.001.

during all homosexual interactions. He was also the only one of the two to display an erection. In early April, the homosexual relationship between the two males, like the aggressive relationship, had flip-flopped completely compared to the previous fall. Now Schooner was the aggressor and also the actor during all homosexual interactions. He also displayed all but one of the observed erections. In August, Bayou again was the aggressor most of the time, and he also made most of the homosexual intromissions and displayed most erections.

DISCUSSION

The relative frequency of attacks and threats between the two males suggests that Bayou was dominant over Schooner in fall 1982, as he was both more aggressive and had a higher proportion of attacks than Schooner. It is thus consistent with a dominance-related function of homosexual interactions that Bayou always was the actor and Schooner always was the recipient during homosexual interactions at that time. The changes in the relationship between the two males also support this conclusion.

Dominance mountings have been described for several species of terrestrial mammals, including giraffes (Coe 1967), mountain sheep (Geist 1971), Japanese macaques (Kawamura 1967), and cattle (Klemm et al. 1983). Homosexual mountings among terrestrial mammals may have functions other than dominance displays, however. In red deer, mounts among females are dominance mounts, while mounts among males are subordinance mounts (Hall 1983). Homosexual mounts among chimpanzees may be a reassurance behavior (de Waal 1982).

M. C. Caldwell and D. K. Caldwell (1977) suggest that bottlenose dolphins "demonstrate penile erection and forceful intromission as a dominance gesture," and they (D. K. Caldwell and M. C. Caldwell 1977) describe a case where a newly introduced, adult male bottlenose dolphin permanently displaced the previously dominant bull in a community tank during a one-hour interaction that "ended with sexual pursuit and two successful intromissions by the victor." McBride and Hebb (1948) observed a community with four males, two large and two smaller, younger, and "subordinate." The two larger males repeatedly attempted intromission with the younger ones. In the homosexual activity, the active (or "male") male swims upside down under the passive (or "female") male. Tavolga (1966) described homosexual behavior among three subadult male bottlenose dolphins (4-5 years old). Two of the males tended to be dominant to the third, "who was relegated to the female role," although the third male occasionally could "assume the male role" in these activities. Amundin and Amundin (1971) observed a captive group of three subadult (about 1 year old) male harbor porpoises, where the most subordinate male generally played the "female role" during homosexual interactions.

Saayman and Tayler (1977) and Würsig and Würsig (1979) suggest that courtshiplike behavior may function in a greeting context among dolphins. In the wild, Irvine et al. (1981) observed "apparent homosexual interactions" repeatedly in a group of four subadult bottlenose dolphins over a six-month period, suggesting that this is not a phenomenon brought on by captivity. Available data on a wide variety of terrestrial mammals, however, suggest that captive mammals tend to mount individuals of the same sex more frequently than their wild conspecifics (Dagg 1982). Additional observations on dolphins in the wild are also consistent with the suggestion that sexual behavior, intra- or intersexually, are used in an aggressive context. Several authors have presented data showing that sexual, "courtshiplike," behavior often is observed among several species of dolphins after several groups of animals meet or coalesce after having been separated for several hours or days. Saayman and Tayler (1979), discussing humpback dolphins (*Sousa* spp.), mentioned that "unusually lively and protracted social interactions were seen when two groups of humpback dolphins coalesced or when single newcomers joined groups already present." The authors also state that these greeting interactions closely resembled courtship activities normally seen in association with sexual behavior. For bottlenose dolphins (*Tursiops truncatus*), behavior patterns that "may be associated with 'play' and copulatory activity," such as belly-up and rubbing, were observed more frequently when two subgroups joined after having separated for several hours (Würsig and Würsig 1979). Among dusky dolphins (*Lagenorhynchus obscurus*), many groups coalesce during feeding bouts, and "much apparent mating" is seen after these bouts (Würsig and Würsig 1980). For spinner dolphins (*Stenella longirostris*), "social behavior, including mating, aerial behavior, sexual play, and aggressive chase, becomes especially evident" as the dolphins move toward the feeding grounds (Norris and Dohl 1980). This is the time when several groups of spinners coalesce into a large group of several hundred animals that later will spend the night feeding together. It should be pointed out that these are occasions in which it would be expected that dominance-subordinance relationships would be established, reestablished, or tested and affiliative relationships strengthened.

This reasoning would also explain why newly captured dolphins often engage in sexual behavior with tank residents within hours after having been introduced into a tank with other dolphins (McBride 1940, Saayman and Tayler 1977). Among spinner dolphins, another sexual behavior, beak-genital propulsion, has been correlated with the dominance rank of the two participants (Bateson 1974).

Homosexual behavior has also been observed among mysticetes. Darling (1977) observed homosexual interactions among three immature male gray whales during the northern migration. Christopher Clark has told me of observing homosexual interactions in southern right whales in both sexes and among all age groups except calves. Female homosexual interactions were seen early in the season and male homosexual interactions in the late season, after the females had left.

Among terrestrial mammals, penis displays are made by dominant animals in squirrel monkeys (Ploog and MacLean 1963) and whiptail wallabies (Kaufmann 1974). It has also been suggested that guard-sitting males among both baboons (*Papio anubis*) and vervet monkeys (*Cercopithecus aethiops*) sit with their penises extended or erect as a warning to foreign conspecifics (Wickler 1966).

It seems probable that, among bottlenose dolphins, displaying an erection in the vicinity of another male is also a dominance-related behavior. When Bayou was dominant over Schooner, during the first and last time periods, he displayed erections more frequently than Schooner did. However, when Shilo was not swimming in the male subgroup in early April and Schooner was dominant over Bayou, Schooner displayed the vast majority of erections.

It is not clear, however, that penis display always was dominance related among the two males. The data is confounded by Shilo probably coming into estrus in late April, as suggested by the intense interest the other three dolphins showed in her genital area. Both males displayed erections when Shilo was present in their subgroup in late April.

This study was carried out on captive dolphins; further studies are needed on wild animals to ascertain what effect the captive environment might have on the behavioral patterns discussed here. However, it is important that we continue to study the social behavior of dolphins in captivity. Captive dolphins, though in a very different situation from their wild counterparts, can still teach us much about their behavior. In such studies, emphasis should be put on quantitative observations, which have largely been absent from studies on dolphin social behavior.

REFERENCES

Altmann, J. 1974. Observational study of behaviour: Sampling methods. Behaviour 49: 227–267.

Amundin, B., and M. Amundin. 1971. Några etologiska iakttagelser över tumlaren, *Phocoena phocoena* (L.), i fångenskap. Zool. Revy 33:51–59.

Bateson, G. 1974. Observations of a cetacean community. In: Mind in the waters, ed. J. McIntyre. New York: Scribner's. 146–165.

Brown, D. H., and K. S. Norris. 1956. Observations of captive and wild cetaceans. J. Mammal. 37:311–326.

Caldwell, M. C., and D. K. Caldwell. 1967. Dolphin community life. Quart. Los Angeles City Mus. Nat. Hist. 5(4):12–15.

———. 1972. Behavior of marine mammals. In: Mammals of the sea: Biology and medicine, ed. S. H. Ridgway. Springfield: C. C. Thomas. 419–465.

———. 1977. Social interactions and reproduction in the Atlantic bottlenose dolphin. In: Breeding dolphins present status, suggestions for the future, ed. S. H. Ridgway and K. Benirschke. 133–142. Final report. Washington, D.C.: U.S. Marine Mammal Commission. 314 pp. Available from NTIS, Springfield, Va.; PB-273/4GA.

Caldwell, D. K., and M. C. Caldwell. 1977. Cetaceans. In: How animals communicate, ed. T. A. Sebeok. Bloomington and London: Indiana University Press. 794–808.

Coe, M. J. 1967. "Necking" behavior in the giraffe. J. Zool. 151:313–321.

Dagg, A. I. 1982. Homosexual behavior and female-male mounting in mammals—a first survey. Mammal Rev. 11:155–185.

Darling, J. 1977. The Vancouver Island gray whales. Waters 2(1):4–19.

de Waal, F. B. M. 1982. Chimpanzee politics: Power and sex among apes. New York: Harper & Row.

Dudok van Heel, W. H., and M. Mettivier. 1974. Birth in dolphins (*Tursiops truncatus*) in the dolphinarium Harderwijk, Netherlands. J. Aquat. Mammal. 2(2):11–22.

Gaskin, D. E. 1982. The ecology of whales and dolphins. London: Heinemann.

Geist, V. 1971. Mountain sheep. Chicago: University of Chicago Press.

Hall, M. J. 1983. Social organization in an enclosed group of red deer (*Cervus elaphus*) on Rhum. II. Social grooming, mounting behavior, spatial organization and their relationship to dominance rank. Z. Tierpsychol. 61:273–292.

Irvine, A. B., M. D. Scott, R. S. Wells, and J. H. Kaufmann. 1981. Movements and activities of the Atlantic bottlenose dolphin, *Tursiops truncatus*, near Sarasota, Florida. Fish. Bull. U.S. 79:671–688.

Kaufmann, J. H. 1974. Social behavior of the whiptail wallaby, *Macropus parryi*, in northeastern New South Wales, Animal Behav. 22:281–369.

Kawamura, S. 1967. Aggression as studied in troops of Japanese monkeys. In: Aggression and defense: Neural mechanisms and social patterns, ed. C. D. Clemente and D. B. Lindsley. Berkeley and Los Angeles: University of California Press. 195–223.

Klemm, W. R., C. J. Sherry, L. M. Schake, and R. F. Sis. 1983. Homosexual behavior in feedlot steers: An aggression hypothesis. Appl. Anim. Ethol. 11: 187–195.

McBride, A. F. 1940. Meet mister porpoise. Nat. Hist. 45: 16–29.

McBride, A. F., and D. O. Hebb. 1948. Behavior of the captive bottlenose dolphin, *Tursiops truncatus*. J. Comp. Physiol. Psychol. 41: 111–123.

McBride, A. F., and H. Kritzler. 1951. Observations on pregnancy, parturition, and post-natal behavior in the bottlenose dolphin. J. Mammal. 32: 251–266.

Norris, K. S., and T. P. Dohl. 1980. Behavior of the Hawaiian spinner dolphin, *Stenella longirostris*. Fish. Bull. U.S. 77: 821–849.

Ploog, D. W., and P. D. MacLean. 1963. Display of penile erection in squirrel monkey (*Saimiri sciureus*). Animal Behav. 11: 32–39.

Puente, A. E., and D. A. Dewsbury. 1976. Courtship and copulatory behavior of bottlenose dolphins (*Tursiops truncatus*). Cetology 21: 1–9.

Saayman, G. S., and C. K. Tayler. 1977. Observations on the sexual behavior of Indian Ocean bottlenose dolphins (*Tursiops aduncus*). In: Breeding dolphins present status, suggestions for the future, ed. S. H. Ridgway and K. Benirschke. 113–129. Final report. Washington, D.C.: U.S. Marine Mammal Commission. 314 pp. Available from NTIS, Springfield, Va.: PB-273/4GA.

Saayman, G. S., and C. K. Tayler. 1979. The socioecology of humpback dolphins (*Sousa* sp.). In: Behavior of marine animals: Current perspectives in research. 4: Cetacea, ed. H. E. Winn and B. L. Olla. New York: Plenum. 165–226.

Tavolga, M. C. 1966. Behavior of the bottlenose dolphin (*Tursiops truncatus*): Social interactions in a captive colony. In: Whales, dolphins and porpoises, ed. K. S. Norris. Berkeley and Los Angeles: University of California Press. 718–730.

Tayler, C. K., and G. S. Saayman. 1972. The social organization and behavior of dolphins (*Tursiops aduncus*) and baboons (*Papio ursinus*): Some comparisons and assessments. Ann. Cape. Prov. Mus. Nat. Hist. 9: 11–49.

Wells, R. S. 1984. Reproductive behavior and hormonal correlates in Hawaiian spinner dolphins, *Stenella longirostris*. Rept. Int. Whal. Comm. Spec. Issue 6. 465–472.

Wickler, W. 1966. Ursprung und biologische deutung des Genitalpräsentirens männlicher Primaten. Z. Tierpsychol. 23:422–437.

Würsig, B., and M. Würsig. 1979. Behavior and ecology of the bottlenose dolphin, *Tursiops truncatus*, in the south Atlantic. Fish. Bull. U.S. 77:399–412.

———. 1980. Behavior and ecology of the dusky dolphin (*Lagenorhynchus obscurus*) in the south Atlantic. Fish. Bull. U.S. 77:871–890.

Dolphin communication research is hampered by physics: human ears cannot hear directionally underwater, so one cannot easily tell which animal is making what sound. Bottlenose dolphins practice wearing "vocalights," small instruments that light up when the wearer whistles, a useful tool in decoding the functions of dolphin vocalizations. (Photo by Mike Greer, Brookfield Zoo.)

CHAPTER ELEVEN

USE OF A TELEMETRY DEVICE
TO IDENTIFY WHICH DOLPHIN PRODUCES
A SOUND

Peter Tyack

INTRODUCTION

Why has progress in the study of social communication been so much slower for marine animals such as whales and dolphins than for terrestrial animals such as birds and primates? A primary obstacle has been the difficulty of identifying which animal within an interacting group produces a sound underwater. Biologists who study terrestrial animals

The advice, engineering assistance, and loans of equipment and space that I received during each phase of development of the vocalight was critical for its success. Thanks to Francis Carey, Frederick Hess, William Watkins, G. Richard Harbison, and Richard Koehler. Ronald Larkin came up with the idea of using light-emitting diodes. George King, the director of Sealand, Keith Wilson, and Dick Gage helped to train and observe the dolphins and provided an environment conducive to research at Sealand. James Bird, Christopher Converse, Eleanor Dorsey, Linda Guinee, Judith Lafler, Victoria Rowntree, and Nancy Wolf helped observe the dolphins to determine when the vocalights lit up. I thank William Watkins for use of the Kay Spectrum Analyzer, hydrophones, tape recorders, and computers for word processing and data management. Francis Carey, Christopher Clark, Donald R. Griffin, Karen Moore, Kenneth S. Norris, Karen Pryor, Laela Sayigh, William Schevill, and William Watkins reviewed the manuscript. This research was performed with financial assistance from a WHOI Postdoctoral Scholar Award and NIH Postdoctoral Fellowship 5-F32-NS07206. This is contribution number 6362 from the Woods Hole Oceanographic Institution.

take it for granted that they can identify which animal is vocalizing. They can use their own ears to locate the source of a sound and to direct their gaze. Most terrestrial animals produce a visible motion associated with the coupling of sound energy to the air medium. Mammals and birds open their mouths when vocalizing, many insects produce visible stridulation motions, and frogs inflate their throat sacs. Once their gaze has been directed to the sound source, terrestrial biologists can watch for the visual correlates of sound production to confirm which animal is vocalizing. The simplicity of this process should not obscure how important it is for ethological research. Without this ability, ethologists could scarcely begin to tease apart the patterns of signal and response that inform us about a system of animal communication.

Humans are not able to locate sounds underwater in the same way they locate airborne sounds. Furthermore, whales and dolphins seldom produce visible motions coordinated with sound production. Some dolphin sounds are coordinated with a visible display, such as the so-called jaw clap that is coordinated with an open mouth display (Wood 1954). Dolphins also occasionally emit a stream of bubbles while vocalizing. But these special cases are not common enough to allow systematic analysis.

The need for some technique to identify which cetacean produces which sound during social interaction has been discussed for over two decades. The following three approaches have emerged: (1) an electronic acoustic link between animals isolated in two tanks; (2) acoustic location of sound sources using an array of hydrophones; and (3) telemetry of information about sound production from each animal in a group.

Several investigators have attempted to study communication between isolated captive dolphins using an electronic acoustic link between two pools (Lang and Smith 1965; Burdin et al. 1975; Gish 1979). This approach has several serious drawbacks. The electronic reproduction of dolphin sounds may be discriminably different from the natural sounds. It is next to impossible to control for this problem in electronic acoustic link experiments. Even if dolphins accept the acoustic quality of the link, the sounds emanate from an underwater sound projector rather than from another dolphin. As soon as an animal approaches the source and can inspect the projector, it is likely to respond differently than if the source were another dolphin. Furthermore, to study the social functions of vocalizations, one must study what roles they play in social interactions; but the isolated dolphins are able to interact only acoustically.

Acoustic localization of sound sources is a promising method for identifying which cetacean is vocalizing. It involves no manipulation of the animals, just placement of hydrophones near them. Analysis is complicated, for one must compare the sound source location to a visual record of the animals' locations in order to identify which animal produced a sound. In some applications, animals may vocalize frequently enough and be sufficiently separated so that source location data alone will suffice to indicate which animal produces a sound. However, questions will remain about the accuracy of location and identification for data not compared to a visual record.

Tracks of continuously vocalizing finback whales (*Balaenoptera physalus*) were obtained in the early 1960s using hydrophones mounted on the sea floor near Bermuda (Patterson and Hamilton 1964). William Watkins and William Schevill of the Woods Hole Oceanographic Institution devised a four-hydrophone array that can rapidly be deployed from a ship (Watkins and Schevill 1972). This array has been used to locate vocalizing finback whales, Northern right whales (*Eubalaena glacialis*), sperm whales (*Physeter catodon*), and several species of dolphins. The array has been used with great success to identify which sperm whale is making a sound, because they click frequently and individuals are often spread out over a wide area. The array has provided important data on the coda repertoires of individual sperm whales and coda exchanges between different sperm whales (Watkins and Schevill 1977). The array has not been as useful in identifying vocalizing dolphins (Watkins and Schevill 1974). These animals move rapidly and swim so close to one another that it has proved difficult to record the location of different individuals in sufficient detail to allow correlation with the acoustic source locations.

Working on Southern right whales (*Eubalaena australis*) off the coast of Patagonia, Christopher Clark developed a device that measures the phase difference between vocalizations arriving at a three-hydrophone array. This device gives bearings to a sound source in real time. It was used successfully to correlate right whale sounds and behavior (Clark 1980). More recently, Clark set up a computer near Point Barrow, Alaska, to obtain source locations of vocalizing bowhead whales (*Balaena mysticetus*), but it proved difficult to associate visual sightings with acoustic source locations in this study (Clark et al. 1986).

The third technique for identifying a vocalizing cetacean, telemetry from the animal itself, eliminates the need for locating each animal within a group. If each animal carries a telemetry device that transmits information about its sound production, then these data can be used directly to determine which animal produces a sound. The problem is deciding how to transmit the information effectively without disturbing the animals. The use of sound for underwater telemetry is common, since it has favorable propagation characteristics underwater. However, there are serious limitations to the use of sonic telemetry with dolphins and possibly other cetaceans. Dolphins of many species can hear above 100 kHz (Popper

1980). Frequencies higher than this attenuate so rapidly in seawater that they are not particularly effective for sonic telemetry. If the sounds of the telemetry device are audible to the animal, they may interfere with normal behavior.

Other researchers have suggested nonacoustic telemetry. Evans and Sutherland (1963) proposed the development of a 27-MHz radiotelemetry device to broadcast sounds from a dolphin's head. This technique is limited to applications where the antenna remains above water, for the propagation of radio signals in seawater is so limited as to render radio impractical for telemetry under the sea.

Visible light propagates favorably in seawater compared to other wavelengths of electromagnetic radiation. I have developed a telemetry device, called a vocalight, that uses light-emitting diodes (LEDs) to indicate which sounds are produced by a dolphin within a social group (Tyack 1985). Observers watch the response of the vocalight directly, and this obviates the need for constructing a telemetry receiver. The vocalight has been successfully used to study the whistle repertoires of two captive bottlenose dolphins (*Tursiops truncatus*) (Tyack 1986a).

Dolphin Signature Whistles

Dolphins produce a large repertoire of complex vocalizations (Popper 1980). Aside from pulsed sounds used for echolocation, the most intensively studied dolphin sounds are frequency modulated narrow band sounds called whistles or squeals (Herman and Tavolga 1980; Watkins and Wartzok 1985). Most authors have categorized dolphin whistles by using variation in the dominant frequency as a function of time. This variation is called a whistle's contour (Dreher 1961). Early work on the function of dolphin whistles attempted to associate particular whistle contours with specific behavioral contexts such as fright or disturbance (Lilly 1963; Dreher and Evans 1964; Dreher 1966). These authors studied whistles recorded from groups of dolphins. They were only able to associate vocalizations with the behavior of the whole group, because they were unable to identify which individual dolphin produced a sound.

David and Melba Caldwell found one way to solve this problem. To determine which animal produced a whistle, the Caldwells recorded dolphins when they were isolated from conspecifics. They recorded whistles from five bottlenose dolphins, all of which had been recently collected from one wild group. These dolphins were housed in the same pool, but each individual was recorded when it was isolated for short periods of time (e.g., for veterinary examination). The Caldwells presented evidence that over 90 percent of the whistles from any one of these dolphins conformed to one contour that was easily distinguished from that of any other dolphin within the same group (Caldwell and Caldwell 1965). Ultimately, the Caldwells studied whistles of over 100 individuals from four species. The Caldwells called these individually distinctive whistles "signature whistles," and they proposed that their function was to broadcast the individual identity of the whistler to other members of its community (M. C. Caldwell and D. K. Caldwell 1968, 1971, 1979; Caldwell et al. 1970, 1973, 1990).

For an animal to use individually distinctive attributes of a signal in this way, it must have learned to associate the distinctive signal with the appropriate individual. This implies something more than simple discrimination of the calls of different individuals. A specific example may clarify this distinction. Playback experiments performed under natural conditions have shown that if the alarm call of a juvenile vervet monkey (*Cercopithecus aethiops*) is played to its group in the field, adult females tend to look at the mother of the juvenile (Cheney and Seyfarth 1980). Since this response occurs in the absence of any apparent cue from the mother, it suggests that the females (1) can discriminate the calls of particular juveniles, (2) have associated specific calls with specific individuals, and (3) have learned the relationships of mothers and juveniles. The ability to recognize the calls of different individuals is necessary but not sufficient to produce the observed response.

Captive dolphins are able to learn to discriminate between whistles recorded from different dolphins. Caldwell et al. (1969) trained a captive bottlenose dolphin to press a paddle after playback of whistles from one dolphin but to ignore whistles from another dolphin. The dolphin quickly learned to discriminate the whistles recorded from different dolphins, and he was able to recall the discrimination after several weeks. Later studies demonstrated this dolphin's ability to discriminate between the whistles of two groups of four conspecifics each or between whistles of individuals of another species (Caldwell, Caldwell, and Hall 1972; Caldwell, Hall, and Caldwell 1972). While these studies show that dolphins can *discriminate* whistles from different individuals, there is no direct evidence that dolphins learn to *associate* each whistle with the individual typically producing it.

The most serious problem with the Caldwells' data in support of the signature whistle hypothesis is that some of their whistle data came from dolphins that were isolated, often out of the water for veterinary attention. In this unusual context, whistling may be abnormal. The data from the electronic acoustic link experiments also indicate that each dolphin tends to produce a distinctive signature whistle. However, the dolphins in these experiments are also physically isolated and cannot interact normally. To study whistle repertoires in more normal circumstances, one must be able to identify which dolphin produces a whistle within an interacting group.

METHODS

I was able to identify which dolphin within a group produced a sound with a telemetry device, a vocalight. The vocalight is attached to a dolphin's head with a suction cup. A scale drawing of the vocalight is presented in figure 11.1. The circuit design and construction of the vocalight is described elsewhere (Tyack 1985). A variable number of LEDs illuminate the vocalight, depending on the loudness of sounds received at a contact hydrophone within the suction cup. The louder the sound, the more LEDs light up. A small piezoelectric ceramic disk is used as a contact hydrophone to pick up the sounds of the dolphin. This signal is amplified with a variable gain usually set at approximately 50 dB. I determined which amplifier gain was suitable by tests with captive dolphins. The signal is also filtered to cut out frequencies below 2 kHz. The 2-kHz cutoff point is a compromise between reducing sensitivity to low-frequency noise while not interfering with sensitivity to whistles. While the device does respond to high-frequency turbulence as a dolphin breaks the water surface, it is not triggered by flow noise underwater or by noise from most water pumps.

This filtered and amplified signal is fed to an LED driver integrated circuit that compares the input signal to a reference voltage. The dynamic range of the circuit is 30 dB. If the input signal is 30 dB less than the reference voltage, no LEDs light up. For every doubling of signal (3 dB) above this level, another LED is turned on. If the signal exceeds the reference voltage, all ten LEDs remain turned on, indicating that the device is saturated.

I used the vocalight to define the whistle repertoires of two captive *Tursiops* individuals that were housed at Sealand, an aquarium in Brewster, Massachusetts, on Cape Cod. These two dolphins, named Spray and Scotty, were caught in Tampa Bay, Florida, in 1977 and were moved directly to Sealand. When caught, the dolphins were 2 m long and were judged to be 5 to 6 years of age (their age at the time of the study was 12 to 13 years). Spray was a female (length at the time of the study was 249 cm), and Scotty was a male (length at the time of the study was 244 cm). After a few days of training, the dolphins showed no obvious differences in behavior whether wearing the vocalight or not. They also showed no obvious response when the LEDs lit up. I noticed no response of one dolphin to a device worn by the other.

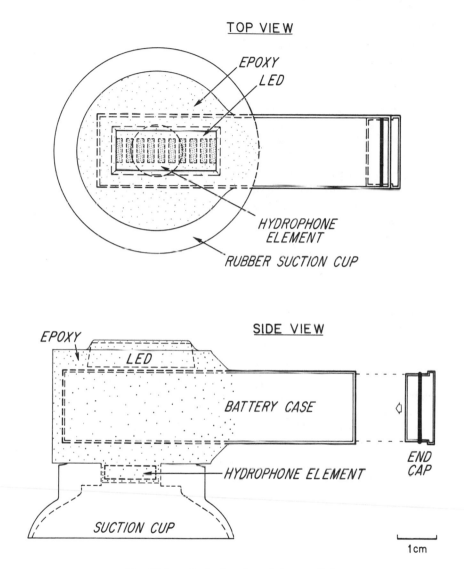

Fig. 11.1.   Scale drawing of the vocalight.

To identify which of the two dolphins produced a sound, I used two vocalights, one with red LEDs, the other with green. A loudspeaker broadcast underwater dolphin sounds picked up by a hydrophone in the pool to four observers around the pool. Each observer watched whichever vocalight was nearest. (If both dolphins were nearby, one observer could sometimes follow both vocalights, but several observers at different locations were usually required to keep both vocalights under observation.) When a whistle was broadcast in air, any observer who saw a vocalight light up in synchrony with the whistle signaled the response of the vocalight by calling out its color and brightness. If observers were watching a vocalight that did not light up when the whistle was produced, they called out the color of the device and stated that no LEDs lit up. If they could not see any vocalights clearly, they remained silent. These comments were recorded on a second channel of the same tape that recorded the underwater dolphin sounds.

It was usually easy to associate observer's comments with the whistles, unless both dolphins produced simultaneous sounds. The vocalights were adjusted to light up several LEDs even for relatively faint whistles. Observers never reported that neither device lit up when a whistle was heard. While the loudness of the whistles varied, the 30-dB range of the vocalight was usually sufficient to identify which animal had whistled. Sometimes both devices lit up; however, there was usually a distinctive difference in how many LEDs lit up, and the dolphin with the brightest vocalight was presumed to have produced the whistle. (Occasionally, when very loud whistles were heard, and when the dolphins were separated by less than approximately 1 m, both devices lit up fully, making it impossible to identify which animal produced the whistle.)

A Sony TC-D5M stereo tape recorder was used to record observers' comments on one channel using a microphone and dolphin sounds on the other using AN/SSQ-41A sonobuoy hydrophones. This system had a frequency response from 30 to 15,000 Hz ($\pm$ 5 dB when recorded at -13 dB VU) with Maxell UDXLII tape. Whistle sounds were analyzed with a Kay Elemetrics Corp. Sonagraph Model 7029A Spectrum Analyzer with a narrow band (90 Hz) filter. The frequency range of all analyses was 160 to 16,000 Hz.

RESULTS

All of the whistles described in this chapter were recorded from the two Sealand dolphins on February 28, 1984, or from Scotty on May 28, 1987. On February 28, I recorded 1,083 whistles in five sessions from 0930 to 1600. During 77.6 minutes of these recordings, the vocalights were removed from the dolphins for control observations; 586 whistles were recorded during this period, yielding a rate of 7.6 whistles/min. The dolphins wore vocalights for the remaining 110.1 min.; 497 whistles were recorded during this period, a rate of 4.5 whistles/min. Observers used the vocalights to identify which dolphin produced 252 of these 497 whistles. Thirty-two whistles were audible in air, and observers could locate the source to identify which dolphin produced them.

A random sample of 50 whistles from the 497 recorded when both dolphins were wearing the vocalights was compared with a random sample of 50 whistles from the 586 recorded when they were not wearing the devices. The whistles were combined into three categories described in the next section: type 1, type 2, and all secondary whistles. A chi-square analysis (Siegel 1956) of these data comparing the kinds of whistles produced under these two conditions indicates no significant differences (chi-square = 0.68, df = 2, p > 0.7). Thus, while the dolphins produced whistles at rates that differed when they were wearing vocalights compared to not wearing them, they produced similar proportions of each kind of whistle in both conditions.

Spray died in April 1985. From this time until Scotty was recorded on May 28, 1987, Scotty was isolated from other dolphins. I analyzed whistles recorded on this day from approximately 1500 to 1545. Scotty produced 70 whistles during this 45-minute period, yielding a rate of 1.6 whistles/min. While I did not measure the absolute level of whistles on either day, the 1987 whistles appeared to be fainter than the 1984 whistles, judging by the level of the whistle relative to the ambient noise in the pool.

Categorization of Whistles

I categorized the spectrograms of whistles by visual inspection of the contours. Figure 11.2 presents sound spectrograms of the most commonly produced whistle, primary contour type 1. The spectrograms on the upper row vary by 18 percent in duration, with little variation in contour. Both dolphins sometimes emitted a truncated type 1 whistle, in which just the first section of this whistle contour was produced with no terminal increase in frequency. Two examples of this truncated type 1 whistle, termed 1A, are shown in the bottom row of figure 11.2. The two sections of the type 1 whistle, termed 1A and 1B, are marked on the upper left spectrogram of the figure, a complete type 1 whistle, termed 1AB. The 1B section of the type 1 contour was only produced as part of 1AB whistles.

Figure 11.3 illustrates the second most commonly produced whistle contour, primary contour type 2. The whistle shown in the spectrogram on the middle left has only 70 percent the duration of that on the upper left, while the contours are otherwise quite similar. Three different sections of whistle type 2 (2A, 2B, and 2C) are labeled on the upper left spectrogram of the figure. All three of these sections were produced sepa-

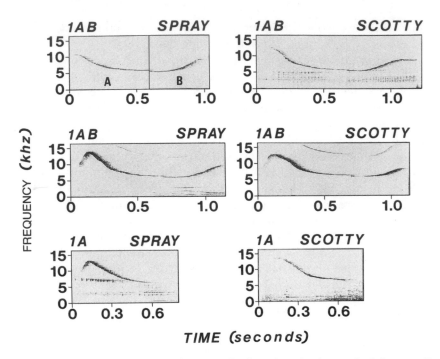

Fig. 11.2. Spectrograms of type 1 whistles. The whistles on the left were all produced by Spray; those on the right were produced by Scotty. Type 1 whistles were formed of two segments, 1A and 1B. These are separated by a vertical line in the spectrogram on the upper left of the figure. The top two rows show complete type 1 whistles, while the bottom row shows whistles that included only the 1A segment.

rately; the most common was 2C, shown on the middle of the bottom row. The bottom row of figure 11.3 also shows examples of sequences of two of these sections—2AB (bottom left) and 2BC (bottom right). The complete type 2 whistle was termed 2ABC.

Secondary whistles were much less common than primary whistles and were not as stereotyped. Any whistle with a monotonic rise in frequency different in structure from 1B or 2C was classed as a RISE whistle (upper right of fig. 11.4). Whistles with little frequency modulation were classed as FLAT whistles (second half of the spectrogram on the upper left of fig. 11.4). DOWN whistles had a monotonic decrease in frequency (middle right of fig. 11.4). SINE whistles were frequency modulated in a

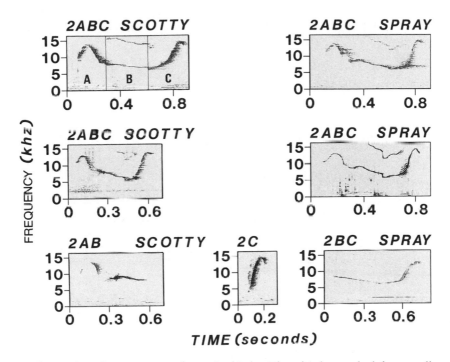

Fig. 11.3.   Spectrograms of type 2 whistles. The whistles on the left were all produced by Scotty; those on the right were produced by Spray. Type 2 whistles were formed of three segments, 2A, 2B, and 2C. These are separated by vertical lines in the spectrogram on the upper left of the figure. The top two rows show complete type 2 whistles; the bottom row shows whistles that included only one or two of these segments.

sine wave pattern (middle left of fig. 11.4). Any whistle that did not match any of the six categories or that could not be sorted into just one category was entered into a seventh VARIANT category.

Whistle Production by Individual Dolphins

Both Spray and Scotty, the two dolphins at Sealand, produced type 1 and type 2 whistles in 1984. Of the 131 whistles from Spray (row 5 of table 11.1), 88 were type 1, 25 were type 2 (19%), and 18 were secondary types (14%). Of the 153 whistles from Scotty (row 10 of table 11.1), 33 were type 1 (21%), 73 were type 2 (48%), and 47 were secondary types

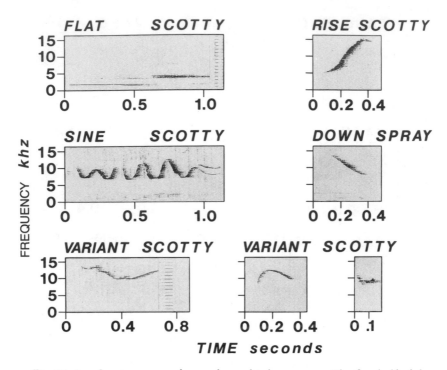

Fig. 11.4.   Spectrograms of secondary whistle contours. The first half of the spectrogram on the upper left shows a 1.5 kHz signal from the trainer's whistle; the second half shows a FLAT whistle of approximately 3 kHz produced by Scotty. The rest of the spectrograms show examples of secondary whistles produced by Scotty, except for the DOWN whistle on the middle right, which was produced by Spray.

(31%). These results, summarized in table 11.2, show that Spray produced 73% of the type 1 whistles while Scotty produced 74% of the type 2 whistles.

Scotty produced more of the secondary whistles than did Spray. He was the only one identified as producing SINE whistles, and he produced many more FLAT whistles than Spray. Two of the FLAT whistles were produced immediately after the trainer blew his training whistle, which was also constant frequency. As can be seen on the upper left of figure 11.4, while the trainer's whistle and Scotty's FLAT whistle differ in frequency, their contours were otherwise very similar.

Scotty and Spray seemed to use different whistles in different behavioral contexts. For example, almost all of Spray's whistles that were

Table 11.1. Whistles Produced by Two Bottlenose Dolphins

ID Data	Primary Whistle Types										Secondary Whistle Types					Total
	1AB	1A	TYPE 1 SUBTOTAL	2ABC	2A	2B	2C	2AB	2BC	TYPE 2 SUBTOTAL	RISE	FLAT	SINE	DOWN	VARIANT	
Spray																
A	0	0	0	11	0	0	0	0	0	11	1	0	0	0	0	12
B	33	1	34	3	0	0	0	0	0	3	0	0	0	0	1	38
C	8	0	8	0	0	0	0	0	0	0	0	0	0	0	1	9
D	38	8	46	9	0	0	1	0	1	11	7	1	0	1	6	72
	79	9	88	23	0	0	1	0	1	25	8	1	0	1	8	131
Scotty																
A	1	2	3	12	0	0	2	0	0	14	0	0	0	0	3	20
B	13	2	15	11	2	0	4	1	1	19	0	5	2	1	7	49
C	2	0	2	3	0	1	2	0	0	6	0	0	0	0	0	8
D	8	5	13	12	4	0	17	1	0	34	9	9	2	2	7	76
Subtotal	24	9	33	38	6	1	25	2	1	73	9	14	4	3	17	153
Grand Total	103	18	121	61	6	1	26	2	2	98	17	15	4	4	25	284

Note: This table includes all whistles from captive dolphins Spray and Scotty recorded on February 28, 1984, for which the dolphin producing the whistle could be identified. A indicates that the whistle could be heard in the air and its source located aurally by observers. B indicates that the LED telemetry device was seen to light up nearly completely for the dolphin originating the whistle, and none or only a few of the LED segments were seen to light up for the other dolphin. C indicates that the observers could not determine the responses of the telemetry device and that none or only a few of the LED segments were seen to light up for the other dolphin. D indicates that the LED telemetry device was seen to light up nearly completely for the dolphin that whistled, and the response of the other dolphin's device could not be determined.

Table 11.2.   Primary Whistle Types Produced by Spray and Scotty

	Kind of Whistle			
	TYPE 1	TYPE 2	OTHER	TOTAL
Spray	88	25	18	131
Scotty	33	73	47	153
Total	121	98	65	284

Note: Any contour consisting of enough of a primary contour to be unambiguously categorized was included as a primary contour. This includes short segments such as the 2C whistle. All secondary contours are here counted as "other."

audible in air were type 2 whistles (11 out of 12, the other was a RISE whistle), whereas 74 percent of her underwater whistles were type 1.

While the type 1 whistles from Scotty and Spray had very similar contours, there tended to be a consistent difference between the type 2 whistles of the two dolphins. Inspection of Spray's type 2 whistles on the upper and middle right of figure 11.3 shows that the signal is not pure tone for a short section of 2C; a series of sidebands are visible on the spectrogram. Figure 11.5 compares spectra and waveforms for two sections of the whistle shown in the middle right of figure 11.3, one section with no sidebands and one with sidebands. The spectrum on the upper left of figure 11.5 is typical of most dolphin whistles. There is one sharp spectral peak with no sidebands. The waveform on the lower left shows part of the waveform from which the spectrum was derived. There is very little amplitude modulation of this waveform. The spectrum on the upper right of figure 11.5 is from the 2C section of the whistle that has sidebands. This spectrum shows the sidebands as three spectral peaks separated by intervals of approximately 500 Hz. The associated waveform on the lower right shows that this section of whistle is amplitude modulated with a period of approximately 2 ms (frequency = 500 Hz). It is this amplitude modulation that produces the spectral sidebands (see Watkins 1967, for analysis of the spectra of similar waveforms).

Scotty's type 2 whistles on the upper and middle left of figure 11.3 do not have sidebands, although the whistle structure is blurred slightly due to reverberation in the pool. A systematic investigation of all whistles from the February 28, 1984 sample made by Spray and containing the 2C segment show that sidebands are present in 16 whistles, absent in 5, and indeterminate in 4. Only 3 of Scotty's whistles, including 2C, have side-

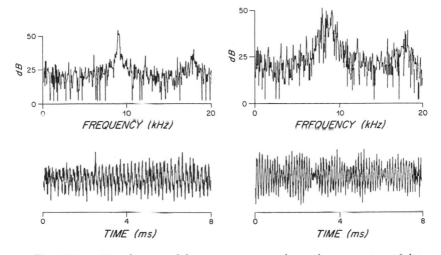

Fig. 11.5. Waveforms and frequency spectra of two short segments of the type 2 whistle produced by Spray and represented in the middle right of figure 11.2. The waveform and spectrum on the left are typical of most whistles, with one strong spectral peak and with little amplitude modulation of the waveform. The waveform and spectrum on the right are taken from the short section of the 2C segment of the whistle. This section of whistle shows sidebands on the spectrogram which are also visible as three spectral peaks on the upper right of this figure. These sidebands are produced by amplitude modulation of the waveform visible on the lower right of this figure. Waveforms and spectra were analyzed on a Rockland FFT/S Real-Time Spectrum Analyzer.

bands, 58 show no such structure, and 3 are indeterminate. Thus, the 2C whistles of Spray and Scotty tend to have predictable differences in acoustic structure.

There are striking differences in the distribution of whistles produced by Scotty when he was with Spray in 1984 compared to those produced after two years of isolation in 1987. Scotty produced no 1AB or 2ABC whistles in 1987. The only primary whistles he produced were 2C whistles and an abbreviated version of 2C, which only included the initial rise in frequency without the inflection and terminal decrease typical of 2C. These abbreviated 2C whistles are called 2C-RISE whistles. This 2C-RISE primary whistle was more stereotyped than the RISE secondary whistles. Not only were RISE secondary whistles highly variable but they also were not recognizable as part of either of the two primary whistles. The 70 whistles analyzed from Scotty when he was isolated were distributed as follows: 2C–20, 2C-RISE–15, FLAT–16, RISE–10, VAR–5, DOWN–3, SINE–1.

DISCUSSION

Categorization of Whistle Contours

Most studies since Dreher (1961) have categorized dolphin whistles on the basis of simple contour parameters such as duration, frequency at the start or end of a whistle, rise or fall in frequency, or number of maxima or minima in the contour frequency. Quantitative analyses of whistle structure (e.g., Graycar 1976; Gish 1979; Taruski 1979; Steiner 1981) have also relied on these absolute acoustic features more than configurational parameters, which are difficult to quantify. Caldwell et al. (1973) report that *Tursiops* and other dolphin genera frequently repeat a distinctive contour for a variable number of times with little break between repetitions. These repeated whistles were called multiloop whistles by the Caldwells. If these multiloop whistles are common, then acoustic measures of duration and number of maxima and minima in a whistle should be accompanied by consideration of the number of loops and the configuration of each loop (Caldwell et al. 1990).

Experiments where dolphins were trained to mimic man-made whistlelike sounds (Evans 1967, Richards et al. 1984) show that *Tursiops* sometimes transpose whistles to a different frequency range while maintaining other aspects of contour configuration. Similarly, as the upper two whistles on the left side of figure 11.3 indicate, Scotty maintained a recognizable type 2 whistle contour while varying whistle duration by 30 percent. Measurements of absolute frequency or duration of a whistle may thus also be misleading unless accompanied by examination of configurational properties of the contour.

In categorizing whistles for this study, I sorted spectrograms by eye. It was particularly easy to categorize primary whistles using contour configuration. Of the 284 whistles identified from a particular dolphin in the 1984 recordings, 77 percent could easily be sorted into two primary contour types. In this sense, these whistles appeared to be stereotyped, but exemplars of the same type were not identical. Deletions, variations in duration and absolute pitch, and even slight variations in contour configuration were common. This is similar to the results of spectrographic analyses of whistles reported by the Caldwells (Caldwell et al. 1990).

It is likely that variations within my categories of contour are discriminable to *Tursiops*. The Caldwells made the same point in their papers. Caldwell et al. (1970, 1990) do not argue that the *only* information broadcast by signature whistles was individual identity but suggest that variation within distinctive contour types might communicate other information (neither in this study nor in those of the Caldwells were data systematically gathered on the contexts in which particular whistle contours were produced).

The major difference between the whistle analyses of the Caldwells and those reported here is that no multiloop whistles were found in this study, except for SINE whistles. This may result from differences in the context in which whistles were recorded, for the Caldwells usually worked with isolated animals that often were removed from the water, while I worked with two dolphins swimming together in a pool. Gish (1979), who worked with physically isolated but undisturbed animals, also reported no multiloop whistles. The Caldwells reported that one individual *Tursiops* was *less* likely to produce multiloop whistles when disturbed. Over 50 percent of the whistles produced by this male dolphin were one loop or less during venipuncture when he was removed from the water, while 93.5 percent of his whistles had more than one loop when he was undisturbed in his isolation pool (Caldwell et al. 1970).

The lack of multiloop whistles reported here could also result, in part, from differences in sound analysis procedures. The Caldwells counted two contours separated by an interval of silence as one whistle if the two contours "carry the suggestion of 'one sound' to the percipient" (Caldwell et al. 1973), while I counted whistles separated by an obvious gap of silence on the spectrogram as two whistles. On the other hand, many other dolphins I have recorded do produce the multiple loop whistles described by the Caldwells. Some bottlenose dolphins may produce whistles with varying numbers of loops, while others may produce a single contour pattern that is not typically repeated without a break.

The Caldwells did not emphasize deletions from signature whistle contours in their earlier papers, but later papers include whistles of less than one loop. Deletions may have been even more common than these results suggest. Short frequency modulated whistles were called "chirps" by the Caldwells, and these were excluded from their whistle analyses. However, chirps might be a short segment of a longer stereotyped whistle contour (Caldwell and Caldwell 1970). Thus, what is called a segment of a stereotyped whistle here, for example 2C-UP, might have been called a chirp by the Caldwells.

Do Dolphins Mimic Each Other's Signature Whistles?

The two Sealand dolphins each produced stereotyped whistle contours similar to signature whistles described by the Caldwells. But the whistle repertoires recorded from these dolphins while they were interacting do not fit a narrow definition of the Caldwells' (M. C. Caldwell and D. K. Caldwell 1968) signature whistle hypothesis, that is, that more than 90 percent of the whistles from a dolphin conform to one individually distinctive contour. While the two dolphins favored different contour types, only 67 percent of Spray's whistles were of type 1 and only 48 percent of

Scotty's whistles were of type 2. Even more important, both dolphins produced both of the distinctive primary contour types. The second most common whistle of both dolphins was the other animal's favored whistle (table 11.2).

Two interpretations of this overlap in Scotty's and Spray's whistle repertoires are possible: either the animals simply shared a repertoire of stereotyped sounds and the signature whistle hypothesis did not hold for them or each dolphin was mimicking the other's signature whistle. Most of the data on *Tursiops* whistles indicates that they do not share a fixed repertoire of stereotyped whistles. Reviews of whistles from over one hundred of these dolphins (Graycar 1976, Caldwell and Caldwell 1979, Caldwell et al. 1990) have indicated that each adult dolphin, when isolated, produced a stereotyped individually distinctive whistle. Caldwell and Caldwell (1979) studied the ontogeny of whistling in captive *Tursiops*. Of the fourteen calves studied, twelve developed stereotyped whistles by the first year of age. Apparently, each calf developed only one stereotyped contour type. Stereotyped whistles in *Tursiops* thus do not appear to be drawn from a fixed repertoire of stereotyped whistles shared by conspecifics.

The second case, that for mimicry, is based on the exceptional abilities of dolphins for vocal mimicry. *Tursiops* has been shown to mimic manmade sounds spontaneously (Lilly 1965; Caldwell and Caldwell 1973; Herman 1980). These dolphins also have been trained to mimic manmade whistlelike sounds on command (Evans 1967, Richards et al. 1984).

Captive bottlenose dolphins appear to imitate each others' signature whistles at varying rates depending on their social context. The Caldwells found that over one hundred *Tursiops* individuals produced only their own signature whistles primarily under conditions when the dolphins were isolated from conspecifics. The electronic acoustic link experiments provide evidence that when *Tursiops* are physically isolated but in acoustic contact, they mimic signature whistles at rates of approximately 1 percent (Burdin et al. 1975, Gish 1979). The results reported here indicate that when captive *Tursiops* are interacting normally, they may imitate each others' signature whistles at rates as high as 20 percent.

Since individual *Tursiops* do not share a large number of distinctive stereotyped whistles, if one animal produced a stereotyped whistle more typically produced by a different individual, this would seem to represent whistle mimicry. However, proof that the overlap in primary whistles represents vocal mimicry of individually distinctive stereotyped whistles will require study of the vocal repertoires of dolphins before and after they come into acoustic contact.

The data on whistles recorded from Scotty two years after Spray's death support the mimicry hypothesis. After two years of isolation, Scotty did not produce any of the type 1 whistles (which were Spray's favored whistle), but half of the whistles he produced were categorized as type 2 whistles (which had been his favored whistle). Scotty's whistle repertoire changed in other ways as well after two years of isolation. When Scotty and Spray were together and were not wearing vocalights, their combined rate of whistling was 7.6 whistles/min. When I could identify which dolphin produced a whistle, Scotty whistled more than Spray. Thus, it would be reasonable to estimate that Scotty produced about half the combined rate, or 3.8 whistles/min., when Scotty and Spray were together. This is more than double the 1.6 whistle/min. rate recorded from Scotty after two years of isolation.

Not only did Scotty whistle less frequently after isolation but he also tended to produce short abbreviated whistles. He did not produce any of the 2ABC or 1AB whistles that dominated his repertoire when he was with Spray in 1984. The most common whistle in 1987 was the 2C whistle, which is only a short fragment of the full 2ABC whistle. By 1987, Scotty even introduced an abbreviated version of the 2C whistle which consisted of just the rapid upsweep at the start of the 2C whistle.

After two years of isolation, Scotty appeared to devote less effort to whistling than when he associated with Spray. He did not imitate her type 1 whistle, he produced quieter whistles, and he whistled less frequently. When he did whistle, he tended to produce short abbreviated whistles. This reduction of whistling in an isolated dolphin is consistent with the notion that the primary function of dolphin whistles is social communication.

Possible Social Functions of Tursiops Whistles

An extensive body of research emphasizes the social functions of dolphin whistles. Those dolphin species known to whistle are more social than nonwhistling species (Herman and Tavolga 1980). Previous research on the social functions of dolphin whistling indicates that whistles function as contact calls: they are used to establish or maintain vocal or physical contact between dolphins. For example, many studies have shown that dolphins whistle when separated from parents, offspring, or other group members (McBride and Hebb 1948; McBride and Kritzler 1951; Wood 1954). Dolphins are reported to respond to whistles either by whistling themselves (Lilly and Miller 1961; Lang and Smith 1965; D. K. Caldwell and M. C. Caldwell 1968; M. C. Caldwell and D. K. Caldwell 1968; Gish 1979) or by approaching the whistler (Lilly 1963, D. K. Caldwell and M. C. Caldwell 1968). When separated, female dolphins and their young calves are reported to whistle frequently until they are reunited (McBride and Kritzler 1951, Sayigh et al. 1989).

Dolphins do not always show an immediate response to a whistle from another member of their group, however. There are often periods when there is a steady background of whistling in a group of dolphins or a sudden chorusing of whistles when many animals whistle simultaneously. If each animal within a group has learned to associate the appropriate signature whistles with each group member, then this frequent repetition of each animal's own signature whistle may function to maintain the spatial coordination of individuals within the group when animals are out of sight of one another. The production of signature whistles in this context might be expected not to evoke immediately obvious responses.

Wild dolphins form individual specific social relationships (Tyack 1986b). *Tursiops truncatus* is the species for which the social structure has been best defined in the wild (Shane et al. 1986). The composition of most groups is characterized by the stability of bonds between particular individuals within more fluid larger groups. For example, *Tursiops* mothers are almost always sighted with their young calves. Some pairs of *Tursiops* males are also sighted together for years (see chap. 6, this vol.). But each mother-calf pair or male-male pair is sighted in different larger groups. How can a dolphin signal to one particular individual in the midst of a large group?

Mimicry of signature whistles is one obvious way in which dolphins might solve this problem. I have suggested that a dolphin may imitate the signature whistle of a specific individual within a social group to initiate social interactions with that individual (Tyack 1986a). This interpretation of whistle mimicry may be called the signature labeling hypothesis. Similar behavior has been reported for birds. Gwinner and Kneutgen (1962) and Thorpe and North (1966) have shown that mated pairs in several monogamous species of birds can signal their mate to approach by imitating vocalizations typical of the mate. No species of dolphin is known to be monogamous, but signature labeling would be equally important in maintaining other individual specific social relationships in dolphins.

If the signature labeling hypothesis is correct, then imitations of signature whistles should evoke specific responses. If dolphins mimic the signature whistles of members of their groups to initiate social interaction with those particular individuals, then one might predict that the animal whose signature whistle has been mimicked would be likely to respond vocally or to approach the animal that imitated its whistle. Other individuals would be expected not to show this response. The frequent reports of dolphins imitating man-made sounds or imitating the movement patterns of other animals (e.g., Tayler and Saayman 1973) suggest that imitation is also used in an open-ended fashion by these animals. While imitation may have a specific role, for example, in signature labeling, it also appears at times to be a form of play. Imitative learning may also have a more general role in the development of behavior in young dolphins.

Is the original signature whistle hypothesis compatible with signature labeling and with the rates of whistle mimicry reported here? If dolphins only produced their own signature whistle, then the task of learning to associate each whistle with the appropriate dolphin would be simpler than if dolphins also imitated each other's whistles. Mimicry of signature whistles need not interfere with the association of these whistles with their hypothesized referent, for example, if the mimicked whistles occur significantly less often than those produced by the "appropriate" animal, or if the mimicked whistles have a similar contour but include some acoustic features that are different from those produced by the "appropriate" animal. In the present study, each dolphin tended to produce an individual variant of each primary whistle. In addition, Spray's type 2 whistles tended to include a section with sidebands, a feature that was very rare in Scotty's type 2 whistles.

The Caldwells suggested that the function of the signature whistle was to broadcast the identity of the whistler to other members of its group (M. C. Caldwell and D. K. Caldwell 1968). For this to occur, each member of a social group would have to learn to associate each signature whistle with the individual that produced it. Do dolphins have the cognitive skills required for signature labeling? The experiments of Richards et al. (1984) demonstrated that dolphins are not only highly skilled at vocal mimicry but also at vocal labeling. After training a dolphin to mimic a variety of man-made whistlelike sounds, Richards et al. showed that a female bottlenose dolphin could learn to associate these man-made sounds with different objects and to produce the appropriate sound when shown one of these different objects. In these experiments, the dolphins were able to label the objects vocally.

Herman et al. (1984) have trained dolphins to recognize these computer-generated sounds and to associate them with objects or actions. While Herman has not specifically tested the dolphins' understanding of whether these acoustic labels function as symbols representing a class of objects (as has been performed with chimpanzees [*Pan troglodytes*]; Savage-Rumbaugh et al. 1980), Herman et al. (p. 207) summarize their results on reference as follows: "The concept that signs stand for referents seems to come easily to the dolphins." It is possible, as Herman (1980) argues, that the training procedures used in these experiments have allowed dolphins to develop cognitive skills not realized in the wild. But as Richards (1986) argues, it seems just as likely that dolphins have evolved these skills because of their functions in natural systems of dolphin communication. My results suggest, for example, that dolphins may imitate signature whistles to establish contact with particular individuals within their social group.

REFERENCES

Burdin, V. I., A. M. Reznik, V. M. Skornyakov, and A. G. Chupakov. 1975. Communication signals of the Black Sea bottlenose dolphin. Sov. Phys. Acous. 20:314–318.

Caldwell, D. K., and M. C. Caldwell. 1968. The dolphin observed. Natural History 77:58–65.

Caldwell, M. C., and D. K. Caldwell. 1965. Individualized whistle contours in bottlenosed dolphins (*Tursiops truncatus*). Nature 207: 434–435.

———. 1968. Vocalization of naive captive dolphins in small groups. Science 159:1121–1123.

———. 1970. Etiology of the chirp sounds emitted by the Atlantic bottlenosed dolphin: A controversial issue. Underwater Naturalist 6:6–9.

———. 1971. Statistical evidence for individual signature whistles in Pacific whitesided dolphins, *Lagenorhynchus obliquidens*. In: Cetology 3:1–9.

———. 1973. Vocal mimicry in the whistle mode by an Atlantic bottlenosed dolphin. Cetology 9:1–8.

———. 1979. The whistle of the Atlantic bottlenosed dolphin (*Tursiops truncatus*)—ontogeny. In: Behavior of marine animals. 3: Cetaceans, ed. H. E. Winn and B. L. Olla. New York: Plenum. 369–401.

Caldwell, M. C., D. K. Caldwell, and N. R. Hall. 1969. Ability of an Atlantic bottlenosed dolphin to discriminate between whistles of other individuals of the same species. Technical Report No. 6. Los Angeles County Museum of Natural History Foundation.

———. 1972. Ability of an Atlantic bottlenosed dolphin (*Tursiops truncatus*) to discriminate between, and potentially identify to individual, the whistles of another species, the common dolphin (*Delphinus delphis*). Technical Report No. 9. Marineland Research Laboratory, St. Augustine, Fla.

Caldwell, M. C., D. K. Caldwell, and J. F. Miller. 1973. Statistical evidence for individual signature whistles in the spotted dolphin, *Stenella plagiodon*. Cetology 16:1–21.

Caldwell, M. C., D. K. Caldwell, and R. H. Turner. 1970. Statistical analysis of the signature whistle of an Atlantic bottlenosed dolphin with correlations between vocal changes and level of arousal. Technical Report No. 8. Los Angeles County Museum of Natural History Foundation.

Caldwell, M. C., D. K. Caldwell, and P. L. Tyack. 1990. A review of the signature whistle hypothesis for the Atlantic bottlenosed dolphin, *Tursiops truncatus*. In: The bottlenose dolphin, eds. S. Leatherwood and R. Reeves. San Diego: Academic Press. 199–234.

Caldwell, M. C., N. R. Hall, and D. K. Caldwell. 1972. Ability of an Atlantic bottlenosed dolphin to discriminate between, and respond differentially to, whistles of eight conspecifics. Technical Report No. 10. Marineland Research Laboratory, St. Augustine, Fla.

Cheney, D. L., and R. M. Seyfarth. 1980. Vocal recognition in free-ranging vervet monkeys. Animal Behav. 28:362–367.

Clark, C. W. 1980. A real-time direction finding device for determining the bearing to the underwater sounds of southern right whales, *Eubalaena australis*. J. Acoust. Soc. Am. 68:508–511.

Clark, C. W., W. T. Ellison, and K. Beeman. 1986. A preliminary account of the acoustic study conducted during the 1985 spring bowhead whale, *Balaena mysticetus*, migration of Point Barrow, Alaska. Rep. Int. Whal. Comm. 36:311–316.

Dreher, J. J. 1961. Linguistic considerations of porpoise sounds. J. Acoust. Soc. Am. 33:1799–1800.

———. 1966. Cetacean communication: Small group experiment. In: Whales, dolphins, and porpoises, ed. K. S. Norris. Berkeley and Los Angeles: University of California Press. 529–543.

Dreher, J. J., and W. E. Evans. 1964. Cetacean communication. In: Marine bioacoustics, vol. 1., ed. W. N. Tavolga. Oxford: Pergamon. 373–393.

Evans, W. E. 1967. Vocalization among marine mammals. In: Marine bioacoustics, vol. 2, ed. W. N. Tavolga. Oxford: Pergamon. 159–186.

Evans, W. E., and W. W. Sutherland. 1963. Potential for telemetry in studies of aquatic animal communication. In: Biotelemetry, ed. L. E. Slater. Oxford: Pergamon. 217–224.

Gish, S. L. 1979. A quantitative description of two-way acoustic communication between captive Atlantic bottlenosed dolphins (*Tursiops truncatus* Montagu). Ph.D. dissertation, University of California, Santa Cruz. University Microfilms, Ann Arbor.

Graycar, P. J. 1976. Whistle dialects of the Atlantic bottlenosed dolphin, *Tursiops truncatus*. Ph.D. dissertation, University of Florida. University Microfilms, Ann Arbor.

Gwinner, E., and J. Kneutgen. 1962. Über die biologische Bedeutung der "zweckdienlichen" Anwendung erlernter Laute bei Vögeln. Zeitschrift für Tierpsychologie 19:692–696.

Herman, L. M. 1980. Cognitive characteristics of dolphins. In: Cetacean behavior: Mechanisms and functions, ed. L. M. Herman. New York: John Wiley and Sons. 363–429.

Herman, L. M., D. G. Richards, and J. P. Wolz. 1984. Comprehension of sentences by bottlenosed dolphins. Cognition 16:129–219.

Herman, L. M., and W. N. Tavolga. 1980. The communication systems of cetaceans. In: Cetacean behavior: Mechanisms and functions, ed. L. M. Herman. New York: John Wiley and Sons. 149–210.

Lang, T. G., and H. A. P. Smith. 1965. Communication between dolphins in separate tanks by way of an electronic acoustic link. Science 150:1839–1844.

Lilly, J. C. 1963. Distress call of the bottlenosed dolphin: Stimuli and evoked behavioral responses. Science 139:116–118.

———. 1965. Vocal mimicry in *Tursiops*: Ability to match numbers and durations of human vocal bursts. Science 147:300–301.

Lilly, J. C., and A. M. Miller. 1961. Vocal exchanges between dolphins. Science 134:1873–1876.

McBride, A. F., and D. O. Hebb. 1948. Behavior of the captive bottlenose dolphin, *Tursiops truncatus*. J. Comp. Physiol. Psych. 41:111–123.

McBride, A. F., and H. Kritzler. 1951. Observations on pregnancy, parturition, and postnatal behavior in the bottlenose dolphin. J. Mammal. 32:251–266.

Patterson, B., and G. R. Hamilton. 1964. Repetitive 20 cycle per second biological hydroacoustic signals at Bermuda. In: Marine Bioacoustics, vol. 1, ed. W. N. Tavolga. Oxford: Pergamon. 125–145.

Popper, A. N. 1980. Sound emission and detection by delphinids. In: Cetacean behavior: Mechanisms and functions, ed. L. M. Herman. New York: John Wiley and Sons. 1–52.

Richards, D. G. 1986. Dolphin vocal mimicry and vocal object labeling. In: Dolphin cognition and behavior: A comparative approach, ed. R. J. Schusterman, J. A. Thomas, and F. G. Wood. Hillsdale, N.J.: Lawrence Erlbaum Associates. 273–288.

Richards, D. G., J. P. Wolz, and L. M. Herman. 1984. Vocal mimicry of computer-generated sounds and vocal labeling of objects by a bottlenosed dolphin, *Tursiops truncatus*. J. Comp. Psychol. 98:10–28.

Savage-Rumbaugh, E. S., D. M. Rumbaugh, S. T. Smith, and J. Lawson. 1980. Reference: The linguistic essential. Science 210:922–925.

Sayigh, L. S., P. Tyack, M. D. Scott, and R. S. Wells. 1989. Signature whistles in free-ranging bottlenosed dolphins, *Tursiops truncatus:* Stability and mother-offspring comparisons. Behav. Ecol. Sociobiol. 26:247–260.

Shane, S. H., R. S. Wells, and B. Würsig. 1986. Ecology, behavior and social organization of the bottlenose dolphin: A review. Marine Mammal Science 2:34–63.

Siegel, S. 1956. Nonparametric statistics for the behavioral sciences. New York: McGraw Hill.

Steiner, W. W. 1981. Species-specific differences in pure tonal whistle vocalizations of five western North Atlantic dolphin species. Behav. Ecol. Sociobiol. 9:241–246.

Taruski, A. G. 1979. The whistle repertoire of the North Atlantic pilot whale (*Globicephala melaena*) and its relationship to behavior and environment. In: Behavior of marine animals. 3: Cetaceans, ed. H. E. Winn and B. L. Olla. New York: Plenum. 345–368.

Tayler, C. K., and G. S. Saayman. 1973. Imitative behaviour by Indian Ocean bottlenosed dolphins (*Tursiops aduncus*) in captivity. Behaviour 44:286–298.

Thorpe, W. H., and M. E. W. North. 1966. Vocal imitation in the tropical boubou shrike, *Lanarius aethiopicus,* major as a means of establishing and maintaining social bonds. Ibis 108:432–435.

Tyack, P. 1985. An optical telemetry device to identify which dolphin produces a sound. J. Acoust. Soc. Am. 78:1892–1895.

———. 1986a. Whistle repertoires of two bottlenosed dolphins, *Tursiops truncatus:* Mimicry of signature whistles? Behav. Ecol. Sociobiol. 18:251–257.

———. 1986b. Population biology, social behavior, and communication in whales and dolphins. Trends in Ecology and Evolution 1:144–150.

Watkins, W. A. 1967. The harmonic interval: Fact or artifact in spectral analysis of pulse trains. In: Marine bioacoustics, vol. 2, ed. W. N. Tavolga. London: Pergamon. 15–43.

Watkins, W. A., and W. E. Schevill. 1972. Sound source location by arrival times on a non-rigid three-dimensional hydrophone array. Deep-Sea Research 19:691–706.

———. 1974. Listening to Hawaiian spinner porpoises, *Stenella longirostris,* with a three-dimensional hydrophone array. J. Mammal. 55:319–328.

———. 1977. Sperm whale codas. J. Acoust. Soc. Am. 62:1485–1490.

Watkins, W. A., and D. Wartzok. 1985. Sensory biophysics of marine mammals. Marine Mammal Science 1:219–260.

Wells, R. S., M. D. Scott, and A. B. Irvine. 1987. The social structure of free-ranging bottlenose dolphins. Current Mammalogy 1:247–305.

Wood, F. G. 1954. Underwater sound production and concurrent behavior of captive porpoises, *Tursiops truncatus* and *Stenella plagiodon*. Bull. Mar. Sci. Gulf and Carib. 3:120–133.

THE DOMESTIC DOLPHIN

Karen Pryor

Whenever we humans have moved into a new habitat, we have tamed and made use of animals living there already, animals whose skills in that environment exceed our own: pigs in the jungle, camels in the desert, Himalayan yaks. Some of these domesticated species, sheep, for example, have been so changed by selective breeding that similar species no longer exist in the wild. Other working animals—the falcon, the elephant—are unchanged and, in fact, are sometimes still captured from the wild and then trained to perform work for us. In the papers of Louis Herman and Patrick Moore, which follow, you will get a close look at wild-caught dolphins that have become willing, even dedicated co-workers in extremely complex research tasks.

Metaphorically, domestication is a trade-off. An animal gives up its freedom and wild companions and contributes produce or work of some sort; in return, it is fed and kept safe from predators, thus escaping the two biggest problems of life in the wild, going hungry and getting eaten. Many species have proved quite willing to make this bargain, and the bottlenose dolphin, in my experience, is one of them. I have even seen the trade-off acted out: for some months, at the Oceanic Institute in Hawaii, we kept a pair of bottlenose dolphins in a pen next to a pier in the ocean, a pen they could jump in and out of at will. They spent most of the day loose, playing at the bows of boats coming in and out of the little harbor, but they spent the nights in the pen, jumping back in at 5:00 P.M. when a trainer showed up with their suppers. And they were once seen to jump hastily into their pen when a large hammerhead shark cruised under the pier.

Living in an environment that to us is so alien, the dolphin can perform tasks that we, even with all our technology, find difficult. The dolphin, for example, is superb at finding lost objects underwater. We taught one of our free-swimming Institute dolphins to look for objects in the waters around the pier and report them to us. Our criterion was "anything man-made and bigger than a breadbox." I do not know how the dolphin defined that, but he soon found engine blocks, a movie camera, quite a lot of fishing equipment, and a World War II airplane.

And apparently the animal enjoyed being asked to do this. Domestication may require an animal to give up the extensive social contact of a large wild group, but working animals, at least, gain the benefit of interesting things to do (gun dogs and cutting horses, for example, appear to be fascinated and exhilarated by their work). Whatever we may think of the merits of what dolphins do in oceanarium shows, for the animals, it can be challenging; I have seen a dolphin, striving to master an athletically difficult trick, actually refuse to eat its "reward" fish until it got the stunt right. When the U.S. Navy released the news that it was using dolphins to search for mines in the Persian Gulf, a reporter asked me if I thought the dolphins would find the work arduous or unpleasant. Knowing the character of the animals and the skills of the navy trainers, I could answer instantly, "Are you kidding? They love it."

It has been suggested in the press that dolphins could be used to sabotage ships; I regard this as unrealistic, if only because the highly streamlined dolphin cannot carry much of a payload. Dolphins could be used, however, to detect the presence of any underwater activity; for example, they could be used to notify swimmers of the presence of sharks. And the echolocation skills have real potential; dolphins would be the partners of choice in searching shallow Caribbean waters for a wrecked treasure ship buried in the sand.

The U.S. Marine Mammal Protection Act prohibits the use of dolphins for any such commercial purpose. Under this carefully administered federal law, dolphins may only be kept for public display, research, or, in the case of the navy, national defense. And a working relationship with dolphins is not something to be entered into lightly.

Dolphins are extremely expensive to maintain. They need a lot of food and room and care. A working dolphin requires a full-time trainer, and not just anyone can be a good trainer. Veterinary supervision alone calls for enormous expertise. But we have developed this expertise. And as Moore describes, we are beginning to breed dolphins successfully in captivity. The dolphin, primarily the bottlenose dolphin, may well prove to be the newest large animal in our history to accept domestication.

Taking a break from the computer, Louis Herman gives his research dolphin, Phoenix, the gestural command for "Jump person." (Photo by Alan Levenson, Kewalo Basin Marine Mammal Laboratory.)

CHAPTER TWELVE

WHAT THE DOLPHIN KNOWS, OR MIGHT KNOW, IN ITS NATURAL WORLD

Louis M. Herman

INTRODUCTION

The major sensory interface of the dolphin with its world is through hearing, both passive listening and echoranging, and through vision (Herman and Tavolga 1980). It is primarily through these two senses that the dolphin creates a representation of the world in which it lives and of the relationships and contingencies within that world. Both hearing and vision are highly developed and specialized for the aquatic medium. Reviews of hearing and sound production can be found in several sources (e.g., Norris 1980; Popper 1980; Ralston and Herman 1989) as can reviews of visual capabilities (e.g., Dral 1977; Dawson 1980; Madsen and Herman 1980). More general reviews of the sensory systems of dolphins are also available (e.g., Caldwell and Caldwell 1977; Kinne 1975; Herman and Tavolga 1980; Nachtigall 1980). A further review of sensory systems is not within the scope of this chapter. Instead, in the following sections, I will briefly review some of what has been revealed about cognition and cognitive processes in dolphins through laboratory studies of auditory and visual information processing and related topics. A concluding section will relate these findings to the life of the wild dolphin.

Preparation of this chapter was supported by Contract N00014-85-K-0210 from the Office of Naval Research and by a grant from the Center for Field Research (Earthwatch).

AUDITORY INFORMATION PROCESSING

Few would be surprised to learn that dolphins are facile at processing auditory information. This expectation is consonant with the remarkable development of the dolphin's hearing and sound production systems. In several earlier reviews (e.g., Herman 1980, 1986), I described some abilities of bottlenose dolphins when carrying out auditory-based cognitive tasks—tasks in which information arrives through the auditory sensory system. These auditory tasks revealed, for example, that the dolphin readily forms and applies rules about relationships among auditory events. Relational rules are rules that generalize to the whole class of problems represented by the particular exemplar or two that may be used in training the animal. For instance, a relational rule governed the correct choices of a dolphin when given pairings of new sounds (Herman and Gordon 1974) or when given a new pairing of old sounds (Herman 1975, Herman and Thompson 1982). In both cases, the correct element of a pair of sounds was determined by a "sample" sound available before the new pair of sounds was presented. In another case, the positive or negative outcome of the immediately preceding choice by the dolphin determined for it which sound to choose on the current test (Herman and Arbeit 1973; Herman 1980, fig. 8.12). The correct responses by the dolphin indicated that the appropriate relational rule for the task at hand had been adopted.

In these same types of tasks, or similar ones, the excellent fidelity of the dolphin's auditory memory was revealed. This included the ability to store new auditory information (Herman and Gordon 1974) as well as to update old information rapidly (Herman 1975, Herman and Thompson 1982). In one particularly demanding task, the dolphin was asked to listen to a list of as many as eight different short sounds and then decide whether a subsequent "probe" sound was or was not a member of that list (Herman 1980, Thompson and Herman 1977). As might be expected, the ability to classify sounds correctly as "old" (a member of the list) or as "new" (not on the list) declined with list length, from well over 90 percent correct for lists of one or two sounds to over 80 percent correct for lists of three or four sounds and to about 70 percent correct for lists of six or eight sounds. The dolphin remembered the sounds near the end of the list best and those nearer the beginning most poorly. This "recency" effect has also been found in analogous memory tasks with humans (e.g., Wickelgren and Norman 1966).

In other auditory work, we found that a dolphin was capable of vocally imitating arbitrary sounds broadcast into its tank through an underwater speaker (Richards, Wolz, and Herman 1984; Richards 1986). Vocal mimicry of arbitrary sounds is rare among nonhuman mammals. The dolphin was capable of reproducing frequency, frequency modulation, amplitude modulation, and pulsed waveforms. The same dolphin was also able to use some of the same sounds it had imitated to label vocally any of six different objects shown it. Thus, if shown a hoop or a ball, it reliably reproduced a learned whistle sound for each, in effect, vocally "naming" these objects.

Perhaps the most impressive auditory accomplishment was the ability of a dolphin to learn to understand sentences expressed within a simple, artificial acoustical language we created (Herman et al. 1984; Herman 1986, 1987). The words of the language were sounds generated by a computer and were broadcast into the dolphin's tank through an underwater speaker. The words referred to objects in the tank, to actions that might be taken to those objects, and to modifiers of place or location. Using a set of syntactic rules governing how words could be combined into "legal" sentences, imperative sentences of two or more words in length were constructed. These imperatives required the dolphin to take named actions relative to named objects and their named modifiers. Understanding was measured by the accuracy and reliability in carrying out the instructions. An important finding, revealing the dolphin's mastery of the sentence forms used, was that understanding was shown for novel instructions as well as for more familiar ones, with only a slight advantage to the latter. Novel instructions consisted of new combinations of words that obeyed the syntactic rules of the language or, in a few cases, new but logical extensions of existing syntactic rules. Except for some anomalous sentence forms deliberately given the dolphin, novel sentences were always meaningful, in that the instructions could be carried out by the dolphin. In most cases, the dolphin was able to distinguish an anomalous form from a correct form. Both the semantic and the syntactic features of sentences were taken into account in making a correct interpretation of a sentence. Syntactic information was used, for example, to distinguish between instructions that used the same words in different sequences, as in the semantic contrasts "take the surfboard to the frisbee" versus "take the frisbee to the surfboard."

These are but a sample of the findings from laboratory studies of auditory information processing, but they illustrate the competency of the dolphin in carrying out complex auditory tasks. With the obvious exception of *Homo sapiens,* few mammals show a facility for performing well on complex tasks when task information is solely auditory. Although hearing is well developed in mammals, the larger-sized, relatively large-brained terrestrial mammals tend to occupy diurnal niches where vision may dominate. The nonhuman primate, a visual specialist, provides a good example. Performance of monkeys or apes on visual tasks is generally much superior to performance on auditory tasks (e.g., D'Amato 1973; cf. D'Amato and Colombo 1986). However, when performance of nonhuman primates (e.g., species of *Cebus*) on visually based cognitive tasks is compared with the performance of *Tursiops* on auditory-based cognitive tasks, the overlaps in capabilities and in limitations are remarkable, suggesting a convergence of information processing skills in these different species, although mediated by different sensory systems (Herman 1980).

VISUAL INFORMATION PROCESSING

The visual system of the dolphin has often been underrated. This is traceable, perhaps, to the early analyses of visual anatomists who, while granting the cetaceans adequate underwater vision, suggested that the dolphin eye and the whale eye were highly myopic in air. Gordon Walls (1942), in his classic text *The Vertebrate Eye*, stated in reference to the behavior of porpoises in the wild, "Their rolling and frequent breaching is mere exuberance, and the eye is probably as completely useless in air as is that of a mysticete" (412). The later findings of an echolocation sense in dolphins (e.g., Schevill and Lawrence 1953, Kellogg 1958) only served to reinforce the view that vision was largely secondary to passive and active hearing in the life of a dolphin. More recent findings, however, suggest that this view needs to be corrected.

We now know that dolphin visual acuity in air is comparable to acuity underwater, although the best viewing distances differ in the two media (Herman et al. 1975). Acuity is best at far distances (ca. 2.5 m or greater) in air and at near distances in water (ca. 1 m or less), exactly as it should be given the viewing conditions in the two media and the availability of discernible visual targets at those ranges. In air, above the ocean's surface, primarily distant scenes intrude on the otherwise featureless environment: flocking birds signaling the presence of fish schools underneath, leaping dolphins splashing at the surface, the topographic features of adjacent landmasses, or the presence of shoaling water. Underwater, the visual scene is limited in range and resolution by the scattering and diffraction of light by small suspended particles, by local refractions arising from thermoclines, and by other factors. Within the limits of the restricted in-water visual range, however, many visual targets of interest may appear: fish, schoolmates, other life forms, and local physiography, to name some.

In the initial visual learning work with dolphins, constraints were noted on visually based performance. In an early review (Herman 1980), I reported the difficulty in teaching a dolphin to match one of two visual alternatives to a visual "sample" when the materials were static, two-dimensional shapes or simple brightness differences. Although learning was possible, it developed slowly at best. Good performance occurred only when the visual materials were mapped onto unique sounds, in effect, giving them auditory names (for a detailed report, see Forestell and Herman 1988). Other investigators (e.g., Chun 1978) encountered similar difficulties with the use of visual materials and found little or no generalization to new problems of the same type as the training problem. At the same time, there was developing evidence from our laboratory of the facility of a dolphin in learning to understand a visually based language in which words were represented by the gestures of a trainer's arms and hands (Herman 1980, Herman et al. 1984). This "gestural" language was the visual analogue to the auditory-based language described earlier. The trainer's gestures were varied, complex, and often contained subtle components. Shyan and Herman (1987) showed that the dolphin was able to extract certain critical features from each prototypical gesture which enabled it to recognize a gesture even when it deviated to a large degree from the prototype. The results of this work led us to speculate that the pattern in time traced by a gesture was the essential characteristic that enabled the dolphin to perform so well (Herman et al. 1984). We reasoned that gestures, like sounds, were temporal patterns—whose form unfolded in time—and that the dolphin might be specialized for the perception and classification of such patterns. The temporal pattern was, in fact, found to be one of several salient features of signs for the dolphin trained in the gestural language (Shyan and Herman 1987).

More recent findings have led to the revision of these views. Initially, we discovered that within the visually based language, an object could be substituted for its symbol, in simple two-word Object + Action sentences. For instance, the two-word sentence instructing the dolphin to jump over a ball is expressed as "Ball Over," with each word represented by a unique gesture. In response, the dolphin almost invariably executes the correct behavior. In the substitution procedure, a real ball is shown (in air) to the dolphin in place of the gestural symbol for "ball" and is followed immediately by the gestural symbol for "over." Again, the dolphin executes the correct behavior. It goes to a ball floating in the tank and jumps over it. The dolphin has used the displayed ball to select another ball. This result is similar to what one finds in a "matching-to-sample" procedure when an animal selects the one alternative, from among two or more offered, that matches a displayed sample stimulus. We therefore carried out a series of matching-to-sample studies in which we demonstrated the ability of the dolphin to recognize almost any three-dimensional object shown it as a sample, including objects new to its experience (Herman et al. 1989). Furthermore, the dolphin was able to select the correct alternative even when delays of over one minute intervened between seeing the sample and seeing the alternatives. Clearly, the dolphin had an excellent short-term memory for things seen, roughly comparable in fidelity to its memory for sounds heard (Herman and Gordon 1974). Further work showed that the method of displaying the sample, whether static or dynamic, affected performance level only marginally. There was only a slight advantage when the sample was moved across the dolphin's field of view (dynamic) during its presentation as contrasted with it being held stationary (static). Hence, there was little support here for the idea that temporal patterning was an important variable in visual performance.

The next step involved retesting the ability to match simple two-dimensional geometric forms similar to those used with little success in earlier studies (Forestell and Herman 1988). We found that as long as such forms were held in air, so that the background was essentially the broad expanse of the sky, that visual matching proceeded successfully. However, when we placed each figure against its own background, consisting of a white square that stood in contrast to the black figure, performance declined to chance levels. It appeared as if the dolphin perceived both figure and background as a unitary percept, and attended to the whole rather than to the black figures alone. In consequence, the different figures, each on its own discrete background identical to other backgrounds, all looked very much alike. The solution we found was to place all figures on one large common background. With that done, the dolphin's performance returned to high levels, and she was able to match geometric forms or abstract forms easily, even when delays were imposed between removal of the sample and the opportunity to select an alternative (Hunter 1988).

To return to the findings with the gestural language itself, performance was in almost all respects comparable to performance on the acoustic language. In fact, over several years of tutoring in the two languages, the dolphin trained in the visual form has shown greater stability in performance than the dolphin trained in the acoustic form. Because dolphins and languages are inseparable statistically, we cannot tell which factor is the major contributor to stability.

Within the gestural language, we went on to test for the understanding of question forms (interrogatives) as well as of imperatives (Herman and Forestell 1985). We developed a "reporting" procedure in which the dolphin was asked binary questions about the presence or absence of named objects in her tank. In reply, the dolphin had only to press one of two paddles, one meaning "Yes" (object is present) and the other "No" (object is absent). For example, the two-word sentence "Ball Question" means "Is there a ball in the tank?" Several different procedures were used to test for understanding. The test took the form of showing the dolphin one, two, or three different objects and placing each in her tank. She was then asked a question referring either to one of those objects or to an object not in the tank. The dolphin reliably reported presence and absence, with comparable accuracy, although in each case performance was better with fewer objects placed in the tank. This suggested that the dolphin was maintaining an inventory in memory of the objects placed in her tank rather than conducting an active search for them. In other types of reporting, however, the dolphin did conduct an active search, resulting in levels of performance that were independent of the number of objects in the tank. A major theoretical result of this study is that the dolphin was able to understand references to absent objects as well as to objects that were present. A practical result is that one can interrogate a dolphin about the contents of its physical world and receive accurate reports of those contents.

SOME OTHER TYPES OF LEARNED CONCEPTS

An important part of the communicative process is the ability to refer to objects or events in the external world. In previous sections, I described the dolphin's understanding of referents given within the acoustic or gestural languages. We also tested the dolphin's ability to understand referents in a simpler context, through the use of indicative gestures. Simply put, could the dolphin understand a pointing gesture by a human? Our informal observations during training the dolphin to go to a particular object or response paddle suggested this was possible. We therefore devised a simple test in which we placed three objects behind the dolphin and to its left, center, or right relative to the dolphin as it faced the trainer. The trainer then pointed to one of the three objects and rewarded the dolphin for going to the indicated object. Our two language-trained dolphins were the subjects. The dolphins were highly accurate in orienting toward the left and right objects, about 94 percent correct for one dolphin and 80 percent correct for the second. The center object yielded only chance performance, suggesting that fine discriminations were difficult. A further finding was that the final destination was much less accurate than the initial orientation, indicating that the dolphin was influenced by factors occurring after the pointing signal. Object preferences may have been one such influence, but the data have not yet been analyzed at that level of detail.

In other work, Pryor, Haag, and O'Reilly (1969) demonstrated the ability of a rough-toothed dolphin to create behaviors in response to a reinforcement contingency that the current behavior should be different from that performed previously. We have verified this capability for our bottlenose dolphins and have placed the creative act under control of a gesture. In response to the gesture, the dolphin "volunteers" behaviors. This is in contrast to other gestures that call for a specific behavior. Many different behaviors are volunteered, although preferences for certain behaviors typically occur. The preferred behaviors tend to be offered first but are followed by other behaviors if the gestural sign is repeated.

Related to this innovative skill is the ability of the dolphins to combine two different gestural action signs to perform a single coordinated behavior. For example, there is a unique gesture for the action of spitting water and another for executing a porpoising leap. If the spit sign is given, followed rapidly by the leap sign, the dolphin will spit while leaping. Many different signs can be combined in this way. Training for this conjoint behavior initially required some instruction by shaping, but once the concept of multiple behaviors was acquired, further combinations were typically carried out without specific training.

Dolphins are well known for their coordinated social behaviors. Swimming, breathing, and leaping in unison are common examples. At our laboratory, we have developed this tendency into coordinated and interrelated behaviors between two dolphins. In these tandem behaviors, the two dolphins are required to carry out a behavior together and on each other. An example is leaping in the air while touching flippers ("holding hands") or leaping in the air while touching snouts ("kissing"). These behaviors are controlled by context and by gestures. Context involves stationing the two dolphins close together, usually with a single trainer present, and asking for a behavior, such as inverted swimming. The dolphins will spontaneously swim together in that posture. To train a mutual hand hold, the dolphins are given the sign for "pec-fin touch" while the trainer signifies through pointing that the target is each other's pectoral fin. Once that is established, then the pectoral touch sign can be followed by the inverted swim sign and the pair will swim in that posture "holding hands." Under these procedures, the dolphins acquired a concept of "do it together" that enabled them to acquire a variety of new tandem behaviors of these combined types without specific training.

These types of behaviors might be termed sensory-motor coordinations and are mediated by both the visual system and the proprioceptive system. Additionally, the auditory system may come into play through vocal communication. We do not know what signals may pass between two dolphins involved in coordinated leaps or other behaviors requiring rapid simultaneous changes in behavior. It may be that some vocal signal is used by one or the other animal to initiate the leap (or other behavior) and serves as a cue for the second animal. The matter deserves further careful study.

IMPLICATIONS FOR PERFORMANCE IN THE NATURAL WORLD

What these findings imply, in general, is that in its natural world, the dolphin is well prepared to perceive, recognize, categorize, and remember the multitude of sounds or sights it receives through its auditory or visual senses. The sources of different sounds, or the targets and target characteristics that yield different echo spectrums, are probably learned through a lengthy period of experience with these sounds and with verification of sources and targets. Verification may come about through visual identification of a source or target, through generalization from what is known of other similar sounding sources or targets, through social observation of the responses of others to the sounds, and, in youngsters, possibly through some degree of more direct social tutoring by adults.

In addition to classification of sounds or sights, the dolphin can use its ability to develop rules to form adaptive strategies for response to detected and classified information. In prey detection, for example, this may include strategies for recruiting schoolmates, for aggregating prey, and for capture. While these are the normal activities of any social predator, the ability of the dolphin to understand semantic information, illustrated in our language studies, and to understand references to objects that are present or absent suggests the possibility of communication among individuals about prey detected or even specific prey species. Norris and Dohl (1980) described how prey detected by a wing of a laterally spread foraging school of spinner dolphins is communicated through at least a portion of the remainder of the spread formation. Würsig and Würsig (1980), studying dusky dolphins, conjectured that the vigorous breaches by the dolphins in the vicinity of the prey might serve to attract more distant school members to the prey patch, to help in capture. Würsig (1986) described additional search and capture strategies in other delphinid species.

Predators, such as large, deep-water sharks, may also be recognized by their sounds or echoes, or through visual detection, and their presence communicated. Reports of predator-specific alarm calls in vervet monkeys (Seyfarth, Cheney, and Marler 1980) or in prairie dogs (Slobodchikoff, Fischer, and Creef 1987) support the idea that either predator-specific or prey-specific calls might be used by dolphin species. It seems plausible, given the ability of dolphins in the laboratory to learn that sounds and visual signals can refer to objects or actions, that they might be able to produce and understand semantic references in their natural world, through sound production and analysis or through visual signals and visually observed behaviors.

In this regard, the findings on vocal mimicry have recently led to a reinterpretation of the function of the whistle sounds of dolphins as not only a self-referral (Caldwell and Caldwell 1965) but as a referral to others (Tyack 1986). Dolphins appear to be able to imitate the whistle sound of their tank mates and may do so in laboratory settings at times of stress (Ralston and Herman 1989). In the wild, an imitation of another's whistle, if it occurs, may be a means to refer to that individual. This may result in a responsive whistle from the intended receiver, locating that individual in space for the sender. It may also serve to attract that individual to the sender or at least to gain the attention of the receiver and allow for the improved transfer of any information that follows the attention call (cf. Richards 1981).

The results of the laboratory studies on visually based performance, together with data from the visually based language studies, affirm the considerable information processing capabilities of the visual system. This system can no longer be considered rudimentary or even secondary. Instead, like the auditory system, it appears to be an important interface between the real world and the cognitive world of the dolphin. What might be some of the functions of vision in the real world? Earlier, C. J. Madsen and I (Madsen and Herman 1980) outlined possible functions of vision in orientation, navigation, group movements, prey detection and capture, predator defense, identification of conspecifics, including individuals and gender and age classes, and in the communication of behavioral states. Earlier in this chapter, I referred to the information potentially available to the dolphin through underwater and aerial vision. Included in that short list were topographic features useful for orientation and navigation and flocking birds as an indicant of prey location. The capture of elusive, fast-swimming prey is a task for which vision is particularly well suited in clear-water conditions or in the final closure on prey in turbid waters. Maintaining visual contact with schoolmates during rapid group movements is a useful function of vision that can be augmented by acoustic contact. Detection and avoidance of predators is a vitally important visual function since a fleeing dolphin cannot maintain contact with a predator through echolocation, but with the wide field of view available to dolphins, vision can direct avoidance maneuvers.

All of these visual functions are basic requirements for survival but may require relatively little cognitive commitment. Other visual functions have more to do with communication and social behavior and may involve higher levels of information processing. Visual markings, body shape and size, coloration patterns, and other visually observable features allow for identification of species and of individuals. Wild bottlenose dolphin schools seem to consist of loosely structured subgroups of individuals that commonly school together but with frequent visits occurring among individuals of different subgroups (Wells et al. 1980, Johnson and Norris 1986). Long-term associations between individuals extending over years may occur. Dominance hierarchies develop among individuals and possibly among subgroups depending on the age or gender classes or the particular individuals comprising the subgroups. All of this places stringent requirements on individuals to learn the identifying characteristics, including behaviors, of many other individuals and how the behaviors of these individuals may be modulated by social and ecological context. Some of this learning may be experiential; other learning may occur more economically through generalizations from these experiences. There are likely to be social rules or conventions within these dolphin societies which may be complex but which govern social relationships, social roles, and social behaviors (cf. Johnson and Norris 1986). In the final analysis, it may be social knowledge that determines the success of the individual dolphin, since the dolphin is dependent on the social matrix for almost all aspects of its life (Norris and Dohl 1980). In all cases, extensive demands would be placed on acquiring and storing social knowledge, including rules, roles, relations, conventions, and contingencies, and on accessing and updating that knowledge as necessary.

Jerison (1986) has described the task of the brain as the construction of a model of reality. The richness of the model developed will depend on the information processing power of the brain, which in turn is a function of relative brain size and enhancements in cortical surface areas. These enhancements are considerable among cetaceans, especially the bottlenose dolphin (*Tursiops truncatus*) and several other members of the delphinid family. By mapping reality, including concepts about objects, events, time, and space, the animal enhances its ability to deal with the real world. Communication involves, in part, an exchange of these models of reality, so that individuals can share objects of attention or make reference to other matters. We have seen in some of the laboratory work the ability of the dolphin to refer or to understand references made by us. This included references made through the language systems and through indicative gestures. In the wild, dolphins may make references to objects of their attention through vocal or visual communication, including the possibility of symbolic references, as was discussed earlier, as well as more concrete references by "pointing" at objects through visual posture or through the direction of the echolocation "searchlight" beam. These illustrate, I believe, perhaps the single most important function of cognition and communication in the wild: the acquisition and use of knowledge to facilitate an exchange of information and of referential indicants within a mutually dependent social network.

REFERENCES

Caldwell, D. K., and M. C. Caldwell. 1977. Cetaceans. In: How animals communicate, ed. T. A. Sebeok. Bloomington: Indiana University Press. 794–808.

Caldwell, M. C., and D. K. Caldwell. 1965. Individualized whistle contours in bottlenosed dolphins, *Tursiops truncatus*. Nature 207: 434–435.

Chun, N. K. W. 1978. Aerial visual shape discrimination and matching-to-sample problem solving ability of an Atlantic bottlenosed dolphin. San Diego: Naval Ocean Systems Center NOSC TR 236, May.

D'Amato, M. R. 1973. Delayed matching and short-term memory in monkeys. In: The psychology of learning and motivation: Advances in research and theory, vol. 7, ed. G. H. Bower. New York: Academic Press. 227–269.

D'Amato, M. R., and M. Colombo. 1986. Auditory delayed matching-to-sample in monkeys (*Cebus apella*). Animal Learning and Behavior 13:375–382.

Dawson, W. W. 1980. The cetacean eye. In: Cetacean behavior: Mechanisms and functions, ed. L. M. Herman. New York: John Wiley and Sons. 53–100.

Dral, A. D. G. 1977. On the retinal anatomy of Cetacea (mainly *Tursiops truncatus*). In: Functional anatomy of marine mammals, vol. 3, ed. R. J. Harrison. New York: Academic Press. 81–134.

Forestell, P. H., and L. M. Herman. 1988. Delayed matching of visual materials by a bottlenosed dolphin aided by auditory symbols. Animal Learning and Behavior 16:137–147.

Herman, L. M. 1975. Interference and auditory short-term memory in the bottlenose dolphin. Animal Learning and Behavior 3:43–48.

———. 1980. Cognitive characteristics of dolphins. In: Cetacean behavior: Mechanisms and functions, ed. L. M. Herman. New York: John Wiley and Sons. 363-429.

———. 1986. Cognition and language competencies of bottlenosed dolphins. In: Dolphin cognition and behavior: A comparative approach, ed. R. J. Schusterman, J. Thomas, and F. G. Wood. Hillsdale: N.J.: Lawrence Erlbaum Associates. 221-252.

———. 1987. Receptive competencies of language-trained animals. In: Advances in the study of behavior, vol. 17, ed. J. S. Rosenblatt; assoc. eds. C. Beer, M. C. Busnel, and P. J. B. Slater. Petaluma: Academic Press. 1–60.

Herman, L. M., and W. R. Arbeit. 1973. Stimulus control and auditory discrimination learning sets in the bottlenosed dolphin. Journal of the Acoustical Society of America 56: 1870–1875.

Herman, L. M., and P. H. Forestell. 1985. Reporting presence or absence of named objects by a language-trained dolphin. Neuroscience and Biobehavioral Reviews 9: 667–681.

Herman, L. M., and J. A. Gordon. 1974. Auditory delayed matching in the bottlenosed dolphin. Journal of the Experimental Analysis of Behavior 21: 19–26.

Herman, L. M., J. R. Hovancik, J. D. Gory, and G. L. Bradshaw. 1989. Generalization of visual matching by a bottlednosed dolphin (*Tursiops truncatus*): Evidence for invariance of cognitive performance with visual or auditory materials. J. Exper. Psych.: Animal Behavior Processes 15: 124–136.

Herman, L. M., M. F. Peacock, M. P. Yunker, and C. J. Madsen. 1975. Bottlenosed dolphin: Double-slit pupil yields equivalent aerial and underwater acuity. Science 139: 650–652.

Herman, L. M., D. G. Richards, and J. P. Wolz. 1984. Comprehension of sentences by the bottlenosed dolphin. Cognition 16: 129–219.

Herman, L. M., and W. N. Tavolga. 1980. The communication systems of cetaceans. In: Cetacean behavior: Mechanisms and functions, ed. L. M. Herman. New York: John Wiley and Sons. 363-429.

Herman, L. M., and R. K. R. Thompson. 1982. Symbolic, identity, and probe delayed matching of sounds by the bottlenosed dolphin. Animal Learning and Behavior 10: 22–34.

Hunter, G. A. 1988. Visual delayed-matching of two-dimensional forms by a bottlenosed dolphin. M.Sc. thesis, University of Hawaii at Manoa, Honolulu.

Jerison, H. J. 1986. The perceptual worlds of dolphins. In: Dolphin cognition and behavior: A comparative approach, ed. R. J. Schusterman, J. A. Thomas, and F. G. Wood. Hillsdale, N.J.: Lawrence Erlbaum Associates. 141–166.

Johnson, C. M., and K. S. Norris. 1986. Delphinid social organization and social behavior. In: Dolphin cognition and behavior: A comparative approach, ed. R. J. Schusterman, J. A. Thomas, and F. G. Wood. Hillsdale, N.J.: Lawrence Erlbaum Associates. 335–346.

Kellogg, W. N. 1958. Echoranging in the porpoise. Science 128: 982–988.

Kinne, O. 1975. Orientation in space. Animals: Marine. In: Marine ecology, II, ed. O. Kinne. London: Wiley. 709–852.

Madsen, C. J., and L. M. Herman. 1980. Social and ecological correlates of vision and visual appearance. In: Cetacean behavior: Mechanisms and functions, ed. L. M. Herman. New York: John Wiley and Sons. 101–147.

Nachtigall, P. E. 1980. Odontocete echolocation performance on object size, shape and material. In: Animal sonar systems, ed. R. G. Busnel and J. F. Fish. New York: Plenum Press. 71–95.

Norris, K. S. 1980. Peripheral sound processing in odontocetes. In: Animal sonar systems, ed. R. G. Busnel and J. F. Fish. New York: Plenum Press. 495–510.

Norris, K. S., and T. P. Dohl. 1980. The structure and functions of cetacean schools. In: Cetacean behavior: Mechanisms and functions, ed. L. M. Herman. New York: John Wiley and Sons. 211–262.

Popper, A. N. 1980. Sound emission and detection by delphinids. In: Cetacean behavior: Mechanisms and functions, ed. L. M. Herman. New York: John Wiley and Sons. 1–52.

Pryor, K., R. Haag, and J. O'Reilly. 1969. The creative porpoise: Training for novel behavior. Journal of the Experimental Analysis of Behavior 12: 653–661.

Ralston, J. V., and L. M. Herman. 1989. Dolphin auditory perception. In: The comparative psychology of audition: Perceiving complex sounds, ed. R. J. Dooling and S. H. Hulse. Hillsdale, N.J.: Lawrence Erlbaum Associates. 295–328.

Richards, D. G. 1981. Alerting and message components in songs of Rufous-sided towhees. Behavior 76: 223–249.

———. 1986. Dolphin vocal mimicry and vocal object labeling. In: Dolphin cognition and behavior: A comparative approach, ed. R. J. Schusterman, J. A. Thomas, and F. G. Wood. Hillsdale, N.J.: Lawrence Erlbaum Associates. 273–288.

Richards, D. G., J. P. Wolz, and L. M. Herman. 1984. Vocal mimicry of computer-generated sounds and vocal labeling of objects by a bottlenosed dolphin, *Tursiops truncatus*. Journal of Comparative Psychology 98: 10–28.

Schevill, W. E., and B. Lawrence. 1953. Auditory response of a bottlenosed porpoise, *Tursiops truncatus*, to frequencies above 100 kc. Journal of Experimental Zoology 124: 147–165.

Seyfarth, R. M., D. L. Cheney, and P. Marler. 1980. Vervet monkey alarm calls: Semantic communication in a free-ranging primate. Animal Behavior 28:1070–1094.

Shyan, M. R., and L. M. Herman. 1987. Determinants of recognition of gestural signs in an artificial language by Atlantic bottle-nosed dolphins (*Tursiops truncatus*) and humans (*Homo sapiens*). Journal of Comparative Psychology 101:112–125.

Slobodchikoff, C. N., C. Fischer, and E. D. Creed. 1987. Alarm calls of prairie dogs identify individual predators. Paper presented at the Annual Meeting of the Animal Behavior Society, Williamstown, Mass., June 21–26, 1987.

Thompson, R. K. R., and L. M. Herman. 1977. Memory for lists of sounds by the bottlenosed dolphin: Convergence of memory processes with humans? Science 195:501–503.

Tyack, P. L. 1986. Whistle repertoires of two bottlenose dolphins, *Tursiops truncatus:* Mimicry of signature whistles? Behavioral Ecology and Sociobiology 18:251–257.

Walls, G. L. 1942. The vertebrate eye. New York: Hafner Publishing Company.

Wells, R. S., A. B. Irvine, and M. D. Scott. 1980. The social ecology of inshore odontocetes. In: Cetacean behavior: Mechanisms and functions, ed. L. M. Herman. New York: John Wiley and Sons. 263–317.

Wickelgren, W. A., and D. A. Norman. 1966. Strength models and serial position in short-term recognition memory. Journal of Mathematical Psychology 3:316–347.

Würsig, B. 1986. Delphinid foraging strategies. In: Dolphin cognition and behavior: A comparative approach, ed. R. J. Schusterman, J. A. Thomas, and F. G. Wood. Hillsdale, N.J.: Lawrence Erlbaum Associates. 347–359.

Würsig, B., and M. Würsig. 1980. Behavior and ecology of dusky porpoises, *Lagenorhynchus obscurus*. Fishery Bulletin 77:871–890.

An aerial view of the Naval Ocean Systems Center labora-
tory facilities in Hawaii. Heptuna and other research dolphins
live in pens visible along the docks and shoreline. (Photo cour-
tesy of U.S. Navy.)

CHAPTER THIRTEEN

DOLPHIN PSYCHOPHYSICS

CONCEPTS FOR THE STUDY OF DOLPHIN ECHOLOCATION

Patrick W. B. Moore

INTRODUCTION

To say that dolphins echolocate is like saying Michelangelo painted church ceilings. The exquisite development of the biological sonar system of the bottlenose dolphin (*Tursiops truncatus*) for shallow water, noisy and reverberant environments such as bays, estuaries and nearshore waterways, is a prime example of evolutionary adaptation. Echolocation is essentially a special extension and adaptation of the mammalian hearing system coupled with an ability to generate sounds. Humans have the ability to judge room size based on reverberation from their own voice. Some blind people can use self-generated sounds to detect reflective objects (Rice 1966). I view echolocation as representing an interval on an acoustic sensory continuum.

The information presented here is by no means exhaustive. Most of the information concerns research methods and procedures with which I have been directly involved. The reader should not assume the work presented comes from one or two individuals, however; it comes, instead, from a team to which each member gave special expertise in each investigation. Attention to the references will guide the reader to a more detailed discussion of specific material.

The discovery of dolphin echolocation is comparatively new, dating back to the late 1940s. The initial suggestion that the bottlenose dolphin used sound for navigation and obstacle avoidance is attributed to Arthur McBride of Marine Studios in Florida (McBride 1956). The study of echolocation began with basic field observational studies. More rigorous experimental studies followed and grew in sophistication as investigators learned more about the dolphin echolocation system.

The first experimental investigations of dolphin echolocation were mainly obstacle avoidance studies. Animals were placed in pools of turbid water and required to navigate through mazes of nets, pipes, and sundry other barriers (for a review of obstacle avoidance studies, see Moore 1980). Early work focused primarily on the animal's ability to use echolocation. However, once Norris et al. (1961) demonstrated echolocation capability in a blindfolded dolphin, the quest to understand dolphin sonar moved from qualifying the dolphin's echolocation skill to quantifying its basic capabilities. Investigators turned to experiments designed to measure such issues as the maximum range of detection, target discrimination ability, and the use of echolocation by other species.

DOLPHIN PSYCHOPHYSICS AND THE NEED FOR CONTROLLED STUDIES

The formalized measurement of sensory abilities is called psychophysics, a branch of experimental psychology involving the study of the relationships of internal sensations associated with specific, measurable external physical stimuli (for a review of dolphin sensory processes, see Nachtigall 1986). Psychophysics began in the 1720s but was not firmly established until the 1850s, when Fechner's law, which relates stimulus sensation to stimulus magnitude, was formulated (Stevens 1975). Psychophysics provides the tools to study dolphin echolocation. The procedures, theories, and even the apparatus from the traditional psychophysics laboratory were adapted to the dolphin experimental setting to measure and analyze the sensory phenomenon of dolphin echolocation.

Human psychophysical studies show that accurate control of the stimulus (i.e., the frequency of a tone to be detected, the spacing of two pinpricks on the skin or any physical event to be perceived) must be precise. Also, control of the subject detecting the stimulus is very important. In a listening experiment, the experimenter requires a quiet environment so as to assure accurate assessment of the subject's ability to hear a tone. The experimenter must also provide clearly understood instructions for the subject to follow. These controls are required to assure accurate and reproducible experimental results.

As the science of psychophysics developed, so did the techniques for controlling and measuring the physical stimulus being presented. With better control of the stimulus, it was possible to study other influences that caused deviating reports of the subjective perceptions of the stimulus. Effects of motivation, for instance reward and punishment, contribute to a dolphin's overall ability to determine and report a stimulus event (Schusterman 1980). For accurate evaluation of the dolphin's hearing, echo detection, and sonar emission capabilities, a controlled experimental setting is required to assure that random influences (which would occur in the natural setting) are controlled so that their effect on the experimental question is minimized.

BEHAVIORAL CONTROL AND PROCEDURAL METHODS

Clearly, animals in captivity have less freedom to roam than animals in a natural setting; however, this limitation does not necessarily mean an uncomfortable existence. Dolphins in captivity for experimental studies are well kept and well maintained for three very important reasons: (1) to assure accurate scientific data, (2) to comply with strict statutory guidelines, and, most important, (3) to fulfill our responsibility to creatures that deserve our respect and humane treatment. The data gathered during experiments I will discuss were all collected using operant conditioning and reward training. The only punishment was an occasional time-out from training.

RESPONSE PARADIGMS

If you intend to investigate stimulus perception, you must provide a method for your subject to respond and thereby to communicate the detection of a stimulus. The methods employed for this communication are called response paradigms (for a review of behavioral methods used in dolphin echolocation studies, see Schusterman 1980).

Our response procedures use a "paddle press" indicator. The animal pushes a ball, hanging rod, disc-shaped object, or other device to indicate the perception of a particular state of stimulus affairs. A response procedure called "go/no-go" is widely used in dolphin psychophysical experiments. This method is particularly adaptable for a variety of experiments because it is composed of simple, easy to train responses. In applying the go/no-go response to dolphins, clearly overt and easily discriminated actions are required to avoid any difficulty in determining the response.

The go/no-go procedure was used to test if the dolphin could taste sweet, sour, bitter, and salty solutions (Nachtigall and Hall 1984, Nachtigall 1986). Nachtigall and Hall trained their animal to swim up and onto a platform just above the water level. The animal then placed its mouth on a thick, hollow bite-plate fitted with holes connected to plastic tubes. A solution of distilled water was gently pumped over the tongue, followed by a second presentation of distilled water or a light solution of a chemical. The chemical used produced one of the four basic tastes. The animal's job was to compare the two solutions. If the second liquid was distilled water, the animal was to remain on the bite-plate (no-go). However, if the second liquid contained the chemical associated with a taste of sweet, sour, bitter, or salty, which the animal could taste, then the dolphin was to release the bite-plate and slide back into the water (the go response). Using this simple method, Nachtigall and Hall demonstrated that the dolphin could indeed taste sweet, sour, bitter, and salty at varying degrees of concentration.

A second response paradigm I have used is the two-alternative forced-choice procedure. This method requires the animal to respond after every presentation of the stimulus. Since we use operant conditioning, the animal must respond correctly to receive reinforcement; incorrect responses result in no reinforcement or a time-out. For example, in a detection task, the animal is trained to use echolocation to detect the presence or absence of a target. It must then strike one of two paddles or balls to show if it detects the target or not (Au and Snyder 1980). (The assignment of the right/left paddle or ball to represent present or absent is arbitrary and is decided during the initial training of the response). The same reinforcement rules apply: the animal must make a correct response to get reinforcement.

It seems intuitively appropriate to use the go/no-go paradigm with a detection task and the forced-choice response with a discrimination task. However, a question can arise. Is the effort required of the animal equal for the two response procedures? A response bias may exist in the go/no-go paradigm because a go response requires more effort than a no-go response. This is a good research question. The effect of the two different response paradigms on dolphin detection bias has yet to be experimentally examined.

STIMULUS PRESENTATION PARADIGMS

Whether investigating the passive hearing system or active echolocation, the stimuli can be presented in various ways. In successive order presentation, stimuli are presented sequentially, on successive trials. For example, a dolphin may be asked to discriminate between stimulus #1 and stimulus #2. Only one stimulus is presented during a trial. The dolphin must observe the stimulus and decide if it is #1 or #2 and then respond differently for each stimulus. Some form of memory for the stimuli must be maintained from trial to trial for the dolphin to discriminate correctly between the stimuli and give the correct response.

With serial order presentation, stimulus #1 is presented for a fixed time and then removed, and stimulus #2 is presented during the same trial. Simultaneous presentation means both stimuli are presented at the same time during a trial.

Let us consider each stimulus presentation method as it might be used in designing an actual experiment; I will select an echolocation detection problem using the go/no-go response. This is a good choice because it is the easiest behavior to train and maintain.

The test dolphin is trained to "station" and to wait in a particular spot, say, to swim to a rubber-covered hoop and put its head through for a reward. Figure 13.1 shows a dolphin stationing in a hoop for an echolocation experiment and a seal stationed in a nose cup for a hearing test. The stimuli are presented in the sequential method. On any given trial, the dolphin will be presented with one of the two stimulus events, either the target or nothing. The animal is to show which event has occurred.

This procedure has a fixed trial time. If a stimulus is presented, the correct response is to leave the station (the go response) before the fixed trial time has elapsed. Conversely, if no stimulus is presented, the animal must remain at its station (emitting the no-go response) until the time limit is up.

The first thing the dolphin must know is when the trial starts. In this experiment, we might put an acoustic screen in front of the animal between trials, so that any emitted echolocation signals will bounce back. The screen has several uses. It acts as a device to keep the dolphins from "peeking" down range as targets are removed or placed into the water, and, since the animal can see the screen, its removal can signal the start of the trial. (In passive hearing tasks, a light works nicely as a trial start and duration indicator; however, the experimenter must be sure that the light or its switches do not introduce noise that can cue the animal to the correct response.)

Testing sessions usually consist of 50 to 100 trials with an equal number of trials in which the stimulus is present or absent. The order of trials in which the stimulus is present or absent is randomized so the animal cannot guess the next trial's stimulus condition based on past trial events. We observe the number of times the correct response is emitted for each stimulus event. At the end of a session, we derive a performance score for

Fig. 13.1.   A dolphin stationed in an underwater hoop station waiting for
an echolocation trial to start. This station is rather large and the hole is lined
with thick, clear plastic tubing. (Photo courtesy of U.S. Navy.)

the subject. The behavior can be adapted to a wide range of experiments;
many different questions can be asked using this simple design (Moore
and Au 1982*a*, 1982*b*; Moore et al. 1984).

Now let us examine an experiment that uses the simultaneous stimulus
presentation method and a forced-choice response procedure. We will
ask a discrimination question. The animal has to discriminate a cylinder
from a sphere to earn a fish reward. With this procedure, we present two

stimuli or targets on each trial. They are presented equidistant from the animal and are clearly distinguishable, one on the right and one on the left. If the targets are the same, the animal must push the "correct" (right or left) paddle; it must push the other paddle if the targets are different (Hammer and Au 1980). When using the simultaneous stimulus presentation paradigm, it is important to eliminate any possible cues for the animal based on target position. This is accomplished by randomizing placement of the targets in the left and right positions.

The same paradigms are used in passive listening experiments. For example, we often study dolphin hearing by using masking sounds. "Masking" means that one sound, presented at the same time as another, can make the less intense sound more difficult to hear (Fletcher 1940). An example of this is being unable to hear what someone is saying when you are working at the kitchen sink and the water is running. That is masking in its basic form of simultaneous presentation.

Masking can also occur when one sound closely follows or precedes a second sound; these paradigms are called backward and forward masking (Moore et al. 1984). Masking is an important paradigm, because it helps us to understand basic auditory processes and signal processing abilities of the dolphin's echolocation system (Johnson 1968, 1980).

STIMULUS CONTROL

Stimulus control is the degree to which a particular stimulus event elicits an observable change in behavior. If, in the initial training, the detection problem is easy, an experimenter can expect the dolphin to learn to recognize the stimulus and give the correct response in a few sessions. The animal soon reaches 100 percent correct performance, thus demonstrating that the animal *understands* the experimental question. This is called stimulus control.

Stimulus control *must* be established before manipulating the stimuli to determine a threshold of hearing or perception. Unlike learning or cognitive studies, psychophysical studies *start* with the assumption that the animal knows the task. The experiment tests only the ability of the animal to detect some specific aspect of the stimulus which is experimentally manipulated. Stimulus control must be continuously scrutinized throughout the experiment. In most of our work, strict stimulus control is demonstrated in every testing session either before the testing session in "warm-up" trials or after testing by "cool-off" trials or both; only then can the psychometric data from the tests be accepted (Au and Moore 1984, Au et al. 1988).

APPARATUS FOR THE STUDY OF DOLPHIN HEARING AND ECHOLOCATION

When I first started working with dolphins in 1972, most of the data collection was accomplished by observation. The researcher recorded responses on paper data sheets with a pen. In echo-detection studies, physical targets were used. In hearing studies, electronic equipment was used only to generate signals and cues. Now most laboratories use computers for almost every aspect of experimental control.

In the study of dolphin echolocation, events take place too rapidly for the experimenter to perceive; also, these events may contain sounds above human hearing. Computers provide the only way investigators can understand the sensory mechanisms dolphins use to extract features from returning echoes. They have the computational speed and accuracy, coupled with programmable flexibility, to detect, capture, analyze, and summarize echolocation signals *as they occur*. A well-developed computer system can schedule target presentations, can detect and record the animal's responses, can generate and deliver tones that tell the animal whether it has made a correct or incorrect response, can count and keep running averages of the animal's performance, and at the end of the experiment, can act as a word processor to help write the manuscript.

Computers are not without problems. In dolphin experiments, they are usually exposed to extreme environmental conditions; salt air and moisture cause many problems. I do not know of any existing computer that can suffer the onslaughts of both nature and experimenter without a failure, usually at the most critical time in the experiment.

Using computers and associated digital systems, one can generate and control acoustic stimuli to a microsecond of accuracy. In the early 1980s we (Moore et al. 1984) conducted a backward masking experiment that would have benefited considerably from computer control. We know that loud noise will mask a signal even if the noise is delayed and occurs after the signal. We developed a system that detected the outgoing echolocation of a dolphin and generated a digital noise that masked the return echo. The target was suspended 150 feet in front of the animal. It took two people to run the study. One person worked in an instrument shelter 25 feet away from the dolphin, to adjust the delay of the noise, position the target in or out of the water, and monitor the power output of the stimulus projector. A second experimenter stayed at the test pen to signal the animal for the start of a trial and to give fish reinforcement for correct responses. Today, the whole experiment, noise stimulus, time delays, reinforcer signals, and all, could be generated and controlled by one small computer, using a "high level language" like BASIC and another, faster, machine language for total control of the experiment.

DOLPHIN ECHOLOCATION SIGNAL

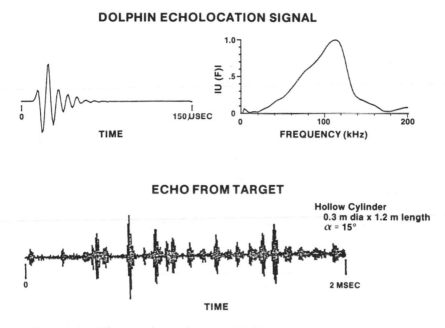

Fig. 13.2.   The waveform of a typical dolphin click and its associated energy spectrum are shown on top. Below is the echo produced from the target from this click. Notice the difference in duration of the click as compared to the echo and the number of amplitude peaks (highlights) contained in the echo (courtesy of Whitlow Au).

The echo returning from a target is a complicated acoustic stimulus (fig. 13.2). Echo characteristics include doppler shifts, extended duration, irregular envelopes, and various modulation effects. The dolphin has an acute ability to extract information from these returns. The animal is capable of judging attributes of size, shape, composition (i.e., if the target is hollow or solid), and even target thickness from the echoes. Computers cannot only control simple sequential experimental events like the masking experiment I described but can also now be programmed to develop and transmit the complex stimuli necessary to investigate the way dolphins obtain information from echoes.

A computer can control the target echo structure by creating a false echo, which allows investigators to examine what part (or parts) of the echo the dolphin uses to distinguish one target from another. New artificial target generators will allow us to ask what internal echo differences are perceivable in time, how large-amplitude echo highlights mask small-amplitude highlights, and what is "enough information" to judge that one target return is different from another (for a description of the recent work with computer-generated targets, see Au, Moore, and Pawloski 1988).

CURRENT DOLPHIN SONAR RESEARCH

One of the most exciting current developments in dolphin echolocation research concerns the dolphin's ability to control the source level of its outgoing signal. Dolphins emit echolocation clicks at levels sufficient to solve a particular task. The source level changes are based on a host of internal and external environmental conditions such as the ambient noise, target range, and target strength. For example, measured source levels (the intensity of the emitted click) for animals echolocating in San Diego harbor, a relatively quiet bay, are lower than source levels of animals in Kaneohe Bay in Hawaii, one of the noisiest. In studying bottlenose dolphins and a beluga whale *(Delphinapterus leucas)* in both locations, we noted both an increased source level (louder clicks) and a shift in peak frequency (higher click tone) for animals in Kaneohe Bay. Peak frequency shifted from 60 to 80 kHz in San Diego to 110 to 120 kHz in Kaneohe (Au et al. 1985). Our curiosity peaked.

The obvious explanation for the change in the echolocation signal was that background noise (mostly from snapping shrimp) is louder and extends to higher frequencies in Kaneohe Bay. The interesting question was, how, and at what cognitive level in the animal's echolocation system, did the shift occur? Was the shift mediated by higher-order centers? Could the animal have voluntary and direct control over the emission level of its click train? Did some sort of learning mechanism exist which allowed the animal to control its emitted source level for particular tasks? Answers to these questions could go a long way toward explaining how echolocation evolved, how it develops in the newborn animal, and how the animal uses echolocation for assessing and learning about targets and their attributes.

My colleagues and I set out to test if, in fact, dolphins could exert voluntary control over the emitted source level of their clicks (Moore and Patterson 1983). If they could do that, we then wanted to find out if the animal could perform a target detection task using clicks that had been altered or controlled. If the animal could be trained to control its emitted signal *at all*, this would verify that an information feedback loop existed. The signals would be open to operant conditioning in actual echolocation detection situations.

Our approach was to use a simple detection task, one that the animal had to learn but that was not so difficult as to limit the level of sound the animal needed to emit to solve the task. We used the go/no-go target response procedure with randomly presented targets (50% target present/50% target absent). A bite-plate tail-rest stationing device kept the animal in a fixed position (fig. 13.3) so we would always be able to measure the emitted click at the same spot in the emitted beam. (Had the animal been allowed to move around, or even just move its head, the measurement of the signal could have been outside the main axis of the emitted sound beam; the source level would have seemed different due simply to the position of the hydrophone in the emitted beam and not to any control developed by the animal.)

Training continued until detection performance was very high (greater than 95%) and stable. This was the animal's baseline performance. This allowed two observations: (1) we could identify the baseline or "natural click" structure the animal preferred to use for solving this detection problem; and (2) we could track the development, over time, of the animal's preferred emitted source level for solving the task. These baseline data would be a basis for comparison of what would happen when we tried to manipulate the source level.

After the animal had solved the detection task and was performing well, we began the training task to establish control over the dolphin's emission level. The first thing an experimenter must be able to do to train a behavior is to know when it occurs. We follow the correct behavior with a primary reinforcer, something the animal likes (fish for a dolphin). When the behavior is very brief in time, we need a signal to help the dolphin understand just exactly when it was doing the behavior that earned that fish.

We began by sounding a short tone and immediately feeding the dolphin a fish. After several repetitions, this coupling of tone and fish makes the tone a secondary reinforcer. The tone indicates to the dolphin "here comes the fish." Then we can use the tone to tell the dolphin, the instant it makes a correct sound, "yes," you are right; here comes the fish. This is the very first training step for *any* dolphin behavior. The "tone" (now a secondary reinforcer) can be a whistle or any event the dolphin can perceive and associate with the primary reinforcer, fish.

To observe the behavior of echolocation, especially a single extremely brief echolocation click, a computer is needed which can collect, analyze, and respond to each outgoing echolocation click the animal emits (Ceruti and Au 1983). Along with this computer, I have discovered, one needs a lot of *functional* software. All of our programs have been developed in-house. The programs we initially used to train changes in source level activated a tone from one of two underwater speakers (to the right of the

Fig. 13.3. Heptuna, a bottlenose dolphin and sonar research expert, stationed in the bite-plate tail-rest station. This type of station device is used to minimize the movement of the animal and afford better sound measurements of the outgoing echolocation signals. (Photo courtesy of U.S. Navy.)

dolphin for increasing level or to the left for decreasing level). The program also detected each emitted click, measured each click's source level (in volts peak-to-peak), and compared it to a criterion source level we entered for each trial. If the emitted click was above the criterion level, the computer interrupted the trial by sounding the reinforcer tone, even if the animal was still emitting clicks. By slowly adjusting the criterion level for reinforcement upward when the tone on the right was on, or downward when the tone on the left was on, we trained the animal to use tones from the right or left underwater speakers as cues to produce high- or low-level click trains (Moore and Patterson 1983).

Then we used similar procedures to bring the *frequency* content of the click train under stimulus control. This was a more difficult training problem, because critical timing was required to analyze the click's frequency as the animal emitted signals. Though the processes were tedious, we successfully trained the behavior. Our expert dolphin can now produce high- and low-frequency click trains *and at the same time* control its source level (Moore and Pawloski 1987).

This kind of training is not easy and must be developed slowly. This task took five years and involved training two animals to control their emitted source level and one to control both source level and frequency content. Heptuna, the animal I have worked with longest (since 1978), is a true professional. We have worked through some tough training and sensory tasks. Several times, he has gone for 50 to 80 trials without a correct response, and still he worked at the problem. He even appears to enjoy the work and rarely refuses to attempt even the most difficult tasks.

The training guidelines we follow are quite simple in many ways. We do not deprive our animals of food. They may not always be correct during training, and they may not earn a fish for each trial, but each day the animals get the full allotted ration required for good health and nutrition. We spend time with the animals, every day of the week, and each animal has a pen-mate to socialize with when not working. Sometimes the animals "socialize" quite a lot; so far, we have had four new baby dolphins from our research family. Our animals receive the best care possible from a staff of two veterinarians with modern medical facilities. The veterinarians conduct daily rounds, and the animals receive well-balanced meals of restaurant-quality fish plus vitamin supplements, along with monthly weighing and semiannual full medical checkups.

One training rule I have always followed: Do not take your "bad day" to a training session. If you are frustrated or angry, you cannot help but let it affect your training. Short-tempered, hot-headed trainers do not make progress; usually, they end up confusing the animal and wasting a training session.

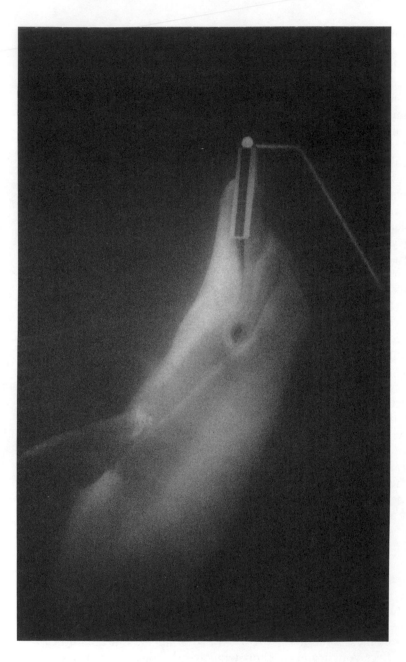

Fig. 13.4. To investigate the up-and-down orientation of the dolphin's receiving beam, we had to turn the bite-plate 90° so that Heptuna would station on his side. The view is straight down into the tank. (Photo courtesy of U.S. Navy.)

OTHER DOLPHIN SENSORY QUESTIONS

Between our computers and our talented dolphins, we are beginning to attempt to understand some other intricacies of dolphin echolocation. We are working on the edge of what we know about dolphin hearing and have to make our way slowly. We are asking the dolphin how it measures target strength (Au, Moore, and Pawloski 1988). Target strength is a sonar engineering term that refers to the amount of energy in the return echo which is measured by the sonar. We want to know how the dolphin attends to the returning energy in an echo (Vel'min and Dubrovskiy 1975, 1976, 1978). We ask the dolphin to tell us the period of time it uses to gather energy from the echo. We also measured the dolphin's hearing band width, the ability of the dolphin ear to separate signals from noise within a range of frequencies (Johnson 1968, Moore and Au 1983). We have found that the dolphin ear is indeed well adapted to the noisy ocean and filters out ocean noise to hear weak signals up to 120 kHz.

We're studying the dolphin's receiving beam, the invisible shape of hearing sensitivity around the animal's hearing apparatus, to discover if it gathers sound from above, below, to the left, and to the right equally well (Au and Moore 1984). Since it was impossible to rotate the floating apparatus in which we mounted the instruments that emitted the test signals, we investigated the up-and-down aspect of the dolphin's beam pattern by rotating the animal 90 degrees. For this experiment, we trained the animal to station on its side (fig. 13.4).

We found that, in fact, the animal is about equally sensitive to sounds from the left and right. We also learned that the hearing sensitivity becomes very narrow and points straight ahead, like a flashlight beam, as the frequency moves higher. When sounds come from above or below, however, best sensitivity is to targets elevated slightly above the head (about 10°). Again, as the frequencies increase, the beam becomes narrower.

FUTURE AREAS OF DOLPHIN ECHOLOCATION RESEARCH

The study of dolphin sonar is fascinating because of the mysteries surrounding the processes. The more we find out about dolphin sonar, the clearer it becomes that nature has created a unique aquatic-adapted biological system that is far better than any system man has yet to devise for short distances in reverberant environments. In spite of all this meticulous testing, however, we still know very little about how the dolphin sonar actually works. Measurements of the outgoing signal indicate no special parameters other than its being a broadband click. We know the click parameters are very flexible, both in frequency and amplitude; but the special part of the dolphin sonar system is the brain behind the signal receiver, the signal processor itself. Determining how this processor works will eventually lead to greater understanding of dolphin sonar.

Learning about the dolphin's natural sonar system helps us build better sonar hardware to study the seas and aid in navigation. Understanding dolphin sonar may also help reduce the problems of dolphin entanglement in fishing gear. The more we understand about these exceptional creatures, the better we can protect and preserve them from the intrusions of human society.

We may never discover all the hidden mysteries of dolphin echolocation. Nature has seldom given up her secrets easily, and the echolocation ability of the dolphin is one of the best kept. But from my point of view, the study of dolphin psychophysics has a most promising outlook, a never-ending series of questions to ask and experiments to design.

REFERENCES

Au, W. W. L., D. A. Carder, R. H. Penner, and B. L. Scronce. 1985. Demonstration of adaptation in beluga whale echolocation signals. J. Acoust. Soc. Am. 77:726.

Au, W. W. L., and P. W. B. Moore. 1984. Receiving beam patterns and directivity indices of the Atlantic bottlenosed dolphin *Tursiops truncatus*. J. Acoust. Soc. Am. 75:255.

Au, W. W. L., P. W. B. Moore, and D. A. Pawloski. 1988. The perception of complex echoes by an echolocating dolphin. J. Acoust. Soc. Am. S1 80:S107.

Au, W. W. L., and K. J. Snyder. 1980. Long-range target detection in open waters by an echolocating bottlenosed dolphin (*Tursiops truncatus*). J. Acoust. Soc. Am. 68:1077.

Ceruti, M. G., and W. W. L. Au. 1983. Microprocessor-based system for monitoring a dolphin's echolocation pulse parameters. J. Acoust. Soc. Am. 73:1390.

Fletcher, H. 1940. Auditory patterns. Rev. Mod. Phys. 12:47.

Hammer, C. E., and W. W. L. Au. 1980. Porpoise echo-recognition: An analysis of controlling target characteristics. J. Acoust. Soc. Am. 68:1285.

Johnson, C. S., 1968. Masked tonal thresholds in the bottlenosed porpoise. J. Acoust. Soc. Am. 44:965.

————. 1980. Important areas for future cetacean auditory study. In: Animal sonar systems, ed. R. G. Busnel and J. F. Fish. New York: Plenum Press.

McBride, A. F. 1956. Evidence for echolocation by cetaceans. Deep-sea Research 3:153.

Moore, P. W. B. 1980. Cetacean obstacle avoidance. In: Animal sonar systems: Biology and bionics, ed. R. G. Busnel and J. F. Fish. Laboratoire de Physiologie Acoustique, INRA-CNRZ, Jouy-en Josas, France.

Moore, P. W. B., and W. W. L. Au. 1982*a*. Masked pure-tone thresholds of the bottlenosed dolphin (*Tursiops truncatus*) at extended frequencies. J. Acoust. Soc. Am. S1:S42.

————. 1982*b*. Directional hearing in the Atlantic bottlenosed dolphin (*Tursiops truncatus*). J. Acoust. Soc. Am. S1:S42.

————. 1983. Critical ratio and bandwidth of the Atlantic bottlenosed dolphin. J. Acoust. Soc. Am. S1:74.

Moore, P. W. B., R. W. Hall, W. A. Friedl, and P. E. Nachtigall. 1984. The critical interval in dolphin echolocation: What is it? J. Acoust. Soc. Am. 76:314.

Moore, P. W. B., and S. A. Patterson. 1983. Behavioral control of echolocation source level in the dolphin (*Tursiops truncatus*). Fifth Biennial Conference on the Biology of Marine Mammals. 70.

Moore, P. W. B., and D. Pawloski. 1987. Voluntary control of peak frequency in echolocation emissions of the dolphin (*Tursiops truncatus*). Seventh Biennial Conference on the Biology of Marine Mammals, Miami, Fla.

Nachtigall, P. E. 1986. Vision, audition, and chemoreception in dolphins and other marine mammals. In: Dolphin cognition and behavior: A comparative approach, ed. R. J. Schusterman, J. A. Thomas, and F. G. Wood. Hillsdale, N.J.: Lawrence Erlbaum Associates.

Nachtigall, P. E., and R. W. Hall. 1984. Taste perception in the bottlenose dolphin. Acta. Zoologica Fennica, 172:147.

Norris, K. S., J. H. Prescott, P. V. Asa-Dorian, and P. Perkins. 1961. An experimental demonstration of echolocation behavior in the porpoise, *Tursiops truncatus* (Montagu). Biol. Bull. 120:163.

Rice, C. E. 1966. The human sonar system. In Animal sonar systems: Biology and bionics, ed. R. G. Busnel and J. F. Fish. Laboratoire de Physiologie Acoustique, INRA-CNRZ, Jouy-en Josas, France.

Schusterman, R. J. 1980. Behavioral methodology in echolocation by marine mammals. In: Animal sonar systems: Biology and bionics, ed. R. G. Busnel and J. A. Fish. Laboratoire de Physiologie Acoustique, INRA-CNRZ, Jouy-en Josas, France.

Schusterman, R. J., D. A. Kersting, and W. W. L. Au. 1980. Stimulus control of echolocation pulses in *Tursiops truncatus*. In: Animal sonar systems: Biology and bionics, ed. R. G. Busnel and J. F. Fish. Laboratoire de Physiologie Acoustique, INRA-CNRZ, Jouy-en Josas, France.

Stevens, S. S. 1975. Psychophysics. New York: John Wiley and Sons.

Vel'min, V. A., and N. A. Dubrovskiy, 1975. On the analysis of pulsed sounds by dolphins. Dokl. Akad. Nauk. SSSR 225:470.

――――. 1976. The critical interval of active hearing in dolphins. Sov. Phys. Acoust. 2:351.

――――. 1978. Auditory perception by the bottlenose dolphin of pulsed signals. In: Marine mammals: Results and methods of study, ed. V. Ye. Sokolov. A. N. Severtsov Inst. Evol. Morphol. Anim. Ecol., Akad. Nauk. SSSR.

AFTERWORD:

DOLPHIN POLITICS AND DOLPHIN SCIENCE

Karen Pryor and Kenneth S. Norris

I

The public attitude toward whales and dolphins has changed enormously since the 1940s, when the routine keeping of dolphins in captivity began. Whales were once mere objects of commerce, to be harvested for profit. Dolphins were simply competition for fishermen. Navy training films portrayed killer whales as dangerous vermin that might attack lifeboats and swimmers; some military fliers reportedly used them for bombing practice, in the belief that they were thus protecting their buddies.

Flipper, the dolphin television star, changed all that. Shamu, the first killer whale in captivity, changed all that. You could not hate an animal that was rolling over to be scratched on its tummy or leaping up to kiss pretty girls on the cheek with its pink tongue. It was not just the fantasies of dolphin-lovers that caught the public fancy; it was the presence of the animals themselves, in oceanariums and on television, first in the United States and then around the world, that enabled us to relate to them. You could *see* that they were friendly and intelligent; their trainability made them great performers; and their innate behavior of making intense (and very readable) eye contact gave the public a sense of connection that to some people continues to seem magical.

What happened? We started saving the whales. And the sea otters, porpoises, and baby harp seals. Perhaps no other wildlife issue has so many vociferous public adherents as the Save the Whales movement and all its spinoffs. Many of the combatants are not scientists but lay members of the public whose motivation is emotional, not necessarily rational: endangered or not, saving these animals is a holy war.

In the United States, this new ethos had a direct political result—the passage by Congress in 1972 of the Marine Mammal Protection Act, an utterly unprecedented piece of environmental legislation. This act provided that the United States would take under its protection all the whales, dolphins, seals, and other marine mammals in its waters (including the polar bear, which was decreed to be a marine mammal by a senator from Alaska). If you run over a manatee in your powerboat, you are now in violation of federal law; and you need a government permit even to approach whales within 300 m to photograph them, which the act defines as "harassment."

The Marine Mammal Commission oversees the management of the act. It is an independent agency, with commissioners appointed by the president and overseen only by Congress. It consists of a Washington staff, three commissioners, and a rotating board of nine scientific advisors. We have both worked closely with the commission: Ken Norris was involved in preparing the act itself and served on the first Scientific Advisory Board, and I was a commissioner from 1984 to 1987.

Small and mobile, with a tiny budget but the ability to make a quick decision, the commission has sparked many major endeavors, such as the international effort now under way to control plastics pollution in the oceans. The Marine Mammal Commission is the only government agency that must, by law, take the advice of its Scientific Advisory Board, or explain to Congress why it has not done so. One consequence is that representatives and senators tend to listen to the commission and to trust its statements, which gives this very small agency some clout. The Marine Mammal Commission is also the only federal agency that *must*, by law, take an ecosystem approach to its problems. The commission cannot be forced from a position by special interest groups, no matter how powerful; staff and commissioners would be breaking the law if they did so. Because this arrangement is proving workable, it is serving as a model for other legislation, such as the International Antarctic Treaty now being developed by a consortium of nations, the first such treaty to be structured around the management not of separate resources but of the polar ecosystem as a whole. To us, this is a profound political change, bearing the promise of long-term and desirable results. It is the product of a momentum that began, perhaps, when people started falling in love with Flipper's smile.   —K.P.

II

Cetology, the study of whales and dolphins, has come a long way in our lifetimes. When I started working with dolphins thirty-eight years ago, scientists had incompletely defined the species of cetaceans and knew precious little about where many of them ranged. In behavioral terms, people regarded them as curious kinds of fish. The story since then is outlined at various places in this volume. We can summarize its parts as follows:

At first, cetologists worked mostly with animals killed or cast up on some beach. With this information we identified species and pieced together where they ranged.

Then, from the first few captives in oceanariums, we began to understand that these cetaceans were complex mammals many of whose behavior patterns bore a startling resemblance to those of terrestrial animals. So a few people, including both of us, began to grapple with learning about dolphins at sea.

This attempt went on at first without much contact with mainstream animal behavior work, which was proceeding apace and rounding into a mature science at many places in the terrestrial world. At first, we cetologists literally did not know whether behavioral studies of dolphins in the wild were possible. And we surely had little information to exchange with our terrestrial colleagues.

In time, we found that we could gather verifiable information. Support for our personal efforts as well as those of others came primarily from two sources at first: fisheries management programs, which faced massive problems involving cetaceans, and the U.S. Navy, which was interested in animals that could perform sensory feats that were difficult or impossible for the military to accomplish with machines. In the early years, these partnerships and funding sources greatly influenced whom we cetologists were talking to and what kinds of questions we were asking.

Meanwhile, the science of terrestrial animal behavior was being built on the long traditions of evolutionary theory, natural history, comparative psychology, and ethology. Its scientists sought out the mainstream sources of governmental and other traditional support. Not a few wondered what the cetologists were up to and if what was happening was decent science. Where were those dolphin people, for heaven's sake, at the national meetings?

Now the two streams have begun to come together. We cetologists have satisfied ourselves that we can learn about animals that live in the very foreign ocean realm. We have developed some sharp observational tools, we have found places where we can meet the oceanic cetaceans to observe them, and we have honed the methods of comparing the behavior of cetaceans in captivity and in the wild. With these basics out of the way, we have turned more and more often to mainstream animal behavior for hard-won theoretical constructs that can shape our next thoughts.

Going beyond these formative events, we now begin to search our experience to understand how complex mammalian societies at sea differ from those ashore. We now begin to learn how living in three-dimensional, open waters has shaped not just the design of cetacean bodies but the patterns of their lives: how behavior, like anatomy, shows the impress of the long journey from that ancient time when they first entered the sea.

—K.S.N.

Chapter 1

V. M. Bel'kovich is Director of the Laboratory of Marine Bioacoustics at the Shirshov Institute of Oceanology, Academy of Science, Moscow State University, Moscow, USSR. He was educated at Moscow State University and has been doing research on morphology, ecology, behavior, and bioacoustics of marine mammals for nearly thirty years. His works have earned him numerous awards and fellowships. The studies reported on here were carried out under the auspices of the Severtzof Institute of Animal Morphology and the Institute of Developmental Biology, with several students and associates. The following persons contributed to chapter 1: *A. V. Agafonov, E. E. Ivanova, S. P. Kharitonov, L. B. Kozarovitsky, F. V. Novikova,* and *O. V. Yefremenkova.*

Chapter 2

Bernd Würsig is director of the Marine Mammal Research Program at Texas A&M University at Galveston. He did his undergraduate work at Ohio State University and received his Ph.D. at the State University of New York at Stony Brook. *Frank Cipriano* obtained a master's degree in oceanography from the Moss Landing Marine Laboratory and is currently a graduate student at the University of Arizona; he has collaborated with the Würsigs on studies of dusky dolphins in New Zealand. *Melany Würsig* has been her husband's research partner since 1972; they have studied the behavior of whales and dolphins in South America, New Zealand, Hawaii, the Arctic, and both coasts of North America. They are authors of many scientific papers and popular articles about dolphins and whales.

Chapter 3

Frederic L. Felleman has a B.S. degree in animal behavior from the University of Michigan and an M.S. degree in fisheries from the University of Washington. He has studied killer whales in Puget Sound since 1980. He is a fisheries biologist for the University of Washington. *James R. Heimlich-Boran* did his undergraduate work at Evergreen State College, in Washington, and completed his master's degree in marine biology at Moss Landing Marine Laboratory in 1987. He is currently a doctoral candidate at Cambridge University and studying pilot whales in the Canary Islands. *Richard W. Osborne* is also an Evergreen alumnus and has a master's degree in primatology from Western Washington State University. He is Research Director of The Whale Museum on San Juan Island in Washington State.

Chapter 4

Susan Kruse conducted the study reported here while an undergraduate at the University of California, Santa Cruz. She has a master's degree in marine science from the University of California at Santa Cruz and is a biologist for the National Marine Fisheries Service Southwest Fisheries Center.

Chapter 5

Karen Pryor was a founder and head dolphin trainer at Hawaii's Sea Life Park from 1963 to 1972. She was educated at Cornell University, with postgraduate work in zoology at the University of Hawaii, New York University, and Rutgers University. She is a writer and consultant and the author of several books as well as many technical and popular articles on human and animal behavior and learning. *Ingrid Kang Shallenberger* has a master's degree in animal behavior from the University of Stockholm. She was curator and head trainer at Sea Life Park from 1972 to 1990. Both authors now live in the Pacific Northwest.

Chapter 6

Randall S. Wells has been the driving force in the twenty-year, ongoing, multiperson study of Florida bottlenose dolphins described in this volume. He received his B.S. degree from the University of South Florida, his M.S. degree from the University of Florida, and his Ph.D. from the University of California at Santa Cruz. He is a behavioral ecologist in the Conservation Biology Department of the Chicago Zoological Society, continuing the Sarasota dolphin research program under the auspices of the Brookfield Zoo. He is the author or coauthor of numerous scientific papers on dolphin social organization.

Chapter 7

Michael D. Scott   is Associate Scientist at the Inter-American Tropical Tuna Commission in La Jolla, California. He did his undergraduate and master's degree work in zoology at the University of California, Los Angeles, and is currently working on his Ph.D. in biology at that university. He has worked on manatees and with R. S. Wells and A. B. Irvine on Florida bottlenose dolphins. He is author or coauthor of numerous scientific papers on cetaceans. *Wayne L. Perryman* has a B.A. in zoology from the University of California, Santa Barbara, and an M.A. in environmental biology from California State University, Hayward. He served as a commissioned officer in the NOAA Corps for seventeen years and commanded the NOAA ships *Chapman* and *Oregon II*. He is presently task leader of the Photogrammetry Group at National Marine Fisheries Service Southwest Fisheries Center.

Chapter 8

Albert C. Myrick, Jr.,   began his scientific career as a marine mammal fossil collector and preparator at the Smithsonian Institution. After eight years, he moved to California to undertake graduate work in marine mammalogy, earning his Ph.D. from the University of California, Los Angeles, in 1979. Since 1977, he has worked as a scientist for the National Marine Fisheries Service Southwest Fisheries Center, La Jolla, California, and has published numerous papers.

Chapter 9

Helene Marsh   is an Associate Professor in Zoology at James Cook University in North Queensland, Australia. She was educated at the University of Queensland and at James Cook University, where she received her Ph.D. in zoology in 1973. She has made extensive studies of the biology of dugongs as well as cetaceans. *Toshio Kasuya* is a senior scientist for the Far Seas Fisheries Research Laboratory in Shimizu, Japan. He was educated at the University of Tokyo and is an authority on the life history and reproductive ecology of small cetaceans.

Chapter 10

Jan Östman   came to the United States from Sweden as an undergraduate. He received his master's degree in psychology from San Francisco State University. He is a doctoral candidate in marine zoology at the University of California at Santa Cruz and is studying spinner dolphins in Hawaii.

Chapter 11

Peter Tyack   is an associate scientist at Woods Hole Oceanographic Institution, Woods Hole, Massachusetts. He graduated from Harvard University and received his Ph.D. in animal behavior in 1982 from Rockefeller University. He has studied humpback, sperm, and gray whales and is the author or coauthor of numerous papers and government reports.

Chapter 12

Louis M. Herman   has been Professor of Psychology at the University of Hawaii since 1966 and Director of the Kewalo Basin Marine Mammal Laboratory, where the work described herein was conducted, since 1972. He received his undergraduate education at the City University of New York, followed by graduate work there and at Emory University. He obtained his Ph.D. in experimental psychology from Pennsylvania State University. He has received many awards and grants for his work with whales and dolphins and is the author of numerous scientific publications.

Chapter 13

Patrick W. B. Moore   is a research scientist and program manager in the marine mammal division of the Naval Ocean Systems Center in Hawaii. He received his B.A. degree at California State University, Hayward, and his master's degree from the University of Hawaii. He began studying marine mammals as a student in 1971 and has been a dolphin researcher since 1978.

Note: The animals are listed under their common names. The scientific names are also listed, with cross-references to the common names.

Acoustic research, 8, 319–22. See also
  Dolphin research; Echolocation research;
  Whale research
    and dolphins, 295–96, 299–300, 302
    hydrophones in, 320–21
Aerial photogrammetry, 226–40
Affiliative behaviors, in spotted dolphins,
  191–92
Affiliative relationships, in captivity, 165
Aggressive behavior. See also Dominance
    in male bottlenose dolphins, 309
    and social aggression, 190–91
Agitation behavior, 178–80
    social activity and, 181–82
Amazon River dolphin (Inia geoffrensis),
  teeth, 252
Andersen, Soren, 300
Animals, domestication of, 345–46
Au, Whitlow, 302
Auditory information processing, dolphins,
  350–52

Balaena mysticetus. See Bowhead whale
Balaenoptera acutorostrata. See Minke
  whale
Balaenoptera physalus. See Finback whale
Beluga whale (white whale; Delphinap-
  terus leucas)
    boats and, 157
    recording sounds of, 8
Bigg, Michael, 14
Boats
    beluga whales and, 157
    bottlenose dolphins and, 157
    bowhead whales and, 157
    humpback whales and, 156–57
    killer whales and, 137–38, 154–57
Bottlenose dolphin (Tursiops truncatus),
  16, 17–19
    aggressive interactions, in male, 309
    benthic foraging by, 57, 58
    in Black Sea, 17–64, 67–75
    boats and, 157
    body abnormalities of, 30
    brains of, 359

calves, 19, 30, 73–74
    circling behavior of, 36
    diving depth of, 301
    dominance behavior in, 21
    feeding behavior
      carousel technique in, 46–48, 49–53
      chasing technique in, 57–58
      kettle technique in, 48–49, 51, 53
      upside down technique in, 60
      wall technique in, 53–57, 59–60
    female-calf pairs, 207–8, 221
    females, groups of, 212–18, 221–22
    in Florida, 14–15, 105, 200–223
    habitat of, 203, 204
  groups of, 18, 19
    composition of, 19, 21–23, 30, 32, 105
    female kinship in, 214–18
    formations in, 28, 34–35
    habitat, movement patterns and, 105
  herds of, 18, 19
    formations in, 19–21, 25, 29
    group structure of, 19, 30, 37
    hunting formations in, 39–41, 46
    sizes of, 24–29
    spatial differentiation in, 19–21
  home ranges of, 202, 204, 205
  homosexual behavior of, 309–12
  hunting behavior
    by groups of, 42–43
    by herds of, 39–42
    jumping during, 60–64
    by lone animals, 59–60
    spiral movement in, 27, 46
  in Indian Ocean, 9, 38, 46
  juveniles, groups of, 208–12
  male subgroups of
    adults, 218–21, 222
    youths, 210–13
  marks and scars in, 25, 31, 32–34
  mortality rates of, 210
  movement patterns, habitat and, 105
  play, 68–69
    acoustic signaling during, 74–75
    by adults, 74
    by calves, 30, 73–74

Bottlenose dolphin (*continued*)
  chasing as, 71
  complex turns in, 70
  fish manipulation in, 71
  free style, 71
  jumps during, 69–70, 71, 72
  in oceanariums, 68
  tactile activity in, 71
  polygamy in, 222–23
  schools of, 18, 24
  scouting behavior in, 19, 39, 43–46
  sex differentiation in, 18–19
  sexual behavior of, as play, 313
  subgroups, in captivity, 308–9
  tooth research on, 259, 274
  vocal mimicry of, 336–37
  vocalizations, social functions of,
    337–40
Bowhead whale (*Balaena mysticetus*)
  boats and, 157
  vocalizations of, 321
British Museum, 245
Bullock, T. H., 297
Busnel, René Guy, 300

Caldwell, David, 297
Caldwell, Melba, 297
Calves. *See also* Female-calf pairs; Juvenile
  groups
  body lengths of, in Central American
    spinner dolphins, 235–36
  play activity of, 30, 73–74
Capture-release research, 205–7
Catching prey. *See* Feeding behavior
Central American spinner dolphin (*Stenella
longirostris centroamericana*). *See also*
Spinner dolphin
  body lengths of, 231–4
  calves, body lengths of, 235–36
  formations of, 239
  schools
    density of, 237–38
    sizes of, 230–31
    structure in, 234–39
Cetacean research. *See* Dolphin research;
  Whale research
Circling behavior, in bottlenose dolphins,
  36
Cochran, William, 10
Common dolphin (*Delphinis delphis*), 37,
  38
  feeding formations of, 67
  jumps of, 67
  migrational movements of, 67
Connor, Richard, 14–15

Dall's porpoise (*Phocoenoides dalli*), 118,
  128, 131
de Haan, F. Reysenbach, 299–300
Delphinapterus leucas. *See* Beluga whale
Delphinis delphis. *See* Common dolphin
Dentition. *See* Teeth
Diurnal movements. *See* Movement
  patterns
Dohl, Tom, 11
Dolphin research, 79–81, 385–86. *See
  also* Acoustic research; Echolocation research; Psychophysics; Teeth: research
  methods; Whale research
  aerial photogrammetry in, 226–40
  anatomical studies in, 297
  behavior codes in, 174–75
  with captive animals
    history of, 293–303
    versus wild animals, 1, 2
  capture-release method, 205–7
  focal animal observation in, 174, 205,
    306–7
  identification tags used in, 80
  long-term studies in, 199–207, 223
  measurement data in, 198
  misconceptions about, 1–2
  using mobile observation chambers,
    11–13
  public interest in, 298, 383–84
  radio tracking in, 9–10, 80, 87–91
  by Russians, 300, 301
  scars and marks analysis in, 13
  telemetry devices in, 321–22, 324–
    26
  theodolite tracking in, 13–14, 80,
    81–87
  and tuna net encirclement, 165–68,
    169, 175–76
Dolphins
  domestication of, 346–47
  misconceptions about, 1–2
  public attitudes toward, 383–84
  and underwater searching for objects,
    346
Dominance, 21. *See also* Aggressive
  behavior
  and disputes, 190–91
  and hierarchies, 165–66, 359
  homosexual interactions and, 312–14
  penis display and, 314
Dusky dolphin (*Lagenorhynchus
obscurus*), 78, 80
  in Argentina, 92–95
  behavioral patterns of
    habitat and, 104

prey availability and, 104
and chimpanzees, compared, 105–6
deep scattering layer and, 96, 104
diurnal movements of, 93–94, 98
feeding behavior of, 78, 92, 93, 96
group composition of, 92–93, 96,
   104–5
habitat, 92, 95–96, 97, 98
   behavioral patterns and, 104
hunting by, 92, 96
killer whales and, 93
in New Zealand, 95–99
prey of, 92, 96
   detection, 357
seasonal movement patterns of, 98–99
sexual behavior of, after feeding, 313
social activity of, 92, 96, 101, 104, 180
spinner dolphins and, 103, 104
Dziedzic, Albin, 300

Echolocation research, 296, 365. *See also*
   Acoustic research
   and hearing sensitivity, dolphins, 379
   history of, 365–66, 379–80
   methods of
      apparatus used in, 372–74
      signal frequency control in, 377–78
      signal intensities in, 374–77
      target strength in, 379
Escherichtius robustus. *See* Gray whale
Essapian, Frank, 294–95
Eubalaena australis. *See* Southern right
   whale
Eubalaena glacialis. *See* Northern right
   whale
Evans, William, 9–10, 11, 298, 300

False killer whale (*Pseudorca crassidens*),
   164
Fear. *See* Agitation behavior
Feeding behavior. *See also* Hunting
   behavior
   of bottlenose dolphins, 46–60
      carousel technique in, 46–48, 49–53
      chasing technique in, 57–58
      kettle technique in, 48–49, 51, 53
      upside down technique in, 60
      wall technique in, 53–57, 59–60
   of dusky dolphins, 78, 92, 96
   of harbor porpoises
      carousel technique in, 64–65
      kettle technique in, 64–65
      wall technique in, 66
   of killer whales, 114–15

resident pods, 124–29, 130–31
transient pods, 133–35
of spinner dolphins, 101
Feeding ecology
   of killer whales, 116
      in resident pods, 123–24, 129–30
      in transient pods, 131–32, 140
Female-calf pairs
   in bottlenose dolphins, 207–8, 221
   in spotted dolphins, 183–85
Females. *See also individual species*
   postreproductive roles of, 282–84,
      287–89
   reproductive cycles of, dentition and,
      268–75
Feresa attenuata. *See* Pigmy killer whale
Finback whale (*Balaenoptera physalus*),
   vocalizations of, 321
Floyd, Robert, 302
Focal animal observation, 174, 205,
   306–7
Foraging. *See* Feeding behavior; Hunting
   behavior
Formations
   of bottlenose dolphins
      in groups, 28, 34–35
      in herds, 19–21, 25, 29
      in hunting behavior, 39–41, 46
   of Central American spinner dolphins,
      239
   of harbor porpoises in hunting behavior,
      64–65
Fraser, F. C., 299

Galler, Sidney, 299
Gaping, 190
Gawain, Elizabeth, 14
Globicephala macrorhynchus. *See* Short-
   finned pilot whale
Globicephala sp. *See* Pilot whale
Gray whale (*Escherichtius robustus*), ho-
   mosexual interactions of, 314
Greeting behavior, 313
   between subgroups, 191–92
Grinnell, A. D., 297

Harbor porpoise (*Phocoena phocoena*),
   36–37, 38
   hunting formations of, 64–65
   jumps by, 66
   migratory behavior of, 66
   sexual behavior of, 66
Hearing. *See also* Acoustic research; Echolocation research

Hearing (*continued*)
  and auditory information processing,
    350–52
  of dolphins, 379
Hebb, D. O., 295
Herding of prey. *See* Feeding behavior;
  Hunting behavior
Herman, Louis, 302, 348
Hohn, Aleta, 247–48
Homosexual behavior. *See* Sexual behavior: homosexual interactions in
Humpback dolphin (*Sousa spp.*), 9
  greeting interactions of, 313
Humpback whale (*Megaptera novangelicae*), boats and, 156–57
Hunting behavior. *See also* Feeding behavior
  of bottlenose dolphins, 39–46
    benthic foraging in, 57, 58
    jumping during, 60–64
  by lone animals, 59–60
  of harbor porpoises, 64–66

Identification
  of bottlenose dolphins by marks and
    scars on, 25, 31, 32–34
  of killer whales, 119
  by marks and scars, 13
    unreliability of, 201
  of spinner dolphins, 169–71
  of spotted dolphins, 169–71
    using color patterns of schools of, 193
  by tagging, 80, 202–3
  by tooth layering patterns, 256–57
Inia geoffrensis. See Amazon River dolphin
Inter-American Tropical Tuna Commission
  (IATTC), 247
International Antarctic Treaty, 384–85
Irvine, Blair, 14

Jumps
  by bottlenose dolphins
    breaching, 62, 63–64
    candles, 61, 64
    during hunting, 60–64
    during play, 69–70, 71, 72
  by Central American spinner dolphins,
    226, 239
  by common dolphins, 67
  by harbor porpoises during hunting, 66
  and porpoising, 60–63, 226–39
Juvenile groups. *See also* Calves
  of bottlenose dolphins, 208–12, 221
    sex ratios in, 209–10
    social activity in, 210
  of spotted dolphins, 185–86

Kanwisher, John, 301
Katsuki, K., 297
Kellogg, Winthrop, 295, 296–97, 299
Killer whale (*Orcinus orca; Orcinus glacialis* [in Antarctica]). *See also* False killer whale; Pigmy killer whale
  in Antarctic Ocean, 115
  behavior categories of, 121
  boats and, 137–38, 154–57
  diurnal movements of, 154
  dusky dolphins and, 93
  feeding behavior
    of resident pods, 124–29, 130–31
    of transient pods, 133–35
  feeding ecology
    of resident pods, 123–24, 129–30,
      140
    of transient pods, 131–32, 140
  foraging, by resident pods, 125–28,
    130–31
  habitats
    of resident pods, 125, 128–29
    of transient pods, 132–33, 138
  hunting behavior of, 114–15
  identification of, 119
  milling index of, 153
  in the Pacific Northwest, 14, 112–40,
    148–58
    habitat of, 116–18, 150–51
  pod composition of, 118
  prey of, 114, 115–16, 118
    marine mammals as, 93, 113–14
    of resident pods, 126–28, 131,
      133–35
    salmon as, 123–24, 130, 139
    of transient pods, 132, 134, 139–40
  resident and transient pods, distinguishing features of, 119, 140
  seasonal movements of, 114
    by resident pods, 140
  and stress from human activities, 158
  surfacing, by transient pods of, 137–38
  and tidal orientation
    by resident pods, 129, 131, 138–39
    by transient pods, 135–36
  travel by resident pods of, 125–26
    formations in, 125–26
    routes in, 124
  and vocal activity, 137
    of resident pods, 129
    of transient pods, 136
Kritzler, Henry, 295

Laboratories d'Acoustique, 300
Lagenorhynchus obscurus. See Dusky dolphin

Lang, Tom, 301
Lawrence, Barbara, 8, 295, 299. *See also* Schevill, Barbara
Leaping. *See* Jumps
Learning. *See also* Training
   acoustic language and, 351–52
   auditory memory and, 350
   auditory relational rules in, 350
   conjoint behaviors and, 356
   gestural language and, 353–55
   pointing gestures by humans and, 355–56
   references to absent objects in, 355
   second-order, 302, 356
   tandem behaviors in, 356–57
Life expectancy. *See* Mortality rates
Lilly, John, 295, 297–98
Long-term studies, 199–207, 223

McBride, Arthur, 294–95
McLean, William, 299
Marineland, Florida, 8, 294
Marine Mammal Commission, 384
Marine Mammal Protection Act of 1972, 202, 346, 384
Marine mammals, as killer whale prey, 93, 113–14, 126–28, 133–35, 139–40
Marine Studios. *See* Marineland, Florida
Marine World Africa USA, California, 304
Marks and scars. *See* Identification: by marks and scars
Marsh, Helene, 246
Mate, Bruce, 10
Megaptera novangelicae. See Humpback whale
Minke whale (*Balaenoptera acutorostrata*), 118, 128
Møhl, Bertel, 300
Moore, Patrick, 302
Mortality rates
   of bottlenose dolphins, 210
   of dolphins, 246, 252, 273–75
   of short-finned pilot whales, 282
Mother-calf pairs. *See* Female-calf pairs
Movement patterns
   of dusky dolphins
      diurnal, 93–94, 98
      seasonal, 98–99
   of killer whales
      diurnal, 154
      seasonal, 114, 140
   of spinner dolphins
      diurnal, 99–103
      seasonal, 103
Murchison, Earl, 302
Myrick, Albert, 248, 250

Nachtigall, Paul, 302
National Marine Fisheries Service (NMFS), 162, 165, 246, 247, 252
Naval Ocean Systems Center Laboratory, Hawaii, 364
Norris, Kenneth S., 10–11, 246, 300–301
Northern right whale (*Eubalaena glacialis*), vocalizations of, 321

Orca. *See* Killer whale
Orcinus glacialis. See Killer whale
Orcinus orca. See Killer whale
Östman, Jan, 12

Payne, Roger, 13
Penner, Ralph, 302
Perrin, William, 247
Perryman, Wayne, 248
Phocoena phocoena. See Harbor porpoise
Phocoenoides dalli. See Dall's porpoise
Photogrammetry. *See* Aerial photogrammetry
Physeter catodon. See Sperm whale
Pigmy killer whale (*Feresa attenuata*), 7–8
   in captivity, 246
Pilot whale (*Globicephala sp.*), 280. *See also* Short-finned pilot whale
   diving depth, 301
Play, in bottlenose dolphins, 68–75
   acoustic signaling during, 74–75
   by adults, 74
   by calves, 73–74
   chasing, 71
   complex turns in, 70
   sexual behavior as, 313
Polygamy, in bottlenose dolphins, 222–23
Powell, Bill, 298
Predator avoidance, by dolphins, 357–58
Prey
   availability of, behavior patterns and, 104
   detection strategies for, 357
   hydroacoustic detection of, 122–23
   salmon as, 123–24, 130, 139
   southern anchovies as, 92
   tuna as, 162
Pryor, Karen, 10, 301
Pseudorca crassidens. See False killer whale
Psychophysics, 366–67
   response paradigms in, 367–68
   stimulus control in, 371
   stimulus response paradigms in, 368–71
Purves, P. E., 299

Radio tracking, 80, 87–91
   and radio receivers, 89–90

Radio tracking (*continued*)
  and tag attachment, 90–91
  transmitter frequencies in, 89
Rafting, 180, 186
Ridgway, Sam, 297, 301
Rough-toothed dolphin (*Steno bredanensis*), second-order learning in, 302, 356

Saayman, Graham, 9
St. Augustine, Florida. *See* Marineland, Florida
Schevill, Barbara, 295–96, 299. *See also* Lawrence, Barbara
Schevill, William, 8, 295–96, 299
Scott, Michael, 14, 248
Sea Life Park, Hawaii, 163, 245–46
Seasonal movements. *See* Movement patterns
Sensory systems, 349
Sex differentiation
  in bottlenose dolphins, 18–19
  in spotted dolphins, 171–74
Sexual behavior, 305–6
  after feeding, 313
  as greeting behavior, 313
  homosexual interactions, 309–312
  as play, 313
Shane, Susan, 14
Short-finned pilot whale (*Globicephala macrorhynchus*). *See also* Pilot whale
  age determination in, 281
  lactation period in, 282–83
  life expectancy of, 282
  ovulation rate in, 282
  post reproductive females
    characteristics of, 283–84
    roles of, 282–84, 287–89
  postreproductive mating, 283
  pregnancy rate of, 281–82
  and rearing of young, 282–84
  school composition of, 283, 287
Signature whistles, 297, 358
  of dolphins, 322–24, 338–39
  mimicry of, 336–37
Smolker, Rachel, 14–15
Social activity
  agitation behaviors and, 181–82
  in captive dolphins, 180
  in juvenile groups, bottlenose dolphins, 210
  levels of, 181
Social aggression, 190–91
Social behavior, vision and, 359–60
Social interactions, vocalizations and, 337–40

Sonar research. *See* Acoustic research; Echolocation research
Sound mimicking, 298
Sousa spp. See Humpback dolphin
Southern right whale (*Eubalaena australis*), 321
  homosexual interactions of, 314
Southwest Fisheries Center, 246–47, 248
Sperm whale (*Physeter catodon*), 95–96
  vocalizations of, 321
Spinner dolphin (*Stenella longirostris*). *See also* Central American spinner dolphin
  in captivity, 163
  chimpanzees compared to, 105–6
  deep scattering layer and, 101, 104
  diurnal movement patterns of, 99–103
  dusky dolphins and, 103, 104
  in the Eastern Tropical Pacific habitat, 161–62
  formations of, 177
  habitat of, 99, 100, 102
  in Hawaii, 99–103
  prey of, 101
    detection of, 357
  schools of, 229
  seasonal movement patterns of, 103
  sexual behavior of, 313–14
  social activity of, 104–5
  spotted dolphins and, 177
  tooth research on, 259–61
  travel by, 193
  tuna fishing and, 193
Spotted dolphin (*Stenella attenuata*)
  affiliative behaviors of, 191–92
  boats and, 192
  in captivity, 163
  in the Eastern Tropical Pacific, 161–93
    habitat, 161–62
  female-calf pairs, subgroups of, 183–85
  formations of, 177–78
  greeting behaviors of, 191–92
  juveniles, subgroups of, 185–86
  male adults, subgroups of, 186–87, 189–90, 192
  maturation stages of, 170–71
  reproduction in, 192–93
  schools of, 229
  sex differentiation in, 171–74
  social activity of, 180, 181–83
  social aggression in, 190–91
  spinner dolphins and, 177
  subgroups of, 183–87
  tooth patterns in, 266
  and transient associations, 187
  travel by, 193
  young adults, subgroups of, 187

Stenella attenuata. See Spotted dolphin
Stenella longirostris. See Spinner dolphin
Stenella longirostris centroamericana. See
  Central American spinner dolphin
Steno bredanensis. See Rough-toothed
  dolphin
Stress. *See* Agitation behavior
Subadult groups. *See* Juvenile groups
Synchronized behavior of spotted dolphins,
  189–90

Tavolga, Margaret, 295
Tayler, C. K., 9
Teeth
  age determination from, 251–52,
    258–63, 275–76, 281
  anatomy of, 254–57
  cemental layers in, 255–56
  dentine layers in, 254–55, 257, 266
  exogenous events and, 266, 275
  female reproductive cycles and, 268–73,
    275
  genetically-based patterns in, 265–68,
    275
  growth layer groups in, 258
  homodont dentition, 252, 253
  individual variability in, 268–75
  monthly layers in, 264–65
  numbers of, in delphinids, 252–54
  research methods
    calibration in, 251, 276
    tetracycline marking in, 258–59,
      262–63
Theodolite tracking, 80, 81–87, 152–53
Threat gestures, 164
Training, 164, 377. *See also* Learning
Tuna fishing
  dolphins and, 160, 162, 193, 252
    agitation behaviors during, 178–80
    population estimates of, 273–74
    purse-seine method of, 166–68
Turl, Wayne, 302
Turner, Ron, 300
Tursiops truncatus. See Bottlenose dolphin

U.S. Navy, 298, 301, 346

Vision
  acuity of, 352
  matching-to-sample studies of, 353–54
  social behavior and, 359–60
  and temporal patterning, 353, 354
  and understanding gestural language,
    353, 354–55
Vocalizations
  in behavioral contexts, 329–333
  of bottlenose dolphins
    social functions of, 334–40
    vocal mimicry, 336–37
  mimicking sounds in, 350–51, 358
  multiloop whistles in, 334–35
  and "naming" objects, 351
  signature whistles, 322–24, 338–39,
    358
  whistle categorization in, 327–29,
    334–35

Watkins, William, 8, 10
Wells, Randall, 14
Whale research. *See also* Acoustic research; Dolphin research; Teeth: research
  methods
  boat encounters in, 119
  fishery capture in, 281
  fishery data correlations in, 119–20
  and habitat use, 121–22
  hydroacoustic detection of prey in,
    122–23
  milling index in, 153
  sightings in, 119
  theodolite tracking in, 152–53
  tidal data in, 122
White whale. *See* Beluga whale
Wood, Forrest, 295–96
Würsig, Bernd, 13–14
Würsig, Melany, 13–14

Compositor:	G & S Typesetters, Inc.
Text:	10/12 Sabon
Display:	Sabon
Printer:	Edwards Brothers, Inc.
Binder:	Edwards Brothers, Inc.